Research and Development Progress in 3D Cadastral Systems

Research and Development Progress in 3D Cadastral Systems

Special Issue Editors

Peter van Oosterom
Efi Dimopoulou

MDPI • Basel • Beijing • Wuhan • Barcelona • Belgrade

MDPI

Special Issue Editors
Peter van Oosterom
Delft University of Technology
The Netherlands

Efi Dimopoulou
National Technical University of Athens
Greece

Editorial Office
MDPI
St. Alban-Anlage 66
4052 Basel, Switzerland

This is a reprint of articles from the Special Issue published online in the open access journal *ISPRS International Journal of Geo-Information* (ISSN 2220-9964) from 2017 to 2018 (available at: https://www.mdpi.com/journal/ijgi/special_issues/3d_Cadastral).

For citation purposes, cite each article independently as indicated on the article page online and as indicated below:

LastName, A.A.; LastName, B.B.; LastName, C.C. Article Title. *Journal Name* **Year**, *Article Number*, Page Range.

ISBN 978-3-03921-056-5 (Pbk)
ISBN 978-3-03921-057-2 (PDF)

Cover image courtesy of Peter Van Oosterom and Efi Dimopoulou.

Contents

About the Special Issue Editors

Peter van Oosterom obtained an M.Sc. in Technical Computer Science in 1985 from Delft University of Technology, the Netherlands. In 1990, he received a Ph.D. from Leiden University. From 1985 to 1995, he worked at the TNO-FEL laboratory in The Hague. From 1995 to 2000, he was a senior information manager at the Dutch Cadastre, where he was involved in the renewal of the Cadastral (Geographic) database. Since 2000, he has been professor at the Delft University of Technology and head of the 'GIS Technology' Section. He is the current chair of the FIG Working Group on '3D Cadastres'. He is coeditor of the International Standard for the Land Administration Domain Model, ISO 19152.

Efi Dimopoulou is Professor at the School of Rural and Surveying Engineering, National Technical University of Athens, in the fields of Cadastre, Spatial Information Management, Land Policy, 3D Cadastres, and Cadastral Modeling. She is the Program Director of the NTUA Inter-Departmental Postgraduate Course (Environment and Development) and President of the Hellenic Society for Geographical Information Systems (HellasGIS).

Preface to "Research and Development Progress in 3D Cadastral Systems"

The content of this Special Issue has its origin at the "5th International FIG Workshop on 3DCadastres", organized in Athens, Greece, 18–20 October 2016. The Workshop provided a further stepping stone to identifying key issues and challenges within 3D Cadastres developments, as well as exchanging ideas and solutions between scientists from academia, industry, and government. After the workshop, authors of selected papers (or in some cases, two teams of authors of related papers) were invited to publish in this Special Issue, which is devoted to the legal, organizational, and technical aspects of 3D Cadastres. Therefore, this Special Issue is where practice and research meet, influence, stimulate, and support each other, showing the progress in the key areas of 3D Cadastres, including the legal framework of 3D Cadastres, initial registration of 3D parcels, 3D data management, visualization, distribution, and delivery of 3D parcels. The 13 papers published in this Special Issue are the end result of the authors' hard work and a rigorous journal's peer review process. We would like to thank the authors for their valuable contributions and of course express thanks to the IJGI Editorial Office and the reviewers for their thoughtful suggestions and constructive criticism. Even though all papers in this Special Issue stand on their own, depending on cultural, geographical, and technical differences, we have tried to produce a rather holistic approach, covering this field of research. Moreover, we have highlighted similarities and dissimilarities in addressing important 3D Cadastral topics, as follows.

Peter van Oosterom, Efi Dimopoulou
Special Issue Editors

International Journal of
Geo-Information

isprs

MDPI

Editorial

Introduction to the Special Issue: "Research and Development Progress in 3D Cadastral Systems"

Peter van Oosterom [1],* and Efi Dimopoulou [2]

[1] Section GIS technology, Department OTB Faculty of Architecture and the Built Environment, Delft University of Technology, Julianalaan 134, 2628 BL Delft, The Netherlands
[2] School of Rural and Surveying Engineering, National Technical University of Athens, Iroon Polytechneiou, 9, Athens 15780, Greece; efi@survey.ntua.gr
*   Correspondence: P.J.M.vanOosterom@tudelft.nl

Received: 5 February 2018; Accepted: 6 February 2018; Published: 9 February 2018

The content of this Special Issue has its origin in the "*5th International FIG Workshop on 3D Cadastres*", organized in Athens, Greece, 18–20 October 2016 [1]; see Figure 1. The Workshop provided a further stepping stone to identifying key issues and challenges within 3D Cadastres developments, as well as exchanging ideas and solutions between scientists from academia, industry and government. After the workshop, authors of selected papers (or in some cases, two teams of authors of quite related papers) were invited to publish in this Special Issue, which is devoted to the legal, organizational and technical aspects of 3D Cadastres. Therefore, this Special Issue is the subject where practice and research meet, influence, stimulate and support each other, showing the progress in the key areas of 3D Cadastres, including the Legal framework of 3D Cadastres, Initial registration of 3D parcels, 3D data management, Visualization, distribution and delivery of 3D parcels.

Figure 1. Participants of the 3D Cadastres workshop in Athens 2016.

Many 3D Cadastres' activities have been conducted during the past two decades: FIG 3D Cadastres workshops, sessions at FIG working weeks and congresses, and two special issues in international scientific journals [2,3] have been published. The lasting interest in 3D Cadastres and their progress become clear when comparing the contents of the special issues: gradually changing from more theoretical into main-stream approach for efficient and effective land administration. This special issue reflects one more step on our journey towards realizing efficient 3D cadastral registration and management.

The 13 papers published in this Special Issue are the end result of the authors' hard work and a rigorous journal's peer review process. We would like to thank the authors for their valuable contribution, and of course express thanks to the IJGI Editorial Office and the reviewers for their thoughtful suggestions and constructive criticism.

Even though all papers in this Special Issue stand on their own, depending on cultural, geographical and technical differences, we have tried to produce a rather holistic approach, covering this field of research. Moreover, we have highlighted similarities and dissimilarities in addressing important 3D Cadastral topics, as follows.

The Special Issue starts with Karel Janecka, and Petr Soucek's paper, which presents a country profile for the Cadastre of the Czech Republic, based on the ISO 19152:2012 Land Administration Domain Model (LADM). The profile considers the requirements of the new Civil Code, which explicitly considers the space above and below the ground as a part of the land. The Czech LADM-based country profile contains all of the classes and code lists required for Level 2 compliance and can be appropriately extended in order to support the registration of 3D parcels and legal components related to utilities.

The second paper, by Jantien Stoter, Hendrik Ploeger, Ruben Roes, Els van der Riet, Filip Biljecki, Hugo Ledoux, Dirco Kok and Sangmin Kim, reports on the first 3D cadastral registration of multi-level property rights in The Netherlands. Within this context, the authors present a methodology applied to two cases, representing legal volumes in an interactive 3D visualization that can be registered in the land registers. Based on the experiences of the 3D registrations of multi-level properties, as implemented in this project, it is concluded that 3D registration should be further developed, and the regulations will be adjusted accordingly.

The third paper, by Rodney James Thompson, Peter van Oosterom and Kean Huat Soon, explores the encoding of spatial units highlighting their 2D extent and topology, while fully defining their extent in the third dimension. The paper presents a conceptual model applied to topology encoding of range of spatial units (2D, simple 3D, complex 3D), expressed in the language of the LADM. Especially in the case of larger apartment buildings, there are many boundary surfaces shared between neighboring 3D parcels, which makes 3D topology very appropriate. Two multi-step real-world examples are given and encoded according to this conceptual model in LandXML for exchange purposes, including the initial registration.

The fourth paper, by Dimitrios Kitsakis and Efi Dimopoulou, investigates public law restrictions (PLR) applying to 3D space along with the current legal framework. A Greek case study concerning the establishment of a subway station is examined, focusing on public utilities, archaeological legislation, and building regulations. Relative legal documentation is compiled and mapped in a 3D PLR model, presenting inefficiencies and malfunctions that can be resolved if PLRs are addressed within a 3D cadastral context.

The fifth paper, by Katerina Athanasiou, Michael Sutherland, Christos Kastrisios, Lysandros Tsoulos, Charisse Griffith-Charles, Dexter Davis and Efi Dimopoulou, addresses the 3D character of marine spaces and proposes the application of the Land Administration Domain Model (LADM conceptual standard ISO 19152) to specific jurisdictional MAS or MC, taking into account the S-121 Maritime Limits and Boundaries (MLB) Standard, which refers to LADM. Several modifications are proposed, including e.g., the introduction of class marine resources into the model, to meet the particular marine and maritime administrative needs of both Greece and the Republic of Trinidad and Tobago.

The sixth paper, by Behnam Atazadeh, Abbas Rajabifard and Mohsen Kalantari, investigates the performance of three BIM-based methods, namely purely legal, purely physical and integrated approaches, for storing, managing and communicating legal interests mainly in multistorey buildings in Victoria, Australia. One finding was that interaction tasks were easier to perform in purely legal or physical models compared to integrated models, while on the other hand, integrated models provide a more visual communication of the location of legal boundaries, indicating that lay users may find it easier to understand their legal rights in integrated BIM models.

The seventh paper, by Nikola Vučić, Miodrag Roić, Mario Mađer, Saša Vranić and Peter van Oosterom, explores Croatian registration related to 3D Cadastral cases, specific interests in strata and rights or restrictions related to spatial planning. Together with the analysis of the current land administration system, this forms the basis for the possible 3D upgrade. Further, the efficiency of implementation and reuse of existing data are considered; e.g., the signs or symbols as used to represent topographic objects on 2D maps (for tunnels, bridges, overpasses, etc.) as these provide a reference context for a 3D Cadastre. The authors propose the establishment of a 3D Multipurpose Land Administration System as the most efficient system of land administration, given the growing 3D data acquisition possibilities and the changing demands, the traditional real estate register is facing.

The eighth paper, by Aleksandra Radulović, Dubravka Sladić and Miro Govedarica describes the development of the Serbian Cadastral Domain Model, as an important step towards the realization of a 3D Cadastre. The proposed model is based on LADM, after analyzing national legislation (and its current incorrect application) and is completely conformant at the medium level (or even higher in some cases). Given the importance of the 3D aspects, such as buildings with sometimes overlapping rights and restrictions, the Serbian country profile uses a 2D representation for simpler situations, and a 3D representation for more complex situations.

The ninth paper, by Eftychia Kalogianni, Efi Dimopoulou, Wilko Quak, Michael Germann, Lorenz Jenni and Peter van Oosterom tries to close the gap between the conceptual model and the actual implementation of the 3D Cadastre, by using the INTERLIS language. This approach allows the inclusion of both the legal 3D spaces, the physical 3D objects (reference objects) and their relationships in a vendor-neutral manner, among others, by using formal constraints and structured code lists. The approach is LADM-based and supports, with the help of INTERLIS tools, the evaluation and validation of 3D legal and physical models. The first results of a Greek case study are presented.

The tenth paper, by Jennifer Oldfield, Peter van Oosterom, Jakob Beetz and Thomas Krijnen, also addresses steps towards implementation of 3D Cadastres. In this paper, the Open Building Information Models (BIM) standards are explored. First of all, by (re)using the available and future data in the BIM Industry Foundation Classes (IFC) as a source for 3D legal spaces. Further, the workflow of the multi-actor processes involved when planning, designing, creating and registering results of new spatial developments are organized according to BIM methodology of the Information Delivery Manual (IDM ISO29481). The proposed BIM IDM approach is used to illustrate how much time and effort are saved (and consistency is enhanced) compared to traditional registration of 3D spatial units in the land administration.

The eleventh paper, by Trent Gulliver, Anselm Haanen and Mark Goodin outlines an approach for turning the 150-years-old New Zealand cadastral survey and tenure systems into a full 3D digital Cadastre supporting 3D RRRs (rights, restrictions and responsibilities). The whole chain of activities is addressed from survey/data capture and validation, via their lodging at the authorities and integration with existing data, to visualization and dissemination. The approach of integration of 2D (default) and 3D (when needed) is promoted by regulators of New Zealand's cadastral survey system. However, given the generic nature of the proposed solution and the involvement of the suppliers of land administration systems, it is believed that other jurisdictions may also benefit from these efforts.

The twelfth paper, by Abdullah Alattas, Sisi Zlatanova, Peter van Oosterom, Efstathia Chatzinikolaou, Christiaan Lemmen and Ki-Joune Li proposes the combined use of IndoorGML and LADM Models. The main application of IndoorGML is to support indoor navigation via 3D primal (actual spaces in building) and dual (connectivity between spaces) representations. However, the actual access rights are not yet included and LADM's capabilities for describing these are applied. The access rights are dependent on the specific user/party (group), spatial unit (location) and time.

Then, together with the 3D model of the building, this allows path computation based on access rights avoiding the non-accessible spaces. The two original standard models remain independent, and their combined use could also support other applications, such as regular building maintenance, facility management work, crisis management/evacuation, etc.

The thirteenth and last paper, by Ruba Jaljolie, Peter van Oosterom and Sagi Dalyot addresses the needed functionalities for 3D Land Management Systems with the capacity to handle various types of data in a uniform way, both above-terrain and below-terrain. Starting with the legal and technical aspects of Survey of Israel's CHANIT (legal set of cadastral work processes' specifications), the authors propose a data structure/model, functionality, and suggestions for closing the regulation gaps for a 3D Cadastre, including those needed for the preparation of 2D and 3D mutation plans (survey plans).

Concluding, we hope the cognitive 3D Cadastres community will find this Special Issue to be a useful collection of papers, to provide an informative foundation to further research in this challenging field. In the meantime, also, the next step on our on-going journey is planned: the 6th International FIG Workshop on 3D Cadastres, to be organized in Delft, The Netherlands, 2–4 October 2018.

Conflicts of Interest: The authors declare no conflicts of interest.

References

1. Van Oosterom, P.; Dimopoulou, E.; Fendel, E. (Eds.) *Proceedings "5th International Workshop on 3D Cadastres"*; FIG: Copenhagen, Denmark, 2016.
2. Lemmen, C.; van Oosterom, P. 3D Cadastres. *Comput. Environ. Urban Syst.* **2002**, *27*, 337–343. [CrossRef]
3. Van Oosterom, P. Research and development in 3D cadastres. *Comput. Environ. Urban Syst.* **2013**, *40*, 1–6. [CrossRef]

International Journal of
Geo-Information

MDPI

Article

A Country Profile of the Czech Republic Based on an LADM for the Development of a 3D Cadastre

Karel Janečka [1],* and Petr Souček [2]

[1] Department of Geomatics, Faculty of Applied Sciences, University of West Bohemia, Pilsen 30614, Czech Republic

[2] Czech Office for Surveying, Mapping and Cadastre, Prague 18211, Czech Republic; petr.soucek@cuzk.cz

* Correspondence: kjanecka@kgm.zcu.cz; Tel.: +420-607-982-581

Academic Editors: Peter van Oosterom and Wolfgang Kainz

Received: 24 February 2017; Accepted: 28 April 2017; Published: 3 May 2017

Abstract: The paper presents a country profile for the cadastre of the Czech Republic based on the ISO 19152:2012 Land Administration Domain Model (LADM). The proposed profile consists of both legal and spatial components and represents an important driving force with which to develop a 3D cadastre for the Czech Republic, which can guide the Strategy for the Development of the Infrastructure for Spatial Information in the Czech Republic to 2020. This government initiative emphasizes the creation of the National Set of Spatial Objects, which is defined as the source of guaranteed and reference 3D geographic data at the highest possible level of detail covering the entire territory of the Czech Republic. This can also be a potential source of data for the 3D cadastre. The abstract test suite stated in ISO 19152:2012—Annex A (Abstract Test Suite) and the LADM conformance requirements were applied in order to explore the conformity of the Czech country profile with this international standard. To test their conformity, a mapping of elements between the LADM and the tested country profile was conducted. The profile is conformant with the LADM at Level 2 (medium level) and can be further modified, especially when legislation is updated with respect to 3D real estate in the future.

Keywords: country profile; LADM; 3D cadastre; GeoInfoStrategy

1. Introduction

The cadastre of real estate (KN) is one of the largest data information systems in the state administration of the Czech Republic. It is composed of data regarding real estate located within the Czech Republic, including a detailed inventory with descriptions of each estate's geometric specifications and location, as well as records of property, other material rights and additional legally-stipulated rights to real estate. The cadastre of real estate also contains a great deal of important data concerning parcels, as well as selected buildings and their owners. The KN, which is the primary information system concerning the territory of the Czech Republic, is administered mainly through computational means, wherein the cadastral unit is the fundamental territorial unit. The cadastre's documentation primarily comprises files of geodetic information encompassing the cadastral map (including its digital representation in given cadastral units), as well as files containing descriptive information regarding cadastral units, parcels, buildings, flats and non-residential premises, owners and other justified persons and legal relations and rights, in addition to other legal facts.

Several objectives have been proposed that drive the creation of a country profile based on the ISO 19152 Land Administration Domain Model (LADM) [1]. In October 2014, the Czech government approved the conception of The Strategy for the Development of the Infrastructure for Spatial Information in the Czech Republic to 2020 (GeoInfoStrategy), which serves as a basis for the National Spatial Data Infrastructure (NSDI). The set of measures to be used to develop the regulatory framework in the field of spatial information was then defined in the GeoInfoStrategy Action Plan [2]. This Action Plan considers the adoption of the ISO 19152 standard within various government initiatives. In particular, using ISO 19152 is recommended for the further development of the register of territorial identification, addresses and real estate and for the feasibility study of the register of passive infrastructure. This indicates that, during the implementation of the GeoInfoStrategy Action Plan, there will be a demand for thorough knowledge and understanding of the LADM concept. Both of the aforementioned registers could be considered as the cornerstone of the Czech National Spatial Data Infrastructure [3].

There is a strong emphasis on the creation of a National Set of Spatial Objects (NSSO) within the GeoInfoStrategy. The NSSO (which includes 3D buildings) is defined as the source of guaranteed, as well as reference (where possible) 3D geographic data at the highest possible level of detail for selected objects in the real world and covers the entire territory of the Czech Republic. It is not explicitly stated within the GeoInfoStrategy that the parcels should be registered in 3D. However, the proposed model based on an LADM could be potentially extended to support the registration of 3D parcels (spatial units) in the future, and it could also serve as the basis for an extension of the current data model of the cadastre in a standardized way.

Furthermore, the Czech version of ISO 19152 has existed since October 2013. The National Mirror Committee 122 of Geographic Information/Geomatics was responsible for the translation. Both of the authors of this paper are members of this committee. During the translation work, the members of the committee discussed the use of ISO 19152 in the Czech Republic. The conclusions reached incorporated a first step wherein a country profile of the Czech Republic based on an LADM should be constructed, after which the country profile should be tested against the LADM concept.

Feedback from professionals (i.e., surveyors) who utilize the cadastre on a day-to-day basis represented an important incentive in this process. This positive feedback was often obtained during presentations regarding a 3D cadastre that were delivered at national events and conferences. For example, the Czech Union of Surveyors and Cartographers (a member of International Federation of Surveyors (FIG)) declared an interest in a 3D cadastre and thereafter demonstrated the need for a 3D cadastre on the grounds of several applications (e.g., the registration of complex buildings, underground constructions, etc.).

Ultimately, the new Civil Code (Act No. 89/2012 Coll.) was enacted within the Czech Republic in 2014. This Act addresses several aspects of the 3D cadastre. The new Civil Code explicitly considers the 3D space both above and below the parcel as a part of the land. This includes the space above and below the surface, buildings established on the land and other facilities (excluding temporary buildings), including what is embedded in the land or fixed in the walls. Furthermore, according to the new Civil Code, the land incorporates real estate and underground construction characterized by separate special-purpose uses, as well as their corresponding property rights. In practice, many underground constructions are not registered in the cadastre. Currently, underground constructions are only registered in the case when some part(s) of the construction is located above ground. Figure 1 illustrates an example of such an underground construction, and the manner in which this construction is displayed on a map is illustrated in Figure 2. Furthermore, even if an underground construction is not considered real estate, it is still a part of the land if it affects (i.e., is located beneath) the parcel. For example, if a building with an underground cellar is standing on Parcel A and the underground cellar is partially located beneath the neighboring Parcel B, then the entire underground cellar is a component of the building standing atop Parcel A (and, according to the superficies solo cedit principle, this building belongs to Parcel A). Consequently, a landowner must also accept the use of space over or under the land if the parcel conforms to these laws and if the owner does not have sufficient cause to oppose it.

Figure 1. (**a**) Visualization of the underground construction of the archeological park in Pavlov, Czech Republic [4]; (**b**) entrance to the archeological park in Pavlov, Czech Republic (photo: Institute of Archeology of the Czech Academy of Sciences (CAS), Brno).

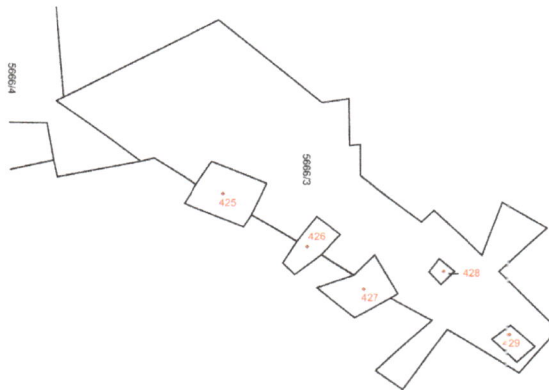

Figure 2. 2D visualization of the boundaries of underground construction for the archeological park in Pavlov. Every component of the construction located above ground has to exist on a separate building parcel (here, a total of five building parcels with bold red numbers are shown, after [4]).

The new Civil Code encompasses numerous provisions from other acts that were recently repealed, e.g., the Flat Ownership Act, the Act on Association of Persons, etc. It re-introduces the former Czech legal terminology, which was gradually abandoned by the Civil Codes of 1950 and 1964. The Property Law component regulates the tenure, possession, ownership and co-ownership, encumbrance, lien and heirship of real estate property. Concordant with the adoption of the New Civil Code, a new Cadastral Act (Law Number 256/2013 Coll., the "New Cadastral Act") was enacted as a basic legal cornerstone of the cadastre of real estate of the Czech Republic. The Cadastral Law defines a parcel as a piece of land that is projected onto the horizontal plane (a 2D cadastral map). The Law does not explicitly state that the 3D space above and below the parcel is a component of the parcel (i.e., it does not constitute a 3D parcel). However, as has been previously discussed [5], though parcels are represented in 2D, someone with a right to a parcel has always been entitled to the 3D space. That is, the right of ownership of a parcel relates to the 3D space, is not solely limited to the flat parcel defined in 2D absent height or depth and can therefore be used by the owner.

Recently, several country profiles based on an LADM have been proposed [6–9]. This paper presents the first version of a Czech country profile [10]. Considering the proposal for an LADM-based country profile, the ambitions of this study are (1) to ensure that the LADM-based country profile reflects the current cadastral registration and the corresponding legal requirements with a possible extension into 3D and (2) to determine the compatibility between the proposed Czech country profile and the LADM. This paper describes a feasible "show case," which could be used as a type of guideline for others conducting or planning a similar application.

2. GeoInfoStrategy and Selected Use Cases Dealing with a 3D Cadastre

The vision of the GeoInfoStrategy is that, in 2020, the Czech Republic is a society educated in the effective use of spatial information. To fulfil this vision, it is necessary to ensure that spatial information and services will be utilized in every aspect of public life. The GeoInfoStrategy represents a conceptual plan that is closely related to other strategic documents of public administration and eGovernment and defines the principles and strategic ambitions for the effective use of spatial information in public administration. The approved GeoInfoStrategy contains a description of several case studies of applications that may employ a 3D cadastre, such as the administration of networks of technical infrastructure or the creation of 3D building models for noise mapping [11].

Considering networks of technical infrastructure, the Czech Republic should establish a register of passive infrastructure that would be intended to contain information regarding the 3D location (using absolute heights) of networks, as well as network owners and administrators. In the approved GeoInfoStrategy Action Plan, the use of ISO 19152 (LADM) is recommended as an input document in order to create a feasibility study of this register. An LADM distinguishes between the physical representation of a spatial object (outside the scope of the LADM) and a legal registration of the space (within the scope of the LADM) required by the physical objects. Such legal spaces could then be registered within the cadastre. Currently, the (legal) information concerning the networks of technical infrastructure is not registered in the cadastre. However, numerous issues could be addressed if this information existed within the cadastre, such as the routes of cable lines beneath certain parcels of real estate, and so on. This is pertinent because many parcel owners are unaware whether cables or pipelines are located beneath their parcel.

This legal registration should then provide a clear overview of the property rights involved, including the rights regarding the network on the one hand (e.g., ownership) and the rights regarding the land established for the benefit of a network on the other hand (e.g., easement). An approach to register the legal space of utility networks could ensure that the geometry of physical utility networks is located within the database of utility companies (or in the established register of passive infrastructure), which can then serve as a dynamic reference within the cadastre. The legal objects for utility networks can then be generated in a controlled (regulated) manner from the 3D descriptions of the physical objects. Because of the permanent nature of the connection, the legal registration can be better maintained. This fits well within the LADM [12]. ISO 19152 offers a subclass LA_LegalSpaceUtilityNetwork (of the class LA_SpatialUnit) in order to support the registration of information (legal spaces) concerning the utilities together with cadastral data. A previous study [12] examined the characteristics of utility networks as 4D (3D space + time) objects and showed that an approach utilizing the 3D space, as well as the separate temporal attributes, is a very promising solution with which to maintain and record temporal changes in utility networks.

Another case study within the GeoInfoStrategy considers the creation of models of 3D buildings primarily for the purpose of strategic noise mapping, the production of which should be accomplished using preexisting spatial data sources (e.g., 2D digital cadastral maps and laser scans). The laser scans could also be used for updating existing 3D models. In this regard, modifications using point clouds can only be detected for visible components (i.e., above ground) of objects or boundaries [13]. The requested level of detail (LOD) for this case study of 3D buildings for strategic noise mapping is LOD1 according to the specifications of CityGML [14]. Furthermore, using the aforementioned spatial data sources, the buildings at LOD2 could also be created. Such models can then serve 3D cadastral purposes [15,16]. If the buildings are originally modeled in 3D, then 3D spatial units can be stored in the cadastral database according to the LADM concepts [17]. However, while CityGML models physical infrastructure, an LADM works with legal spaces, since it is a legal model with support for 3D objects. Within this standard, a spatial unit entity (LA_SpatialUnit) provides various spatial representations of ownership interests defined within the jurisdiction, e.g., areal and volumetric 3D objects. The CityGML standard can be expanded with legal objects and ownership attributes by leveraging its Application Domain Extension (ADE) capability [18]. Previous studies [19–22] have thoroughly examined issues regarding how the LADM conceptual model, and more importantly the representation of legal spaces, can be mapped to and encoded as a CityGML ADE [19–22].

An LADM has been developed heretofore in order to provide an international framework for the most effective development of a 3D cadastre. While this generic framework encompasses a wide range of eventualities, it does not stipulate the requisite data format. Other existing sources from which data could be obtained are 3D building information models (BIMs). For instance, a BIM geometry could be reprocessed for 3D cadastral parcels [23]. The open data model used for BIMs is the Industry Foundation Classes (IFC) standard, which provides a hierarchical spatial structure in order to store building information. The employment of the IFC's virtual spaces and zones, designed for energy analysis, could be a manner in which virtual cadastral legal spaces are defined within BIMs [23].

In the current version of GeoInfoStrategy, there is no explicitly-stated necessity for 3D models of buildings for cadastral purposes. However, this situation could change in the near future, especially since newly-built apartment buildings and other construction projects in the Czech Republic are often too complex for 2D registration and because the new buildings are modeled using BIM. If the buildings are modeled in more detail, including with respect to the interior, then such 3D models can serve as a basis for the modeling of legal spaces of apartments and building units [24]. To accomplish this, the LODs that are to be accepted within the national cadastre must be described, and the manner in which they are to be captured and registered must be ascertained. An important attribute of this plan to consider is that it is not entirely necessary to produce a 3D model for every building registered within the cadastre; it is only necessary for buildings wherein the current 2D registration is insufficient [25].

3. The Czech Profile Based on the LADM

The cadastre of the Czech Republic was designed as a multipurpose land information system that was foremost intended for legal and fiscal purposes, as well as for land management and the provision of a database for other information systems. The cadastre covers the entire territory of the Czech Republic. Since 1993, the information system has integrated the former Land Cadastre (technical instrument) and Land Registry (legal instrument). The sole authority responsible for the Czech cadastre is the Czech Office for Surveying, Mapping and Cadastre.

The LADM specifies a conceptual model. To test the compliance between the LADM and the national data cadastre model, it is necessary that an application scheme (i.e., a country profile) is developed. To accomplish this, reverse engineering was applied, i.e., the first step was to explore the physical model and thereafter create the logical model. Subsequently, the conceptual model of the cadastre was created based on the logical model, after which a mapping of the Czech tables (classes) onto the LADM classes was applied where possible. In some cases, the Czech tables (classes) could be inherited from the LADM classes. It was also necessary to consider the code lists and to compare them with the ones stated in the LADM.

In the Czech Republic, there is a compulsory title registration. The core of the Czech LADM profile consists of the following four fundamental classes derived from the basic classes of the LADM:

- Class CZ_Party (with LA_Party as a superclass). An instance of this class is a party.
- Class CZ_RRR (with LA_RRR as a superclass). An instance of a subclass of CZ_RRR is a right (CZ_Right) or a restriction (CZ_Restriction).
- Class CZ_BAUnit (with LA_BAUnit as a superclass). An instance of this class is a basic administrative unit.
- Class CZ_SpatialUnit (with LA_SpatialUnit as a superclass). An instance of this class is a spatial unit.

The relationships among these four main classes are illustrated in Figure 3.

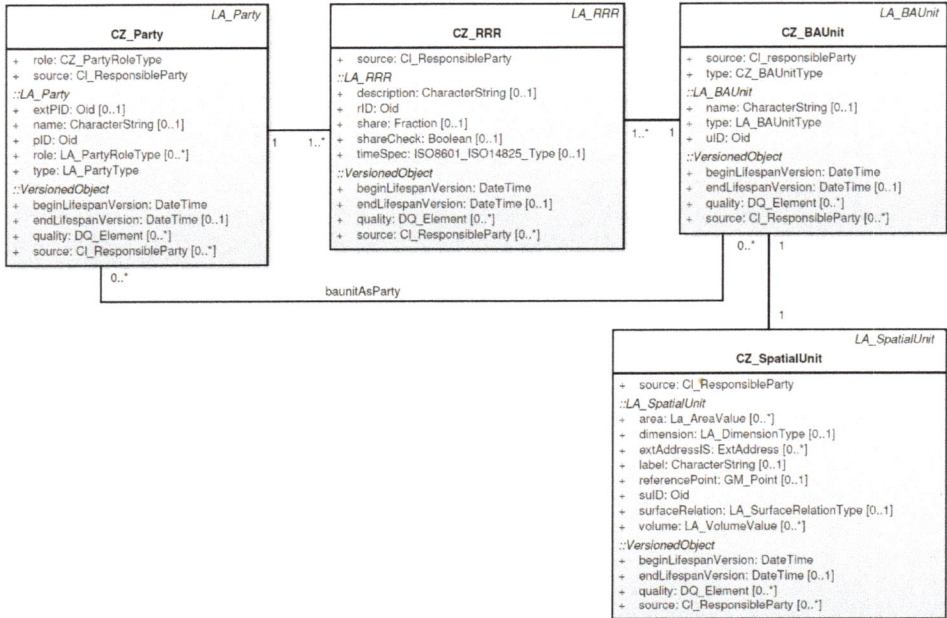

Figure 3. The four fundamental classes of the Czech LADM-based profile.

Two or more parties can be grouped into a group party (CZ_GroupParty). In the Czech profile, the following three types of group parties are distinguished: family, association and baunitGroup. These values are enumerated within the CZ_GroupPartyType code list. There is also an association class CZ_PartyMember between CZ_Party and CZ_GroupParty. An instance of the class CZ_PartyMember is a party member. The sum of the shares of the group party members has to equal unity, wherein an attribute share (of the class CZ_PartyMember) is utilized in order to model the size of the share.

The class CZ_RRR consists of the subclasses CZ_Right and CZ_Restriction. As depicted in Figure 4, the class CZ_Right has a single subclass, CZ_RightOfBuilding. Based on this new instrument within Czech law, it will be possible to construct a building on land belonging to a third party. The right of building is a temporary right (with a 99 year maximum) established by an agreement between the land owner and the developer. From a legal point of view, the right of building as a whole is considered to be real estate and is subject to registration within the cadastre of real estate. The right of building can be subject to a transfer, mortgage, heritage or easement and is independent of the existence of the structure and can therefore be established even if the construction procedure has not yet begun. After expiry of the right of building, it can be theoretically prolonged, otherwise the building becomes a part of the land. The mandatory attribute *validTo* (of the class CZ_RightOfBuilding) is used for storing this information until the right of building is valid.

The class CZ_Restriction has a single subclass, CZ_Mortgage. The Czech cadastre was open to the public from the very beginning. This principle of openness proved to be very useful after mortgages were introduced in the cadastre. Figure 5 shows the principle of mortgage (modeled as CZ_Mortgage). Against the LA_Mortgage class, the attribute *type* (of the class CZ_Mortgage) is mandatory.

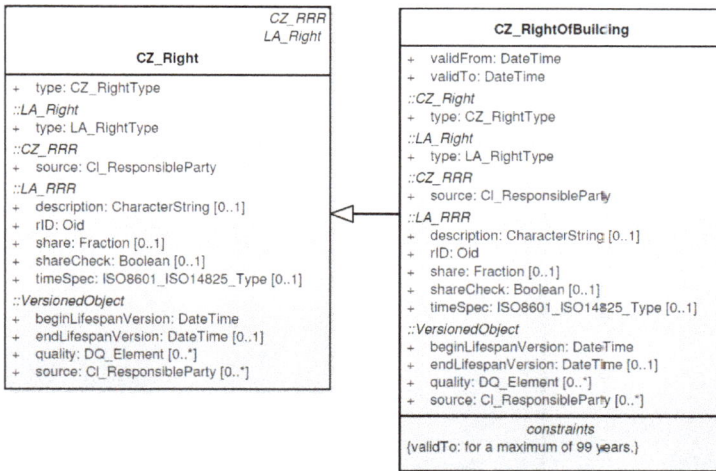

Figure 4. Right of building. The right of building can be established for a maximum of 99 years.

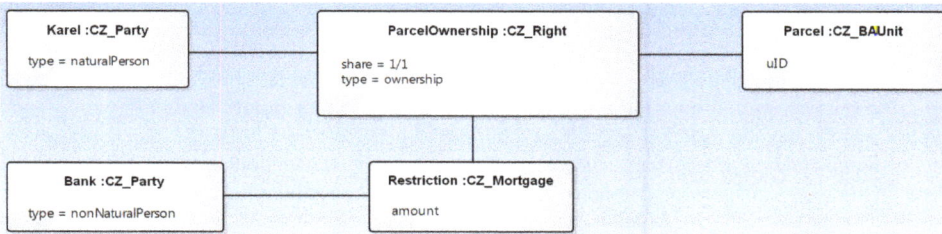

Figure 5. An example of a mortgage.

In the Czech LADM-based profile, administrative and spatial sources are modeled. For this purpose, there is an abstract class CZ_Source and its two subclasses CZ_AdministrativeSource and CZ_SpatialSource. The different real estate rights are based in certain transaction documents. This fact is represented by the association between CZ_RRR and CZ_AdministrativeSource. Furthermore, the mandatory attribute *changeOfRRR* (of the CZ_Source class) indicates whether the source document alters the registered right (i.e., restriction).

Regarding the relationship between the classes CZ_BAUnit and CZ_SpatialUnit, one instance of CZ_BAUnit (with the same RRRs attached) is associated with exactly one instance of CZ_SpatialUnit. CZ_BAUnit can also represent a party (where baunit represents a party). This approach is employed to model the supplementary co-ownership, which is a special type of ownership in that owners of their own property are allowed to utilize that property only in association with some shared property. The ownership of this shared property is known as supplementary co-ownership (according to the new Civil Code). Ownership of property as supplementary co-ownership is inextricably linked with the ownership of properties for whose use the property with supplementary co-ownership serves. The separate transfer of such property is not possible. One example includes a serving parcel as demonstrated in Figure 6, wherein the serving parcel provides access to neighboring parcels, and the serving parcel is not public, but is instead commonly owned by the neighboring parcels (modeled as baunit as party).

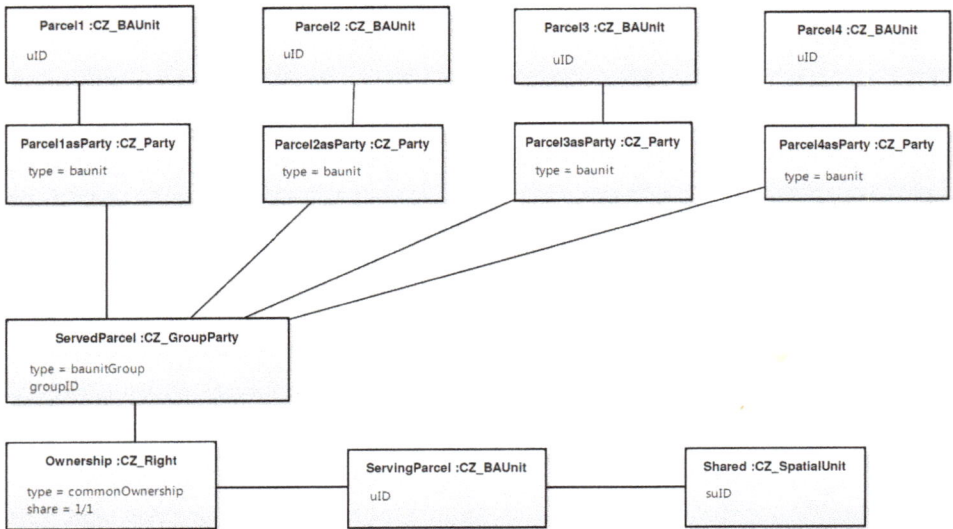

Figure 6. An example of a serving parcel owned by neighbors.

3.1. Ownership of a Building Unit

The ownership of a building unit is one of the most common situations within the cadastre. Figure 7 illustrates the principle of owning a building unit. The new Civil Code also regulates the ownership of building units, which was previously contained under a separate act (the Unit Ownership Act). To model the ownership of a building unit, the CZ_LegalSpaceBuildingUnit class was proposed, which is a subclass of the LA_LegalSpaceBuildingUnit class. A building unit, as defined by the former Unit Ownership Act, remains a separate piece of real estate and does not constitute a part of the land even in the case wherein the owner of the building unit and the owner of the land parcel is the same person. The former Unit Ownership Act and the new Civil Code provide a different view on the subject of the right of ownership to that building unit. According to the Unit Ownership Act, the right of ownership (de iure) consists of the following three separate components: a unit, a share of common parts of the building and a share of the land parcel. The new Civil Code considers the ownership of the unit and the share of common parts of the building as one inextricably linked element. Regarding the land parcel (from the new Civil Code point of view), in the case that the owner of both the building (containing the unit) and the land parcel is the same person, the land parcel then constitutes a part of the common parts of the building. The size of the share is determined by the size of the building unit in relation to the total area of all of the units. However, this is not the only method with which to compute the size of the share. The new Civil Code also makes it possible to consider, for example, the height of the ceiling. The greater the height of the ceiling, the more space that the owner would own, and thus, the share would be greater.

The code list CZ_BuildingUnitType helps to distinguish between types of building units. In the Czech cadastre, there are still many registered building units that are defined by the Unit Ownership Act. Thus, the CZ_BuildingUnitType code list also comprises the value of unitDefinedAccordingToTheUnitOwnershipAct, and the second possible value involved is unitDefinedAccordingToTheCivilCode.

The digital cadastral map displays neither the flat structure nor the spatial distribution of use rights. Rather, it only displays a schematic drawing illustrating the floor plans and a textual description of the flats (as required by legislation).

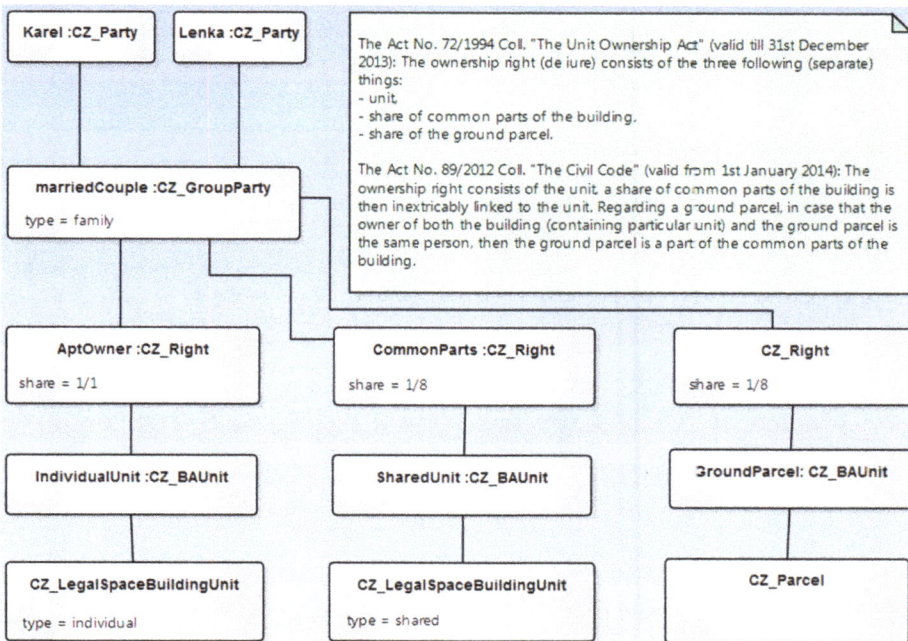

Figure 7. Ownership of a building unit. In principle, the owner has a right to the individual unit, a share of common components and a land parcel (but the owner of the unit and the owner of the land parcel could be different people).

3.2. Building as a Component of the Land

After 1 January 2014, any building under ownership became a component of the land upon which it stands (i.e., Czech real estate law returned to the principle wherein structures are a part of the land upon which they are built, which is a superficies solo cedit principle). Buildings constructed on land (except for temporary buildings, utility lines and some other exemptions) are no longer the objects of law and only comprise a component of the land.

If the land owner and the building owner were two different persons at this time (1 January 2014), the building remained as real estate status, but the land owner was provided a pre-emptive right to the building and the building owner provided a pre-emptive right to the land. The building then became a component of the land when the building and the land were adjoined within the hands of a single owner. The building would not become part of the land if the building or the land was encumbered by a right in rem (i.e., a right associated with a property that is not based on any personal relationship). The CZ_Building class (as a subclass of the CZ_SpatialUnit) serves to model the buildings prior to the effectiveness of the new Civil Code, which means that such buildings are considered as separate entities (real estates) and not parts of the land. After 1 January 2014, when the new Civil Code was enacted, the buildings were not (due to the superficies solo cedit principle) further registered as separate real estates. Newly constructed buildings were not registered in the case when the owner of the land parcel and the owner of the building was the same person. However, a building was registered in the case when the owner of the parcel and the owner of the building were two different persons. For example, this situation could occur when a newly-constructed building is established with the right of building. The owner of the land parcel is then different from the owner of the building.

Figure 8 gives an overview of the legal component of the Czech country profile.

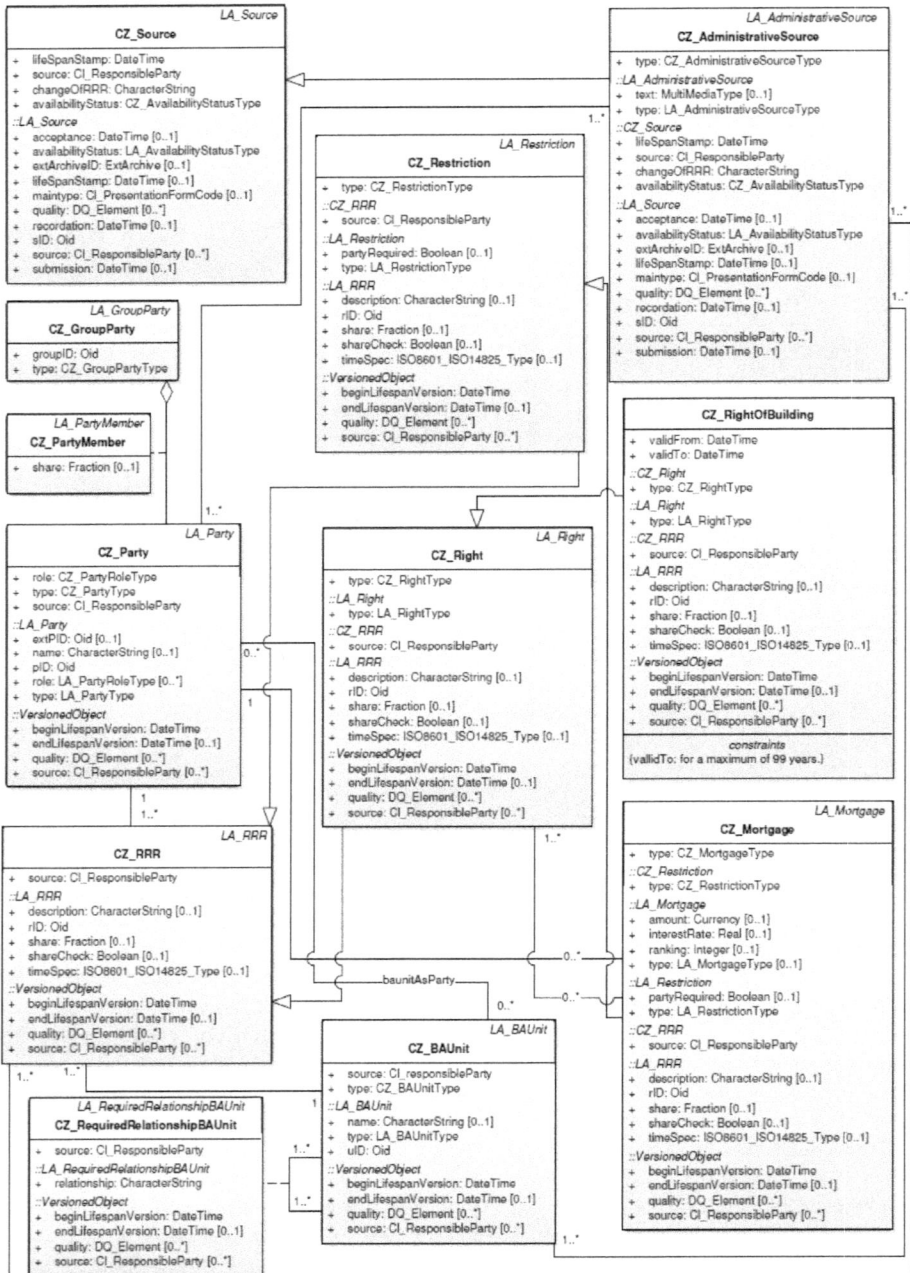

Figure 8. Overview of the legal component of the Czech country profile.

3.3. Spatial Units

Spatial units are modeled using the CZ_SpatialUnit class and its three subclasses: CZ_Parcel, CZ_LegalSpaceBuildingUnits and CZ_Building. Regarding the geometry of spatial units, the Czech LADM country profile uses 2D topological-based spatial units. All topological boundaries (which are instances of the CZ_BoundaryFaceString class) are applied once in the positive and once in the negative direction, except in the event that the boundary is situated on the edge of the domain, in which case either the positive or the negative direction is used once (and the other is not used). All associated boundaries together form one or more non-intersecting rings that define at least one outer ring (with a counter-clockwise orientation) and optionally one or more inner rings (with a specified orientation). CZ_BoundaryFaceString is associated with the class CZ_Point and the class CZ_SpatialSource to document the origin of the geometry.

The Czech LADM profile also contains the CZ_Level class. Level is defined here as a set of spatial units characterized by geometric and topologic coherence. A spatial unit cannot be associated with more than one level. The different types of spatial units are indicated by the value of the structure attribute (i.e., of the CZ_Level class). At the conceptual level, there is a base level (Level 1) with topologically-defined spatial units (structure: CZ_StructureType = topological) and a Level 2 with polygon-based spatial units (structure: CZ_StructureType = polygon) that represent the protection zones.

The spatial units can be grouped into spatial unit groups (wherein the corresponding class for spatial unit groups is CZ_SpatialUnitGroup). An example of a spatial unit group is a cadastral unit consisting of parcels. The boundaries of parcels and territories (similar to a cadastral unit) are based on the topology and are hierarchically structured. In the case that a boundary of a parcel is identical to the boundary of another territory, the boundary shall be classified according to the following priorities (where the top values have the highest priority):

- state border,
- regional border,
- district border,
- municipality border,
- boundary of cadastral unit,
- boundary of a parcel.

These values are also included in the code list CZ_TypeOfBoundary (attribute *boundary* of the class CZ_BoundaryFaceString). Figure 9 gives an overview of the spatial component of the Czech country profile.

Table 1 contains the proposed CZ classes and the corresponding LADM classes. The relationships between all of the classes are illustrated in Figure 10. Table 2 provides an overview of the newly-defined CZ subclasses (CZ_Parcel, CZ_Building and CZ_RightOfBuilding). The CZ attributes and their corresponding CZ code lists are displayed in Table 3.

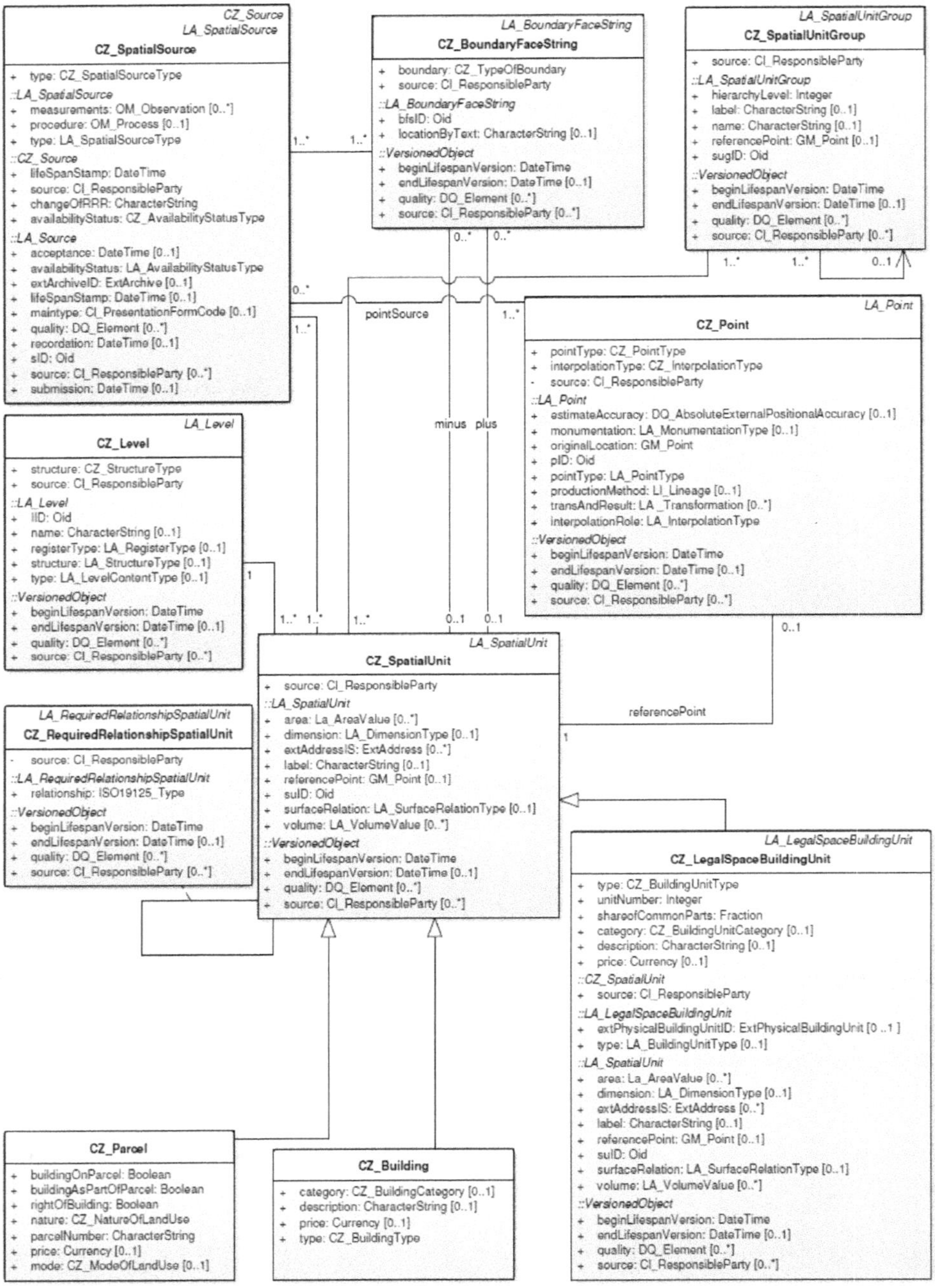

Figure 9. Overview of the spatial component of the Czech country profile.

Table 1. The proposed CZ classes for the Czech cadastral profile and the corresponding LADM classes. The CZ classes with their attributes are illustrated in Figures 8 and 9.

LADM Package	CZ Class	Corresponding LADM Class
Party	CZ_Party	LA_Party
	CZ_GroupParty	LA_GroupParty
	CZ_PartyMember	LA_PartyMember
Administrative	CZ_RRR	LA_RRR
	CZ_Right	LA_Right
	CZ_Restriction	LA_Restriction
	CZ_Mortgage	LA_Mortgage
	CZ_BAUnit	LA_BAUnit
	CZ_RequiredRelationshipBAUnit	LA_RequiredRelationshipBAUnit
	CZ_Source	LA_Source
	CZ_AdministrativeSource	LA_AdministrativeSource
Spatial Unit	CZ_SpatialUnit	LA_SpatialUnit
	CZ_LegalSpaceBuildingUnit	LA_LegalSpaceBuildingUnit
	CZ_SpatialUnitGroup	LA_SpatialUnitGroup
	CZ_Level	LA_Level
	CZ_RequiredRelationshipSpatialUnit	LA_RequiredRelationshipSpatialUnit
Surveying andRepresentation	CZ_Point	LA_Point
	CZ_SpatialSource	LA_SpatialSource
	CZ_BoundaryFaceString	LA_BoundaryFaceString

Table 2. The newly-added CZ subclasses for the Czech cadastral profile.

CZ Class	New CZ Subclass
CZ_SpatialUnit	CZ_Parcel
	CZ_Building
CZ_Right	CZ_RightOfBuilding

Table 3. The CZ attributes and corresponding CZ code lists.

CZ Class	CZ Attribute	CZ Code List
CZ_Party	type	CZ_PartyType
	role	CZ_PartyRoleType
CZ_GroupParty	type	CZ_GroupPartyType
CZ_Right	type	CZ_RightType
CZ_Restriction	type	CZ_RestrictionType
CZ_Mortgage	type	CZ_MortgageType
CZ_BAUnit	type	CZ_BAUnitType
CZ_Source	availabilityStatus	CZ_AvalabilityStatusType
CZ_AdministrativeSource	type	CZ_AdministrativeSourceType
CZ_LegalSpaceBuildingUnit	type	CZ_BuildingUnitType
	category	CZ_BuildingUnitCategory
CZ_Parcel	nature	CZ_NatureOfLandUse
	mode	CZ_ModeOfLandUse
CZ_Building	type	CZ_BuildingType
	category	CZ_BuildingCategory
CZ_Level	structure	CZ_StructureType
CZ_Point	pointType	CZ_PointType
	interpolationType	CZ_InterpolationType
CZ_SpatialSourceType	type	CZ_SpatialSourceType
CZ_BoundaryFaceString	boundary	CZ_TypeOfBoundary

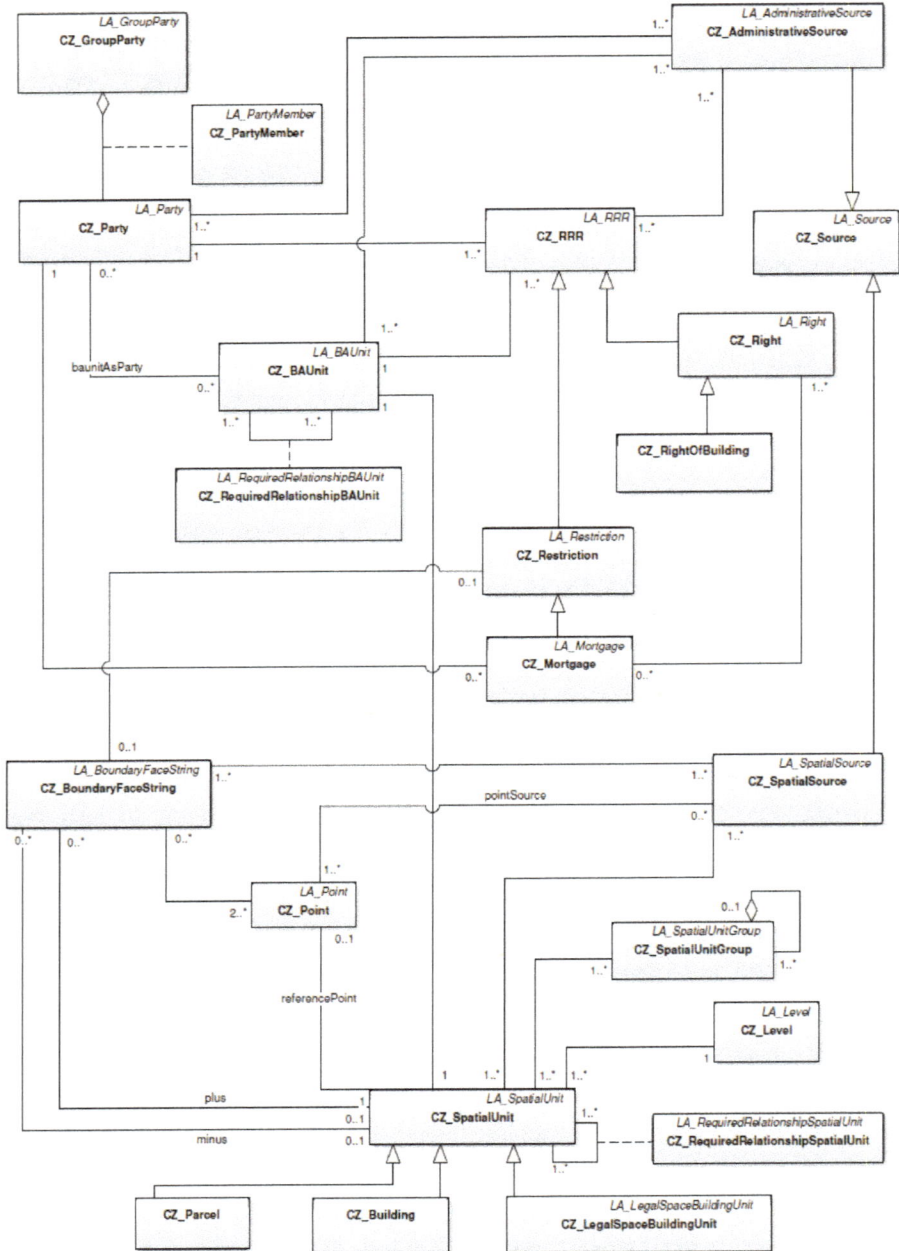

Figure 10. The Czech LADM-based country profile. Due to the complexity of the model, only the names of classes are mentioned here. The classes with their attributes are illustrated in Figure 8 (legal component) and Figure 9 (spatial component).

4. Conformity of the Czech Profile with ISO 19152

The LADM consists of three packages and one sub-package, each of which are specified with a conformance test in Annex A—Abstract Test Suite [1]. The following three conformance levels are specified per (sub)package: Level 1 (low level), Level 2 (medium level) and Level 3 (high level). Level 1 tests include the most basic classes per package, and Level 2 further includes the more common classes. Level 3 includes all of the classes. Any LADM-based profile claiming conformance to the ISO 19152 standard must satisfy the requirements of Annex A.

The possible manners in which to test this conformity are as follows:

(1) show an inherited structure between the LADM and the tested model (elements) or

(2) show a mapping of elements between the LADM and the tested model.

Figure 11 provides an overview of each package to check for LADM compliancy. The proposed country profile for the Czech Republic contains all of the necessary classes for conformance Level 2 (medium level) and meets all of the required dependencies as mentioned in Figure 11.

LADM package	LADM class	CI[a]	Dependencies
-	VersionedObject	1	
	LA_Source	1	Oid, (as a minimum one of the specializations must be implemented [LA_AdministrativeSource or LA_SpatialSource]), LA_AvailabilityStatusType
Party Package			Exist only if Administrative Package is implemented
	LA_Party	1	VersionedObject, Oid, LA_PartyType
	LA_GroupParty	2	Oid, LA_Party, LA_GroupPartyType
	LA_PartyMember	2	VersionedObject, LA_Party, LA_GroupParty
Administrative Package			Exist only if Party Package is implemented
	LA_RRR	1	VersionedObject, Oid, LA_Party, LA_BAUnit, LA_Right (as a minimum, this specialization shall be implemented), LA_AdministrativeSource
	LA_Right	1	LA_RRR, LA_RightType
	LA_Restriction	2	LA_RRR, LA_RestrictionType
	LA_Responsibility	3	LA_RRR, LA_ResponsibilityType
	LA_BAUnit	1	VersionedObject, Oid, LA_RRR, LA_BAUnitType
	LA_Mortgage	2	LA_Restriction
	LA_AdministrativeSource	1	LA_Source, LA_Party, LA_AdministrativeSourceType, LA_AvailabilityStatusType
	LA_RequiredRelationshipBAUnit	3	VersionedObject, LA_BAUnit
Spatial Unit Package			
	LA_SpatialUnit	1	VersionedObject, Oid
	LA_SpatialUnitGroup	2	VersionedObject, Oid, LA_SpatialUnit
	LA_LegalSpaceBuildingUnit	3	LA_SpatialUnit
	LA_LegalSpaceUtilityNetwork	3	LA_SpatialUnit
	LA_Level	2	VersionedObject, Oid
	LA_RequiredRelationshipSpatialUnit	3	VersionedObject, LA_SpatialUnit
Surveying and Representation Subpackage			
	LA_Point	2	VersionedObject, Oid, LA_SpatialSource, LA_PointType, LA_InterpolationType
	LA_SpatialSource	2	LA_Source, LA_Point, LA_Party, LA_SpatialSourceType
	LA_BoundaryFaceString	2	VersionedObject, Oid, LA_Point (if using geometry)
	LA_BoundaryFace	3	VersionedObject, Oid, LA_Point (if using geometry)

[a] CI = Conformance level

Figure 11. The LADM conformance requirements table [1]. The classes for conformance Level 2 are highlighted in red rectangles. The Czech profile also contains some Level 3 classes (highlighted in blue rectangles).

Furthermore, all necessary code lists (for Level 2) are designed within the Czech profile as depicted in Figure 12.

Code lists for Party Package

«CodeList»
CZ_PartyType
+ baunit
+ group
+ naturalPerson
+ nonNaturalPerson

«CodeList»
CZ_GroupPartyType
+ association
+ baunitGroup
+ family

«CodeList»
CZ_PartyRoleType
+ communityPropertyOfSpouses
+ citizen
+ stateCzechRepublic
+ organisationalEntityOfTheState
+ entityOfNonNaturalPerson
+ landFund
+ fundOfChildrenAndYouth
+ municipalityDistrict
+ otherNonNaturalPerson
+ foreignCountry
+ entityOfRegion
+ commonFund
+ investmentCompany
+ enforcementAuthority
+ nonNaturalPersonSetUpByCzechRepublic
+ entityOfStateEntity
+ trustFund
+ assocationOfHeirs
+ investmentCompartment
+ trustAdministrator

Code lists for Administrative Package

«CodeList»
CZ_RightType
+ ownership
+ pre-emptive
+ rightToManageStateProperty
+ supplementaryCoownership
+ administrationOfTrustFund
+ reservationOfOwnershipRight
+ reservationOfTheRightToPurchaseBack
+ reservationOfTheRightOfSaleBack
+ reservationOfTheRightOfBetterPurchaser
+ trialPurchaseArrangement
+ lease
+ tenure
+ surrenderTheRightForDamageCompensationOnTheEstate
+ distributionOfRightToRealEstateIntoSingleOwnershipRightsToUnits

«CodeList»
CZ_MortgageType
+ administrative
+ contract
+ executors
+ future
+ judicials
+ legal
+ mortgage
+ submortgage

«CodeList»
CZ_AdministrativeSourceType
+ agreement
+ agreementAboutCreationOfMortgage
+ courtJudgment
+ decisionAboutExpropriation
+ contractOnTransferOfRealEstate
+ purchaseContract
+ contractOfDonation
+ notarialRecord
+ distressWarrant
+ buildingPermit
+ swornStatement

«CodeList»
CZ_AvailabilityStatusType
+ docAvailable

«CodeList»
CZ_BAUnitType
+ apartment
+ building
+ parcel

«CodeList»
CZ_RestrictionType
+ easement
+ futurePossibilityOfUsingThePropertyAfterItsTransfer
+ prohibitionOfAlienationOrEncumbrance

Code lists for Spatial Unit Package

«CodeList»
CZ_BuildingUnitCategory
+ apartment
+ atelier
+ garage
+ otherNonResidentialPremise
+ unitUnderConstruction
+ workshop
+ groupOfApartments
+ groupOfNonResidentialPremises
+ groupOfApartmentsAndNonResidentialPremises

«CodeList»
CZ_BuildingUnitType
+ unitDefinedAccordingToTheUnitOwnershipAct
+ unitDefinedAccordingToTheCivilCode

«CodeList»
CZ_BuildingCategory
+ factoryUnit
+ farmhouse
+ blockOfFlats
+ familyHouse
+ garage
+ glasshouse
+ dam
+ weir

«CodeList»
CZ_NatureOfLandUse
+ arableLand
+ bodyOfWater
+ builtUpArea
+ forestLot
+ garden
+ hopGarden
+ orchard
+ permanentGrassland
+ vineyard

«CodeList»
CZ_ModeOfLandUse
+ cemetery
+ dump
+ forest
+ glasshouse
+ highway
+ infertileSoil
+ nurserySchool
+ pond
+ road
+ sportsAndRecreationGrounds

«CodeList»
CZ_StructureType
+ polygon
+ topological

«CodeList»
CZ_BuildingType
+ building
+ buildingUnderConstruction
+ hydraulicStructure
+ buildingWithUnitsUnderConstruction

Code lists for Surveying and Representation Package

«CodeList»
CZ_TypeOfBoundary
+ boundaryOfCadastreUnit
+ boundaryOfParcel
+ building
+ districtBorder
+ easement
+ municipalityBorder
+ protectedZones
+ regionalBorder
+ stateBorder

«CodeList»
CZ_SpatialSourceType
+ surveySketch
+ documentationOfDetailedSurveyOfChanges
+ rasterMap

«CodeList»
CZ_InterpolationType
+ end
+ isolated
+ mid
+ midArc
+ start

«CodeList»
CZ_PointType
+ control
+ noSource
+ source

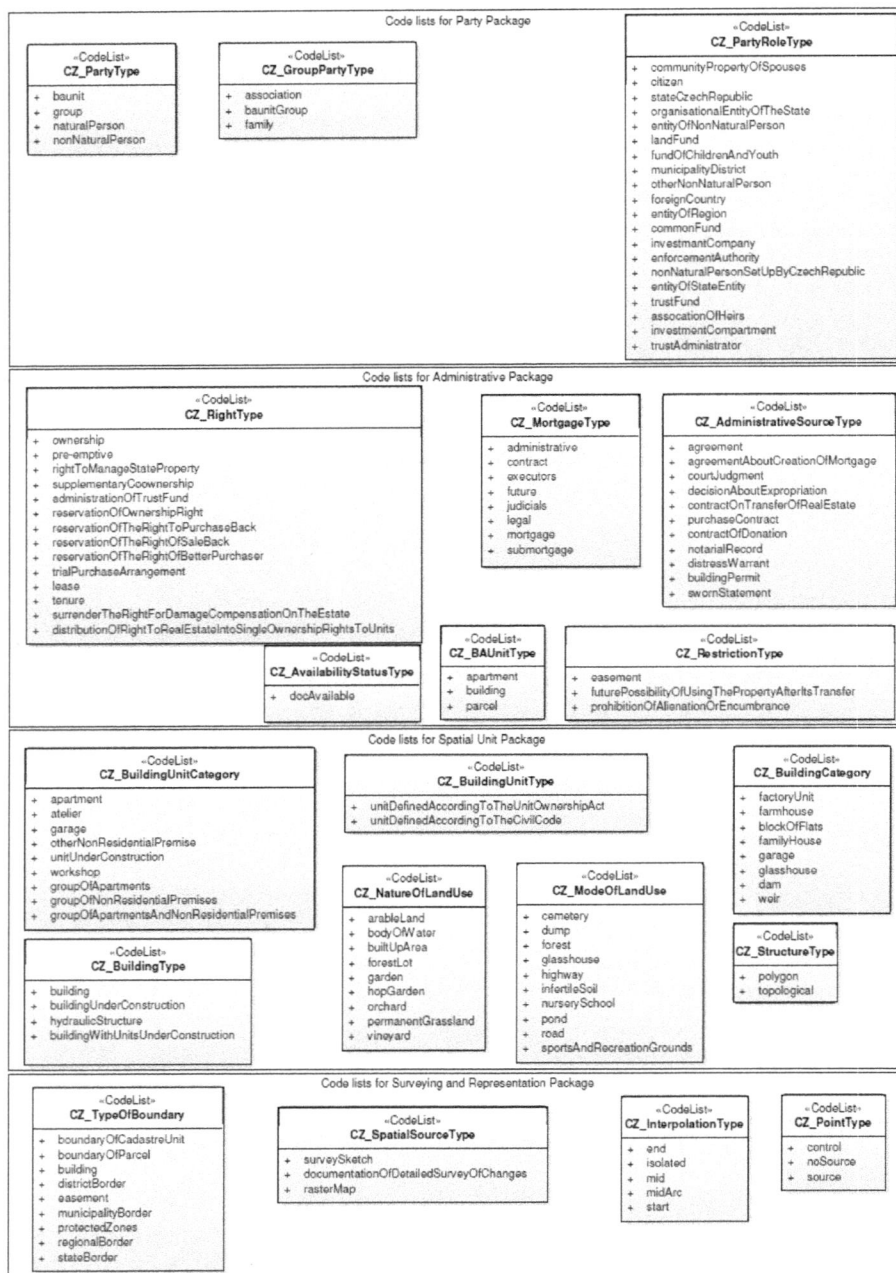

Figure 12. The code lists used in the Czech LADM-based country profile. In some cases, the presented code list contains only a subset of all of the possible values. For example, in reality, the LA_RightType code list contains more than 160 types of different types of right.

It should be noted that the Czech LADM-based profile also contains some Level 3 classes, e.g., the CZ_RequiredRelationshipSpatialUnit class (a subclass of LA_RequiredRelationshipSpatialUnit). Due to historical reasons, in some areas, the geometry of the spatial units is not sufficiently accurate, and therefore, there is a need for explicit spatial relationships. The union of the geometry of all parcels should cover the entire area. Neighboring parcels must not overlap, and no gaps are allowed between adjacent parcels. Furthermore, the boundary of a building must be inside the boundary of the parcel upon which the building is located, and the boundary of a building must not touch the boundary of the parcel.

Another Level 3 class contained within the Czech LADM-based profile is CZ_RequiredRelationshipBAUnit (a subclass of LA_RequiredRelationshipBAUnit). This class serves to model relationships (of a legal, temporal or spatial nature) between instances of LA_BAUnit. For example, only one building can be constructed on a parcel; however, that building can be built on one or many parcels. The next Level 3 class included in the Czech LADM-based profile is the LA_LegalSpaceBuildingUnit class, which serves to model building units. The remaining LADM Level 3 classes are not contained within the Czech LADM-based profile, e.g., the LADM class LA_Responsibility. Currently, there are no responsibilities registered, and therefore, these responsibilities are not included in the proposed country profile.

5. Conclusions

3D geospatial data are becoming an important attribute of geographical information systems. In the Czech Republic public administration spectrum, such 3D geospatial data are primarily utilized at the municipality level and mostly cover the area of a particular city. However, the Strategy for the Development of the Infrastructure for Spatial Information in the Czech Republic to 2020 was recently approved. This represents a government initiative that emphasizes the creation of a National Set of Spatial Objects, which is defined as the source of both guaranteed and reference 3D geographic data at the highest possible level of detail for selected objects covering the entire territory of the Czech Republic. This can also be a potential source of data for a 3D cadastre, especially regarding the legal components of buildings. Furthermore, the GeoInfoStrategy Action Plan recommends the use of the ISO 19152 standard as one of the source documents for obtaining the selected measurements. For example, ISO 19152 should be used for buildings in the register of passive infrastructure.

To enable a comparison between the ISO 19152 standard and the current cadastre data model, an LADM-based country profile was proposed. The profile considers the requirements of the new Civil Code and related legislation. Despite the fact that the new Civil Code explicitly considers the space both above and below the ground as a part of the land, the Czech cadastre still retains the 2D paradigm. The Czech LADM-based country profile contains all of the classes and code lists required for Level 2 compliance. Moreover, it also contains some Level 3 classes. The Czech cadastre is based on the registration of 2D parcels, and therefore (for example), the Level 3 class LA_BoundaryFace is not integrated into the Czech profile. If the necessity for the further development of 3D systems arises within the Czech cadastre, then the profile can be expanded precisely to support 3D parcels modeled using LA_BoundaryFace. Furthermore, with regard to the discussed utilities, the LADM offers the LA_LegalSpaceUtilityNetwork class, and consequently, the Czech LADM-based profile can be appropriately extended in a standardized manner in order to support the registration of legal components related to utilities.

Acknowledgments: The first author of the publication was supported by the project Sustainability support of the centre NTIS—New Technologies for the Information Society (LO1506) of the Czech Ministry of Education, Youth and Sports.

Author Contributions: Karel Janečka wrote an outline of the entire article and is the first author of Sections 1–5. Petr Souček is a co-author for Sections 1–5.

Conflicts of Interest: The authors declare no conflicts of interest.

References

1. ISO Online Browsing Platform (OBP). Available online: https://www.iso.org/obp/ui/#iso:std:iso:19152: ed-1:v1:en (accessed on 24 February 2017).
2. Čada, V.; Janečka, K. The Strategy for the Development of the Infrastructure for Spatial Information in the Czech Republic. *ISPRS Inter. J. Geo-Inf.* **2016**, *5*, 33. [CrossRef]
3. Čada, V.; Janečka, K. The Fundamental Spatial Data in the Public Administration Registers. *Int. Arch. Photogramm. Remote Sens. Spatial Inf. Sci.* **2016**, *41*. [CrossRef]
4. Zobrazení Netypických Staveb v Katastru Nemovitostí. Available online: http://csgk.fce.vutbr.cz/Oakce/A98/prezentace/Olivova_KN16.pdf (accessed on 20 February 2017).
5. Stoter, J.; van Oosterom, P. *3D Cadastre in an International Context. Legal, Organizational, and Technological Aspects*; CRC Press: Boca Raton, FL, USA, 2006.
6. Bydłosz, J. The application of the Land Administration Domain Model in building a country profile for the Polish cadastre. *Land Use Policy* **2015**, *49*, 598–605. [CrossRef]
7. Vučić, N.; Markovinović, D.; Mičević, B. LADM in the Republic of Croatia-making and testing country profile. In Proceedings of the 5th FIG International Land Administration Domain Model Workshop 2013, Kuala Lumpur, Malaysia, 24–25 September 2013.
8. Zulkifli, N.A.; Rahman, A.A.; Van Oosterom, P.; Choon, T.L.; Jamil, H.; Hua, T.Ch.; Seng, L.K.; Lim, Ch.K. The importance of Malaysian Land Administration Domain Model country profile in land policy. *Land Use Policy* **2015**, *49*, 649–659. [CrossRef]
9. Lee, B.M.; Kim, T.J.; Kwak, B.Y.; Lee, Y.; Choi, J. Improvement of the Korean LADM country profile to build 3D cadastre model. *Land Use Policy* **2015**, *49*, 660–667. [CrossRef]
10. Janečka, K.; Souček, P. Country profile for the cadastre of the Czech Republic based on LADM. In Proceedings of the 5th International FIG 3D Cadastre Workshop 2016, Athens, Greece, 18–20 October 2016.
11. Ministry of the Interior of the Czech Republic. *The Strategy for the Development of the Infrastructure for Spatial Information in the Czech Republic to 2020—Annex 3: Services Working with Spatial Data—Use Cases*; Ministry of the Interior of the Czech Republic: Prague, Czech Republic, 2012.
12. Döner, F.; Thompson, R.; Stoter, J.; Lemmen, Ch.; Ploeger, H.; van Oosterom, P.; Zlatanova, S. Solutions for 4D cadastre—With a case study on utility networks. *Inter. J. Geograph. Inf. Sci.* **2011**, *25*, 1173–1189. [CrossRef]
13. Koeva, M.; Elberink, S.O. Challenges for updating 3D cadastral objects using LiDAR and image-based point clouds. In Proceedings of the 5th International FIG 3D Cadastre Workshop 2016, Athens, Greece, 18–20 October 2016.
14. OGC City Geography Markup Language (CityGML) Encoding Standard. Version 2.0. Available online: http://www.opengis.net/spec/citygml/2.0 (accessed on 24 February 2017).
15. Góźdź, K.; Pachelski, W.; van Oosterom, P.; Coors, V. The possibilities of using CityGML for 3D Representation of buildings in the cadastre. In Proceedings of the 4th International Workshop on 3D Cadastres 2014, Dubai, UAE, 9–11 November 2014.
16. Seifert, M.; Gruber, U.; Riecken, J. Multidimensional cadastral system in germany. In Proceedings of the FIG Working Week 2016, Christchurch, New Zealand, 2–6 May 2016.
17. Thompson, R.; van Oosterom, P.; Soon, K.H.; Priebbenow, R. A conceptual model supporting a range of 3D parcel representations through all stages: Data capture, transfer and storage. In Proceedings of the FIG Working Week 2016, Christchurch, New Zealand, 2–6 May 2016.
18. Atazadeh, B.; Kalantari, M.; Rajabifard, A. Comparing three types of BIM-based models for managing 3D ownership interests in multi-level buildings. In Proceedings of the 5th International FIG 3D Cadastre Workshop 2016, Athens, Greece, 18–20 October 2016.
19. Li, L.; Wu, J.; Zhu, H.; Duan, X.; Luo, F. 3D modeling of the ownership structure of condominium units. *Comp. Environ. Urban Sys.* **2016**, *59*, 50–63. [CrossRef]
20. Rönsdorf, C.; Wilson, D.; Stoter, J. Integration of land administration domain model with CityGML for 3D cadastre. In Proceedings of the 4th International Workshop on 3D Cadastres 2014, Dubai, UAE, 9–11 November 2014.
21. Çağdaş, V. An application domain extension to CityGML for immovable property taxation: A Turkish case study. *Inter. J. Appl. Earth Observ. Geoinf.* **2013**, *21*, 545–555. [CrossRef]

22. Dsilva, M.G. *A Feasibility Study on CityGML for Cadastral Purposes*; Eindhoven University of Technology: Eindhoven, The Netherlands, 2009.
23. Oldfield, J.; Van Oosterom, P.; Quak, W.; Van der Veen, J.; Beetz, J. Can data from BIMs be used as input for a 3D Cadastre? In Proceedings of the 5th International FIG 3D Cadastre Workshop 2016, Athens, Greece, 18–20 October 2016.
24. Atazadeh, B.; Kalantari, M.; Rajabifard, A.; Ho, S.; Ngo, T. Building information modelling for high-rise land administration. *Trans. GI.* **2016**, *21*. [CrossRef]
25. Kalantari, M.; Rajabifard, A. A roadmap to accomplish 3D cadastres. In Proceedings of the 4th International Workshop on 3D Cadastres 2014, Dubai, UAE, 9–11 November 2014.

International Journal of
Geo-Information

isprs

MDPI

Article

Registration of Multi-Level Property Rights in 3D in The Netherlands: Two Cases and Next Steps in Further Implementation [†]

Jantien Stoter [1,2,*], Hendrik Ploeger [3,4], Ruben Roes [2], Els van der Riet [5], Filip Biljecki [1], Hugo Ledoux [1], Dirco Kok [1,2] and Sangmin Kim [1]

1   3D Geoinformation, Delft University of Technology, 2628 BL Delft, The Netherlands;
f.biljecki@tudelft.nl (F.B.); h.ledoux@tudelft.nl (H.L.); dircokok@live.nl (D.K.); S.Kim-2@tudelft.nl (S.K.)
2   Kadaster, 7311 KZ Apeldoorn, The Netherlands; ruben.roes@kadaster.nl
3   Geo-information and Land Development, Delft University of Technology, 2628 BL Delft, The Netherlands;
h.d.ploeger@tudelft.nl
4   Faculty of Law, Vrije Universiteit Amsterdam, 1081 HV Amsterdam, The Netherlands
5   Municipality of Delft, 2627 BM Delft, The Netherlands; evdriet@Delft.nl
*   Correspondence: j.e.stoter@tudelft.nl
†   This article is an expanded version of the previously published workshop paper: Stoter, J.; Ploeger, H.;
Roes, R.; van der Riet, E.; Biljecki, F.; Ledoux, H. (2016): First 3D Cadastral Registration of Multi-level
Ownerships Rights in The Netherlands. Proceedings of the 5th International FIG Workshop on 3D Cadastres,
Athens, Greece, pp. 491–504.

Academic Editors: Peter van Oosterom, Efi Dimopoulou and Wolfgang Kainz
Received: 31 March 2017; Accepted: 24 May 2017; Published: 31 May 2017

Abstract: This article reports on the first 3D cadastral registration in The Netherlands, accomplished in March 2016. The solution was sought within the current cadastral, organisational, and technical frameworks to obtain a deeper knowledge on the optimal way of implementing 3D registration, while avoiding discussions between experts from different domains. The article presents the developed methodology to represent legal volumes in an interactive 3D visualisation that can be registered in the land registers. The source data is the 3D Building Information Model (BIM). The methodology is applied to two cases: (1) the case of the railway station in Delft, resulting in the actual 3D registration in 2016; and (2) a building complex in Amsterdam, improving the Delft-case and providing the possibility to describe a general workflow from design data to a legal document. An evaluation provides insights for an improved cadastral registration of multi-level property rights. The main conclusion is that in specific situations, a 3D approach has important advantages for cadastral registration over a 2D approach. Further study is needed to implement the solution in a standardised and uniform way, from registration to querying and updating in the future, and to develop a formal registration process accordingly.

Keywords: 3D cadastral registration; architectural models (BIM); 3D deed; 3D land administration

1. Introduction

In many jurisdictions, property rights in relation to land are registered on 2D parcels. As long as there is only one owner of the land, this mode of registration is sufficient to represent the legal situation. However, providing a clear insight is challenging in multi-level property situations, i.e., if there are more land owners, for example, in the case of a tunnel or an underground parking garage [1].

In such cases, it is almost always possible to entitle a person a right to a volume, whether it is for the ownership of a (part of a) physical construction (such as an apartment right), or for air space (for example, a wind right relating to a wind mill or wind turbines; an easement to protect a right of view) [2]. However, the legal and technical practices used to create such 3D legal entities and to register them in the land registers and cadastral systems vary from jurisdiction to jurisdiction.

In most jurisdictions, a 2D parcel is the main entity of property situations, registration. To be able to establish the legal situation of multi-level property rights via the 2D parcel, limited real rights are established on the land parcel, such as easement, right of superficies (*opstalrecht*), and right of ground lease (*erfpacht*) [3,4].

As shown in many studies, this mode of registration (i.e., registering the 3D legal situation via 2D parcels) does not provide proper insight into the legal situation. This is often not a problem at the moment in which the legal situation is created. At that moment, all stakeholders have to agree on the registration and therefore, for them, the legal situation is clear. The main challenges arise in future transfers of multi-level property rights, when the involved parties (buyer, seller, and others, such as a mortgage bank) need to reconstruct the existing 3D property situation from both the 2D cadastral map and the deeds registered in the land register.

For several years, jurisdictions have provided the possibility to describe the 3D spatial extent of such legal volumes in registered deeds or title documents. Examples of this practice can be found in Sweden [5], Greece [6], Croatia [7], Australia [8], and China [9]. However, until now, these 3D descriptions have been mainly paper-based or static: e.g., volumetric plans that show isometric views to depict 3D spatial units or building plans that display sketches of building numbers, locations, levels, and layout [10]. These traditional, analogue systems of registration do not support the storage of 3D data, the automated validation of 3D data, or the interactive visualisation of the 3D entities. 3D solutions have recently been developed for apartment units, e.g., in Spain [11]. However, these do not provide solutions for multi-level properties unrelated to apartments or condominiums.

This article presents the first registration of an interactive 3D visualisation of legal volumes in the land registers. This registration was realised in The Netherlands to address both the limitations of a 2D parcel-based land administration, as well of solutions that do enable one to register 3D legal entities (i.e., legal volumes), but which do not include a proper way to represent this multi-level property situation in the cadastral registration.

The developed registration consists of an interactive 3D visualisation of all legal volumes of one multi-level property situation. This 3D visualisation is registered as a legal document in the land registers.

This article describes the background, methodology, and the results of the 3D registration that was implemented.

Overview of This Article

To put the research and developments in The Netherlands in an international perspective, the article starts with an overview of 3D cadastre implementations worldwide and provides more background relating to the Dutch situation (Section 2).

Sections 3 and 4 present the methodology that was developed to create the 3D visualisation of legal volumes in 3D PDFs that can be registered as legal documents in the land registers. These sections also show how the methodology was applied to two real world cases of multi-level property. Both cases had first been registered in a "traditional" 2D way (also described for both cases). The 3D PDF of the first case has additionally been registered as a legal deed in cadastral registration.

Section 5 presents the resulting workflow and it evaluates the implemented 3D registration.

The lessons learned from both cases led to guidelines for further developments for the second phase, as described in Section 6. We finish with questions for further research in Section 7 and conclusions in Section 8.

2. Background

This section presents previous research on 3D Cadastre worldwide (Section 2.1) and describes the specific cadastre situation of The Netherlands in Section 2.2 (to provide the context of this paper).

2.1. 3D Cadastre Developments Worldwide

Cadastral organisations around the world are taking steps to register multi-level property rights in such a way that the registration provides a clearer insight of the legal situation. There are many examples of partial (or prototype) implementations. Kitsakis et al. [12] and Dimopoulou et al. [6] provide an overview of these developments. Kitsakis et al. [12], discussing 3D real property legal concepts, conclude that several issues are reported when developing 3D cadastres in the studied countries (i.e., Austria, Brazil, Croatia, Greece, Poland, and Sweden). These issues concern the status of 3D real property, the national legal definition of 3D objects, and the types of rights that can be registered in 3D. However, issues such as real property recording in 3D or the management of cross boundary objects within cadastral databases still remain unsolved. Dimopoulou et al. [6] examine many examples of partial implementations of a 3D cadastre, but the conclusion is that the functionalities are always limited in some way. Most progress has been achieved in providing legal provisions for the registration of multi-level property. Many cadastres have started to show these properties on cadastral plans such as isometric views, vertical profiles, or textual information. However, real 3D solutions (i.e., implementing interactive 3D visualisations) hardly exist.

One of the costliest phases of the implementation of a 3D cadastre is the 3D data capture. According to Dimopoulou et al. [6], for representing multi-level properties in 3D, new ways of 3D data capture are being studied, such as BIM, lidar data, 3D topographic data, and laser surveys of individual units. However, linking these data to existing cadastral frameworks is still challenging, as is the validation of such data. Research on the validation of 3D data geometries related to physical constructions has been done [13]. Applying these solutions to 3D geometries representing legal volumes is a next step which will be addressed in our research.

For the registration of legal spaces in 3D, there are two options for 3D data capture; i.e., (1) surveying in 3D, or (2) using data sources such as construction and design models (as explored in this article). The first option is only possible for already existing constructions. This article will explore the registration of legal volumes that still need to be built and will therefore show the reconstruction of legal volumes from design data (i.e., option 2).

2.2. Multi-Level Property Rights in The Netherlands

Multi-level property rights in The Netherlands have been in existence since long before the start of the Dutch cadastre (1832). Historical examples are cellars under the public street leading from canals to houses in the city of Utrecht. More recent examples are complex combinations of commercial areas, public transport hubs, and parking garages, or buildings over highways [14–16].

As in many countries, these multi-level property situations are established via limited rights on 2D parcels.

To be able to sufficiently represent the spatial extent of each property right in the cadastre, the "speciality principle" is enforced, meaning that in land registration, and consequently in the documents submitted for registration, the concerned subject and object (i.e., real property) must be unambiguously identified [17]. This means that, in the event that a real right is only established on a part of a cadastral parcel, this parcel will be subdivided according to the boundaries of the real right. For example, a right of superficies for a building that is constructed in, on, or over a part of the land. In such a case of 3D property rights, the objects (i.e., buildings or other constructions) are projected on a 2D parcel map and 2D parcels are divided into smaller parcels to be able to register the property of objects above and below the surface via an accumulation of limited rights. Applying this principle may result in an unclear fragmentation of parcels, especially when objects above and under the ground are all projected on the same parcel.

The difficulties (sometimes impossibility) of reconstructing the 3D legal situation from only the description in the deed (in words and accompanying 2D maps), in combination with the 2D parcel boundaries on the cadastral map, have already been experienced in The Netherlands, as can be understood when looking at the complex parcel patterns in Figure 1.

The very small parcels in this Figure are the result of the projection of the outer-boundary of 3D objects (both above and below ground level) in the 2D map. In this way, existing parcels are subdivided in order to ensure that only the land intersecting with the 3D objects is burdened with a limited right for the object.

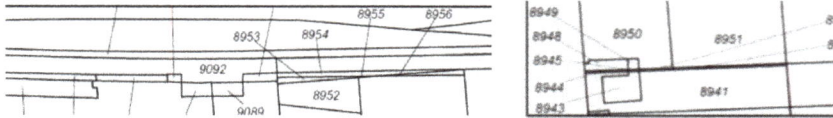

Figure 1. Registration of multi-level ownership on 2D parcels results in fragmented parcel patterns.

To address the issue of subdivision resulting in complex parcel patterns, The Netherlands Kadaster has started a study to improve the registration in such cases [14].

The implementation of the proposed improvements consists of two phases.

The solution of the first phase was sought within the limitations and possibilities of the existing legal and cadastral frameworks, and aimed at gaining experience in the challenging domain of 3D cadastre where technical possibilities, on the one hand, and legal and cadastral needs, on the other, interact. The aim was to provide a solution for the problem that the rights for 3D volumes can be established, but the dimensional aspect cannot be made visible in the cadastre. Therefore, a procedure was developed to accept 3D representations of legal volumes in a 3D PDF format, as part of the deed. Because of the recently established acceptance of the digital registration of deeds in the Dutch land registers, the registration of a 3D visualisation of multi-level rights in the form of a 3D PDF was possible, without a change of the laws. This method was introduced in practice in 2016, where a 3D visualisation of multi-level rights was registered in the public registers (see the Delft case in Section 3).

The second phase consists of research in progress and builds on the lessons learned from the first phase. It comprises the study of how to accommodate the explicit registration of rights and restrictions limited in 3D and the actual inclusion of these volumes in the registration, to enable complete validation and to better support 3D data management and dissemination.

This article reports on the results of the first phase (which is the input for the second phase).

3. Multi-Level Property Rights Case I (Delft)

The first multi-level property rights case of this article is the new combined structure of the city hall and underground railway station in Delft.

3.1. Description of the Factual Situation

This building is located in the heart of the Delft Railway Zone project. The Delft Railway Zone covers an area of 24 hectares in total and consists of:

- a 23 km long railway tunnel, replacing a partly elevated railway, running along the historic city centre of Delft;
- underground railway platforms and station;
- station hall with shops at ground level;
- the city hall: municipal offices for the city of Delft on four levels;
- two underground bicycle parkings (5000 spaces) (and 2700 spaces, under construction at the time of writing);
- underground car parking (650 spaces);
- new public space, including a park on top of the tunnel and a non-navigable canal for water storage above the car parking; and
- urban redevelopment, mainly of former railway yards.

The 3D cadastre case covers only a small part of the total project: the combined new Railway Station and the new City Hall, together with the underground platforms and railway tunnel, several technical installations, and the large underground bicycle parking (see Figure 2).

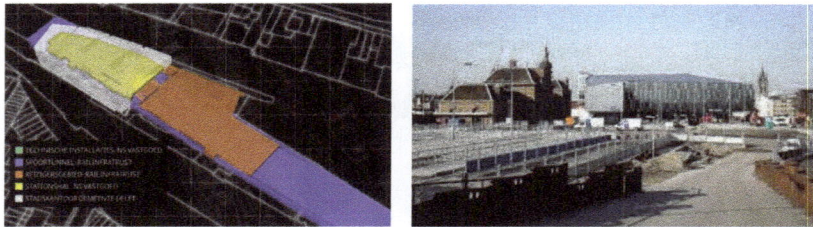

Figure 2. The building complex of Delft station, with (left) the property rights situation and (right) a picture of the situation above ground.

The multi-layered construction combines the property rights of three parties: the municipality of Delft as the owner of the land and the City Hall, NS Real Estate (Dutch railroad company for passenger transportation) as the owner of the Station Hall with shops and installations, and ProRail/Railinfratrust (Dutch railroad infrastructure company) as the owner of the traveller's area, the tunnel, and the platforms. ProRail/Railinfratrust and NS are separate legal entities.

3.2. Registration of Property Rights in 2D

For this multi-level property rights situation, six property rights related to volumes (i.e., legal volumes) have been established:

- Station hall (NS Vastgoed);
- Traveller's area (Railinfratrust B.V.);
- Elevator and stairs (NS Vastgoed);
- Technical installations (NS Vastgoed);
- Tunnels (Railinfratrust B.V.);
- Municipality of Delft: everything that remains.

The first five property rights were established with a right of superficies. Each of these rights can contain more than one object (space) and can also overlap several ground parcels.

The municipality (sixth property) owns the land and therefore it owns all the space that is left after subtracting the volumes for the rights of superficies (and all buildings that belong to this space, like the office building of the municipality).

The stakeholders in this project agreed that they would register the involved 3D property rights via a 3D deed, based on the model for 3D registration that was developed in Stoter et al. [14]. However, because the 3D registration was the first one to ever be accomplished, the stakeholders did not want to run a risk of any delay due to the new mode of registration. Any delay in the registration process would have an impact on the moment when the new owners could make use of the new building complex, because the complex can only be used when the rights are secured. Therefore, it was decided that the real rights would first be established via traditional (2D) registration.

For 2D registration, the property rights were established for the different parts of a deed that was recorded in the Land Register. In this deed, the legal volumes were described in wordings, accompanied by 2D maps. Figure 3 (left) shows the 2D cadastral map at the location and Figure 3 (right) is an example of one of the 2D cross sections that was added to the deeds to clarify the complex property situation.

In this registration process, new ground parcels were formed by Kadaster, i.e., the original parcels (still reflecting the historical ownership situation before the existing houses and other constructions that have been demolished for the Delft Railway Zone project) were consolidated and were subsequently subdivided in order to specify the different accumulation of rights of the new complex.

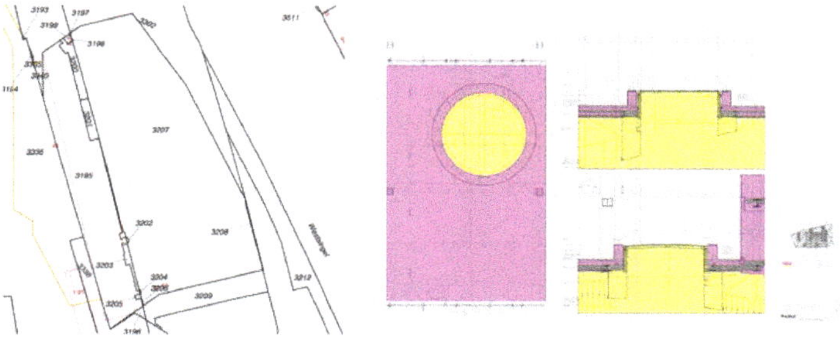

Figure 3. 2D cadastral map of the building complex in Delft (**left**); and registration of rights for a volume by a 2D description in the deed (**right**).

3.3. Methodology for 3D Registration

This section describes the procedure that was followed to create the 3D representations of the legal volumes and to register it via a legal 3D deed (i.e., an interactive 3D PDF registered as an official document in the Land Register), as alternative registration.

For the 3D registration, the architect of the building complex (Mecanco) converted the 3D data of the construction itself (the building information model, BIM) into 3D geometries representing the six legal volumes as described above, based on: (a) the design data of the complex; (b) the already registered deed with 2D maps of the complex, and (c) the input of all stakeholders (collected via four work sessions). Several researches have studied how to use BIM for modelling 3D property rights. See, for example, EL-Mekawy [5] and Atazadeh et al. [18]. Furthermore, Stoter et al. [14] have demonstrated the possibility to extract legal volumes from a BIM.

It was decided that the office building of the municipality would be represented, even though the total space owned by the municipality is much larger. That is, the total space owned by the municipality contains all of the space left after the space for the other constructions is subtracted from that. From a legal point of view, this includes the subsurface not occupied by the railway's tunnel and the underground part of the railway station, as well as the exclusive use rights of the air space above the building. Consequently, the city hall is included in these rights. However, for orientation reasons, it was deemed better to include the building owned by the Municipality in the PDF, be it in a transparent way, to show the difference in status compared to the other parts (legal volumes based on limited real rights).

In a next step, the 3D representations of the rights were translated into 3D PDF. Besides the 3D representation of the rights, some elements were added to improve the usability: a legend of the rights; the 2D cadastral map in which the parcels of concern are identified; and the x, y, and z coordinates of the national reference system that show the location of the complex in the real world.

Finally, the notary firm of Houthoff Buruma has issued a certificate for the deposit of the 3D PDF in the Land Registry as an official deed by the notary (see Figure 4). This was done supplementary to an earlier deed in which the rights were established and only described in 2D. The actual registration of the document by Kadaster took place in March 2016, setting a milestone in the development of a 3D Cadastre.

In cadastral registration, a 3D complex ID was generated and the different rights were assigned unique indices. These IDs are both registered in the drawing as in the textual part of the deed. However, they cannot be considered to be 'formal' parcel numbers as on the traditional 2D cadastral map because this requires a change in the registration framework. Additionally, a reference was made in the cadastral registration to the interactive, 3D visualisation of property rights in 3D.

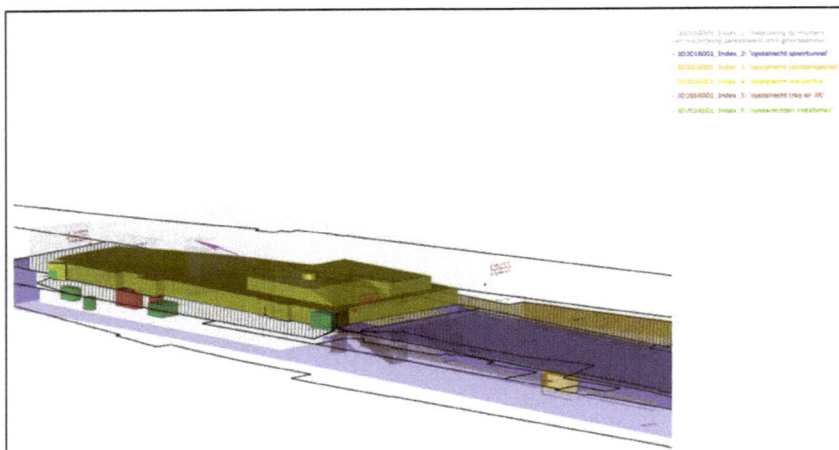

Figure 4. 3D PDF, official document that visualises the multi-level property rights in 3D (case Delft station).

Finally, the 3D data itself (describing the 3D geometries of the legal volumes) are stored by Kadaster in view of future needs that may require the juridical situation to be adjusted. The 3D data is stored (and maintained) by the public registers.

3.4. Accessing the 3D Registration

The multi-level property rights can be queried in 3D via cadastral registration: the registration shows a parcel complex on the cadastral map with a notification of "3D visualisation". This notification refers to the deed with the 3D drawing. This 3D PDF is publicly available, not only from the public registers, but also from Kadaster [19], and can be viewed in any PDF viewer that supports 3D. In the viewer, the 3D situation can be interactively viewed, and one can see the relationship between the different legal volumes (which cannot be done if they are registered via separate 3D surveys), generating individual volumes that are visible (or invisible) for further inspection. When "clicking" on objects, one sees the 3D-indices and owners of the volumes.

4. Multi-Level Property Rights Case II (Amsterdam)

After the 3D registration of the Delft case, another multi-level property rights case was considered. The second case of this article is the new combined structure of the large Congress hotel Maritim, a residential building, and an underground parking garage in the northern part of Amsterdam. The creation of the 3D PDF visualising the legal volumes is a collaboration of Kadaster, Bentley, and Delft University of Technology. The 3D PDF was created to study an alternative method to for the 2D registration, but more importantly, to use the experiences of this case (together with the Delft case) as an input to describe a standardised workflow for future 3D registrations. The actual registration in the cadastral registration of this case is future work.

4.1. Description of the Factual Situation

The congress centre can host 5000 visitors on the first three floors and has a hotel on top with about 580 rooms, with a number of bars and restaurants. It also contains a fitness area that can be used by both hotel guests and the inhabitants of the residential building. The buildings were designed by Team V Architectuur.

4.2. Registration of Property Rights in 2D

For this building complex, three legal volumes have been established:

- Underground parking garage with a right of superficies (Figure 5, left)
- Congress building and hotel with a right of ground lease (Figure 5, middle)
- Residential building, with a right of ground lease (Figure 5, right)

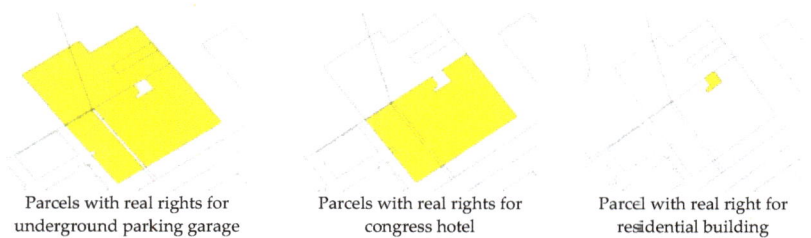

Parcels with real rights for underground parking garage   Parcels with real rights for congress hotel   Parcel with real right for residential building

Figure 5. 2D parcel map around the Congress hotel in the Amsterdam case

The municipality of Amsterdam is the land owner and therefore owns all of the space that is left after subtracting the volumes that comprise the rights of superficies and the two rights of ground lease.

The parcels that are affected by the legal rights of the three respective property rights are shown in Figure 5.

To show the extent of the rights in the deed, an 2D overview of every floor was generated and added to all three deeds (underground parking garage, congress hotel, and residential building). As can be seen from Figure 6, it is a mental challenge to combine those floor plans into a (mental) 3D view that shows the extent of all the legal volumes involved.

Figure 6. Thumbnails of floor plans that have been added to the 2D deed to clarify the complex property registration in the Amsterdam case.

4.3. Methodology for 3D Cadastral Registration

To prepare for alternative 3D registration, we collaborated with the architect of the building complex, Team V Architectuur. From the 3D design data, volumes were created, representing the legal entities. These were used to create the 3D PDF.

A few improvements were made based on the evaluation of the 3D visualisation of the Delft case. This first user evaluation showed that navigating in a 3D PDF is not straightforward. Therefore, different views on the 3D building complex were generated so that users can easily switch between different viewpoints in the 3D PDF to understand the situation (e.g., top view, different side views, etc.).

Additionally, an exploded view was prepared from the data and added to the 3D PDF. This view gives a clear impression of how different legal entities on different heights interfere, with little user interaction needed (see Figure 7).

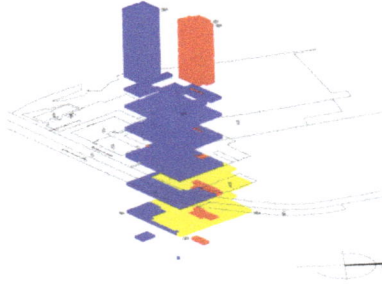

Figure 7. Exploded view of the legal volumes of the Amsterdam case in 3D PDF. Red = residential building; Blue = congress centre; Yellow = underground parking garage.

Finally, we performed data quality checks. This is important because the 3D geometries are generated by the architect, with no involvement from Kadaster. Kadaster has to check that the generated geometries indeed represent the legal volumes as agreed by the stakeholders. Therefore, the first step included integrating the legal volumes created with the original design data.

This integrated view (see Figures 8 and 9) showed that both the balconies of the residential building, as well as the façade of the Maritim Hotel tower, had been excluded from the initial legal volumes. To repair this, a new boundary had to be reconstructed, enclosing the complete physical model for the final version of the legal volumes.

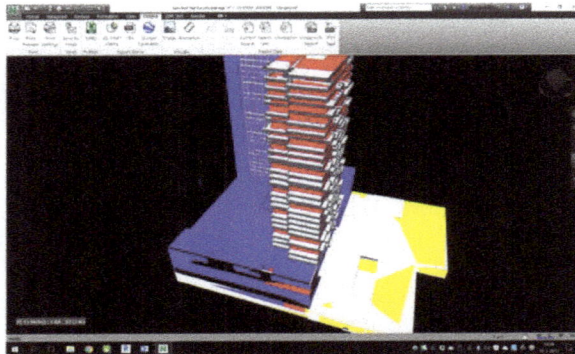

Figure 8. Integration of the legal model and the physical model (Amsterdam case) shows that these are not consistent.

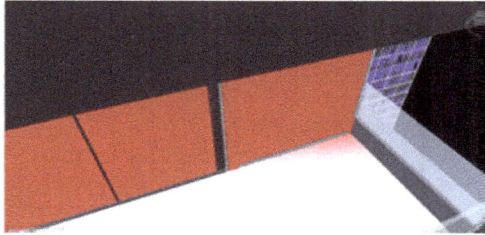

Figure 9. Balcony of the residential building in the Amsterdam case is not enclosed by legal volume (**red**).

Another data quality check that we performed was the tightness of legal volumes, i.e., there should be no overlap or gaps, by using tools in Bentley Map Enterprise. We identified all volumes to be watertight (i.e., closed), with no overlap or gaps. These requirements are basic for cadastral applications and are in 2D. However, in 3D, checking the validity of 3D geometries is not trivial. See Karki et al. [8], Thompson and van Oosterom [20], and Ledoux [13] for examples.

A final data quality check included matching with the 2D cadastral map that is included in the 3D PDF for reference purposes (to identify which parcels are affected by the 3D registration). This required georeferencing the legal volumes that were generated from the design data

Georeferencing is a manual process (it requires localisation and scaling) and can therefore easily lead to errors. It is, however, important that any mismatches between the 3D legal volumes and the 2D parcel boundaries are avoided.

5. Results of 3D Registration

In this section, the 3D registration is evaluated (Section 5.2). First, the developed workflow is presented (Section 5.1).

5.1. Developed Workflow

Based on the experiences of these two cases, we have defined the workflow to create a 3D PDF that acts as an interactive visualisation of the legal volumes in the case of a multi-level property situation. This workflow can be used in future situations and is shown in Figure 10.

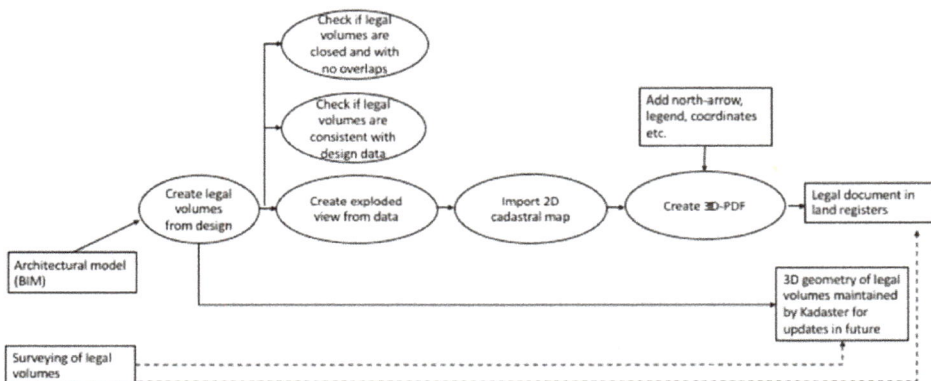

Figure 10. Schematic overview of a workflow to create a 3D PDF visualising a multi-property situation to be registered as a legal document in the land registers.

The workflow has two possibilities for creating the geometries of the legal volumes: (1) surveying (which is only possible in the case of a physical construction that already exists) and (2) from the architectural model (this article). The first option (surveying legal volumes) represents future work. Checking whether legal volumes are consistent with their physical counterpart, as well as if they are watertight with no overlap, is also part of the workflow.

The 3D PDF contains the 3D geometries representing legal volumes, the 2D cadastral map of the situation, and some additional aspects such as a north-arrow, legend, and real world coordinates. An exploded view is also created to better present the situation (see Figure 7). Finally, 3D data relating to the geometry of the legal volumes is maintained to be able to update the document in the future.

5.2. Evaluation of Implemented 3D Registration

A comparison of the two types of registration (i.e., 2D and 3D) for both cases shows the significant advantages of 3D registration in the case of multi-level property situations.

At first, complex multi-level property rights can be made more (and sometimes 'only') clear in 3D.

An example of this is a revolving door in the Delft case (Figure 11): the revolving door between the station hall (owned by NS) and the city hall (owned by the municipality). The construction extends into the property of NS, but does not extend to the ceiling. Therefore, the question arises: to whom does the space belong from the top of the revolving door to the ceiling? This situation can only be clarified in 3D.

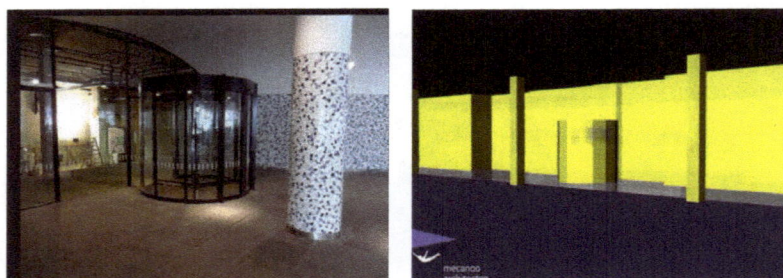

Figure 11. Revolving door of the municipality of Delft extending into the property of another party.

Moreover, for the building complex of the Amsterdam case study, the 3D approach provided more insights than the 2D drawings used for the registration. All legal volumes are integrated in one environment, and therefore, one view is sufficient to understand the ownership situation, instead of mentally combining 14 floor plans in one 3D situation.

Another example of why 2D cannot sufficiently represent the property rights in complex situations demonstrates that the conversions of the 3D data relating to the physical construction into legal volumes revealed some errors in the 2D registration.

An error that appeared in both cases is that not all parcels intersecting with the 3D objects were encumbered with limited property rights. Furthermore, tiny parcels were created because of a mismatch between the location of 2D cadastral boundaries and the projection of the 3D reconstruction. The mismatch was not due to a wrong location of the involved objects (parcels on the one hand and 3D (BIM) objects on the other hand), but due to a difference in accuracy. The design data has a higher accuracy than the cadastral map and integrating both resulted in unneeded, and as we might call them, "artificial", parcels (Figure 12). The 2D cadastral map was therefore improved with this new information.

Another advantage of 3D above 2D, is that—at least in the Dutch context—the registration is more cost effective.

First, this is because the cumbersome process of describing the complex situation in a 2D description required for the registration can be skipped. This also applies vice versa: 3D reconstruction of the legal situation from a sub-optimal registration required for future transactions is not needed. However, the costs of surveying can also be saved because the surveying of ground parcels within complexes is no longer necessary to assure that the 2D parcel map reflects the 3D situation. These costs may be significant in cases of multi-level property rights where boundaries are not exactly on top of each other when projected on a 2D plane. Specifically, those situations produced issues among the stakeholders in the first case: who has a right to the very tiny parcels that were created by projecting objects above and below the surface in a 2D plane? This also led to insights that the 2D map does not reflect the correct situation and it was therefore improved (see before).

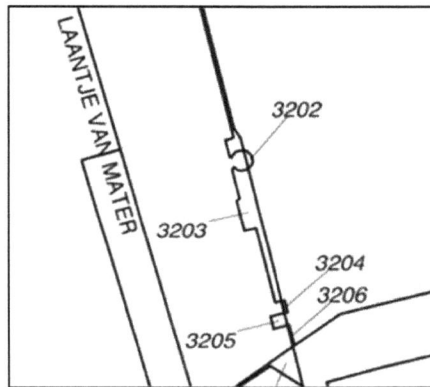

Figure 12. Tiny parcels due to the division of ground parcels in the Delft case to permit the 2D map to reflect objects above and below the surface.

In conclusion, a 3D cadastral registration of multi-level property rights has many advantages. However, the fact that the 2D cadastre map no longer reflects the legal situation, but refers to a 3D registration that does, requires a change in the cadastral legislation and procedures, as will be elaborated in the next section.

6. Next Phase of 3D Cadastre Registration

The optimal solution for adequately registering multi-level property rights in the cadastral registration lies somewhere between current registration (with lots of additional information needed such as text and 2D drawings to explain the situation) and the implementation of a full 3D cadastral registration [21].

In the full 3D cadastre, the 3D space (universe) is subdivided into volume parcels partitioning the 3D space [21]. The legal basis, real estate transaction protocols, and cadastral registration support the establishment and conveyance of 3D rights. There is no longer a 2D cadastral map that lays down restrictions on 3D rights. Every property right is treated as a volume and these are not related to the surface configuration of parcels.

Although a full 3D cadastre would in theory meet all needs to establish and register multi-level property rights, it is not the aim of our research. At first, a full 3D cadastre would require a change of the existing framework in cadastral law (*Kadasterwet*), as well as the land law, as laid down in the Dutch Civil Code. Changing the legal framework is a lengthy process. In addition, cadastral registration has been based on 2D parcels for a very long time and it is hard to predict how a full 3D cadastre would work in this well-established practice. Changing from a 2D parcel-based concept to a 3D parcel-based concept would also mean re-surveying and re-registering millions of existing situations. Finally, the current 2D cadastral registration still serves single property situations very well. Therefore, the optimal solution for practice is an interchange of juridical, organisational, and technical aspects.

To avoid lengthy discussions between experts from different disciplines on the optimal implementation of a 3D cadastre, the solutions presented in this article were sought within existing legal and cadastral frameworks, with the aim of further developing the solution after an evaluation of these first 3D cadastral cases.

This section evaluates the above described 3D registrations for future developments.

6.1. Juridical Expertise Is Required to Create (or Validate) Legal Spaces from Physical Model

In both cases described in this article, the technical drawer of the architect's office actually created the boundaries of the legal volumes based on the technical 3D design of the construction. The lack of juridical knowledge of the people involved may result in an incorrect reconstruction of the legal volumes, as occurred in the Amsterdam case. Therefore, for future registration, Kadaster needs to request additional information to validate the legal volumes, for example, visualisations that show that the physical objects are consistent with the legal volumes and how these (spatially) relate to each other.

6.2. 3D Drawing to Support the Complete Registration Chain

The advantages of a 3D drawing can only be fully realised if the 3D drawing is part of the registration process from the start.

As was mentioned above, in both cases, the stakeholders decided to first register the property rights in a "traditional" (2D) way. This was necessary to secure the rights before the construction of the building was completed. The 3D registration was new and therefore would take (much) longer than the traditional mode of registration in which all of the steps to follow are clear (although resulting in an unclear registration).

To realise the traditional 2D registration, the legal reality was first made "flat" (i.e., described in 2D) and then had to be reconstructed again in 3D from the existing documents. As could have been expected, this resulted in grey areas without an explicit ownership in the deed. A 3D approach from the start would have avoided these grey areas. However, more importantly, it would have saved substantial time in the process (a significant advantage of 3D above 2D), because the 3D reconstruction of complex situations from 2D descriptions would not have been needed.

6.3. Reconsidering the "Speciality Principle"

Another lesson learnt relates to the "speciality principle" followed by Dutch land administration (which is also applicable in many another countries). This principle prescribes that the original parcel needs to be subdivided if a limited right is only established on part of a parcel, in order to assure that parcels that do not intersect with (the projection of) other properties are not affected by a limited right. Because of this principle, the building complex of the Delft case (Section 3) needed to be divided into 15 cadastral parcels and in the Amsterdam case (Section 4), 14 parcels were required. However, the initial idea was not to divide the ground parcel. The speciality principle, as observed in the traditional 2D registration, is a significant problem for multi-level property situations, since considerable time is needed to understand the property situation of these small parcels, specifically at those locations where the tunnel of the first case crosses small objects above ground. As we saw in Section 3.2, errors were even detected in the 2D parcel boundaries when the 2D cadastral map was confronted with the much more accurate BIM data. Maintaining the original parcel and not dividing it into small parcels, but instead referring it for the exact location to the 3D volumes in the 3D PDF, avoids these problems.

Current techniques show that there are better ways to provide insight into multi-level property rights than dividing parcels into smaller parcels. Therefore, it is indeed recommended that only one parcel is registered in such cases where the 3D visualisation of the 3D property rights of the whole complex is registered. For Dutch law, this requires a change of the *Kadasterbesluit* (Cadastre Decree) stating that surveying within a building complex is not needed, if all rights within the complex are represented in 3D. As argued before, this change will also save costs.

Apart from the registration of property via parcels, it is already possible in The Netherlands (as in many other countries) to register specific object types as independent objects with their own geometry, like cables and pipelines (see [22,23]).

6.4. 3D Parcels

The legal volumes formed with real rights overlapping with several ground parcels, can be considered as 3D parcels according to the definition of (Van Oosterom et al. [24]):

"A 3D parcel is defined as the spatial unit against which (one or more) unique and homogeneous rights (e.g. ownership right or land use right), responsibilities or restrictions (RRRs) are associated to the whole entity. Homogeneous means that the same combination of rights equally apply within the whole 3D spatial unit. Unique means that this is the largest spatial unit for which this is true. Making the unit any larger would result in the combination of rights not being homogenous. Making the unit smaller would result in at least two neighbouring 3D parcels with the same combinations of rights."

The legal volumes meet this condition and are registered with their own ID. A next study should address the following questions. What is that status of the 3D legal volumes? Can they be treated as single (new) 3D parcels, i.e., as individual objects? Or, do they always need to be linked to a 2D parcel?

6.5. Legal Boundaries Versus Physical Boundaries

Another lesson learned is that juridical boundaries in 3D are not always bound by physical boundaries. For example, we had to agree on how to demarcate the property of the space that accommodates the (use of a) staircase; only the visualisation of a constructed building is not sufficient. The division between the different rights in relation to the physical objects needs to be unambiguously clear to assure legal certainty.

6.6. Registration of the 3D Data

There is also a lesson to be learned from the registration of the 3D data needed for validation and updating in the future.

The legal volumes have been drawn originally in a specialised CAD software (Rhino 3D in the first case; and Autodesk Revit in the second case) and exported to a 3D PDF for everyone to inspect and visualise. In the specialised software, we ensured that the volumes were closed (i.e., "watertight") and that they were valid. This could be done with functions available in the software.

However, since our ultimate aim is to register these rights and allow everyone to manipulate the geometries, we did some experiments to store the geometries of the legal volumes in a format that is open and easily readable.

Here, we explain our experiments for the first case. The first case has six property rights, and each of them can consists of multiple volumes. The legal volumes for one of the rights can be seen in Figure 13a; in total, there are seven of them and they refer to the green ones in Figure 4. Most are rather simple volumes, but some are more complex, such as the one on the right in Figure 13b (which is the green volume in Figure 4).

Several challenges were faced when trying to obtain the legal volumes in an open standard. First, the export function of the software modified the boundaries of the volumes, by adding several vertices on the surfaces and edges (presumably to translate the parameterised geometries of the CAD software into explicit geometries), as shown in Figure 10 on the left. While in theory these challenges should not be an issue, the software inserted these very close to each other at certain locations (at the sub-millimetre level), which can be problematic when imported in different software. Second, the definition of what is a valid volume is different in different software and in different disciplines, as Ledoux [13] explains. In a GIS context, the ISO 19107 standard defines a solid (a 3D volumetric primitive) as a closed volume, and there should, for instance, not be any self-intersections in the bounding surfaces and no duplicated vertices. Unfortunately, in CAD software, self-intersections are sometimes allowed. This means that when we validated the volumes against the ISO 19107 rules, some of the volumes were invalid. The solution to this problem is to decompose the volumes into sub-parts, for example, if GML (Geography Markup Language) was used, then CompositeSolid would be used. Part of the future work will define a workflow so that practitioners can export their legal rights and store them in valid GML.

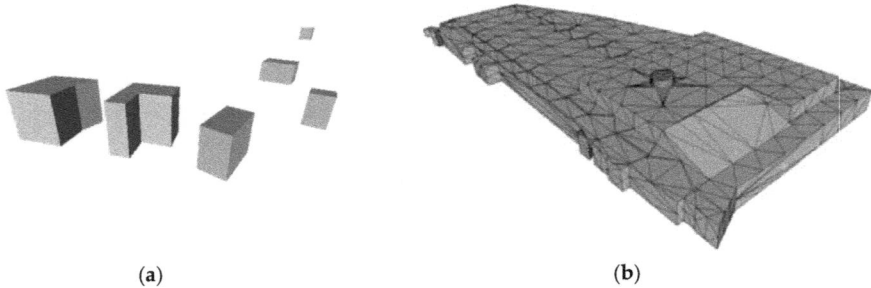

(a)   (b)

Figure 13. Examples of legal volumes in the Delft case: (**a**) seven different sub volumes forming one legal volume; (**b**) one complex legal volume that has been triangulated.

7. Questions for Further Research

Based on experiences from the first phase, investigations focusing on how to establish a more formal procedure for the 3D registration of multi-level property rights are currently underway. In these investigations, the following issues need to be addressed:

- Can the Kadaster enforce such a 3D registration in certain situations and if so, in which cases?
- What is required if 3D needs to be part of the registration from the start? Based on the experiences of these two cases, we have defined a workflow that explains all of the steps required to convert a BIM model into legal volumes and to write these to a 3D PDF with specific requirements. This also covers the export from legal volumes constructed in CAD software to valid GML.
- How does one change the legal rule so that one parcel can be registered for one multi-level property rights situation instead of dividing the 2D parcels to reflect constructions above and below the surface?
- What is the legal status of the 3D visualisation? Is the visualisation additional to the deed (3D is leading) or is the visualisation only meant to clarify the 2D deed only (2D is leading). If the 3D deed is leading, then the 3D data need to be validated via an official procedure. How does one implement these procedures? Asking for additional information to show the consistency between legal volumes and their physical counterpart can contribute to the quality procedure. In addition, Kadaster can require a proof of the geometric validity of each volume (as can be tested in different software).
- What are the (minimum) requirements for the 3D visualisation?
- What does one do if there is a difference between the deed and the 3D visualisation?
- What is the relationship between 2D and 3D (should the 3D visualisation always fit within 2D registration)?
- How does one maintain the underlying 3D data?
- The 3D PDF is registered as a legal document. Does this document need to be updated if a (small) part of the multi-level property case is changed?

8. Conclusions

In this article, we presented the first 3D cadastral registration of multi-level property rights in The Netherlands. Although it has been possible to legally establish 3D rights for centuries, until recently, it was impossible to visualise these 3D rights. Since, on the one hand, technologies required to handle 3D information have matured, and on the other hand, multi-level property rights situations are encountered that cannot be unambiguously registered in the current 2D based system, Kadaster (in collaboration with the Delft University of Technology) has developed a methodology to improve the registration in multi-level property rights situations through the use of a 3D PDF. This article presented the methodology for two real-world cases. The first case was also officially registered in the land registers in March 2016, setting a milestone in the development of a 3D Cadastre.

As can be concluded from the results, a 3D registration of multi-level properties, as implemented in this project, provides better insight into the case of multi-level ownership. It is no longer required to infer the 3D legal situation from 2D maps and verbal descriptions. Some legal situations can even only be shown in 3D.

The 3D PDF can be registered as a legal document in the land registers, as was done for one of the two cases presented.

Based on the experiences of the 3D registrations presented in this article, the 3D registration will be further developed and the regulations will be adjusted accordingly.

The step from 2D to 3D is not only providing insight into 3D via an available 3D technique such as a 3D PDF. Instead, the most challenging aspect is to assign a 3D geometry to legal volumes with possibly the same juridical value as parcel boundaries in 2D (or not). For cadastral registration, this requires a new way of defining, validating, and maintaining information about property rights in the cadastral registration.

The optimal 3D cadastre solution is a trade-off between juridical, cadastral, and technical aspects and we cannot predict how these aspects best come together. Therefore, the further development of 3D registration will be researched and the results of each step will be evaluated with cadastral, technical, and legal experts before the next step is made.

Acknowledgments: We thank all partners of the 3D cadastre study case for their contributions. These are for the first case study: The legal entities of the 3D cadastre case: ProRail/Railinfratrust, NS Vastgoed, and the Municipality of Delft. The architecture firm Mecanoo that designed the municipal offices and the station of Delft and converted the design into geometries representing legal volumes. Houthoff Buruma civil notaries that have issued the certificate for the deposit in the Land Registry. For the second case study, we thank Team V Architectuur for creating the legal volumes from the 3D design data and Bentley for providing help and test licenses for their software to develop the workflow from 3D CAD data to the official document in a 3D PDF format. This work is part of the research programme Innovational Research Incentives Scheme with project number 11300, which is financed by The Netherlands Organisation for Scientific Research (NWO). This project has received funding from the European Research Council (ERC) under the European Union's Horizon 2020 research and innovation programme (grant agreement No. 677312 UMnD).

Author Contributions: Jantien Stoter led both case studies and was the overall editor of the article. Ruben Roes and Hendrik Ploeger wrote all of the parts about the legal aspects of 3D cadastral registration. Els van der Riet contributed to the Delft case. Dirco Kok and Sangmin Kim contributed to the workflow for Amsterdam. Hugo Ledoux and Filip Biljecki performed the data experiments for the Delft case.

Conflicts of Interest: The authors declare no conflict of interest.

References

1. Van Oosterom, P.J.M.; Stoter, J.E.; Ploeger, H.D.; Lemmen, C.; Thompson, R.; Karki, S. Initial analysis of the second FIG 3D cadastres questionnaire: Status in 2014 and expectations for 2018. In Proceedings of the 4th International Workshop on 3D Cadastres, Dubai, United Arab Emirates, 9–11 November 2014; van Oosterom, P., Fendel, E., Eds.; International Federation of Surveyors (FIG): Copenhagen, Denmark, 2014.; pp. 55–74.

2.   Ho, S.; Rajabifard, A.; Stoter, J.; Kalantari, M. Legal barriers to 3D cadastre implementation: What is the issue? *Land Use Policy* **2013**, *35*, 379–385. [CrossRef]

3.   Groetelaers, D.A.; Ploeger, H. Management of redeveloped industrial areas with mixed use in The Netherlands. *J. Leg. Aff. Disput. Resolut. Eng. Constr.* **2010**, *2*, 73–81. [CrossRef]

4.   Ploeger, H.; Bounjouh, H. The Dutch urban ground lease: A valuable tool for land policy? *Land Use Policy* **2017**, *63*, 78–85. [CrossRef]

5.   EL-Mekawy, M.; Paasch, J.M.; Paulsson, J. Integration of 3D Cadastre, 3D Property Formation and BIM in Sweden. In Proceedings of the 3D Cadastre Workshop, Dubai, UAE, 9–11 November 2014.

6.   Dimopoulou, E.; Karki, S.; Roić, M.; De Almeida, J.P.D.; Griffith-Charles, D.; Thompson, R.; Ying, S.; Van Oosterom, P. Initial Registration of 3D Parcels. In Proceedings of the 3D Cadastre Workshop 2016, Athens, Greece, 18–20 October 2016.

7.   Vučić, N.; Roić, M.; Tomić, H. Registration of 3D Situations in Croatian Land Administration System. In Proceedings of the International Symposium & Exhibition on Geoinformation ISG 2013, Kuala Lumpur, Malaysia, 24–25 September 2013.

8.   Karki, S.; Thompson, R.; McDougall, K. Development of validation rules to support digital lodgement of 3D cadastral plans. *Comput. Environ. Urban Syst.* **2013**, *40*, 34–45. [CrossRef]

9.   Guo, R.; Luo, F.; Zhao, Z.; He, B.; Li, L.; Luo, P.; Ying, S. The Applications and Practices of 3D Cadastre in Shenzhen. In Proceedings of the 3D Cadastre Workshop, Dubai, UAE, 9–11 November 2014.

10.   Karki, S.; Thompson, R. Position article on "Initial registration of 3D parcels". In Proceedings of the 3D Cadastre Workshop, Dubai, UAE, 9–11 November 2014; Available online: http://www.gdmc.nl/3DCadastres/workshop2014/programme/Workshop2014_29.pdf (accessed on 20 May 2017).

11.   Olivares García, J.M.; Virgós Soriano, L.I.; Velasco Martín-Varés, A. 3D Modeling and Representation of the Spanish Cadastral Cartography. In Proceedings of the 2nd International Workshop on 3D Cadastres, Delft, The Netherlands, 16–18 November 2011.

12.   Kitsakis, D.; Paasch, J.; Paulsson, J.; Navratil, G.; Vucic, N.; Karabin, M.; Carneiro, A.T.; El-Mekawy, M. 3D Real Property Legal Concepts and Cadastre: A Comparative Study of Selected Countries to Propose a Way Forward (Overview Report). In Proceedings of the 3D Cadastre workshop 2016, Athens, Greece, 18–20 October 2016.

13.   Ledoux, H. On the validation of solids represented with the international standards for geographic information. *Comput. Aided Civ. Infrastruct. Eng.* **2013**, *28*, 693–706. [CrossRef]

14.   Stoter, J.; Ploeger, H.; van Oosterom, P. 3D cadastre in The Netherlands: Developments and international applicability. *Comput. Environ. Urban Syst.* **2013**, *40*, 56–67. [CrossRef]

15.   Stoter, J.E.; Ploeger, H.D. Property in 3D-registration of multiple use of space: Current practice in Holland and the need for a 3D cadastre. *Comput. Environ. Urban Syst.* **2003**, *27*, 553–570. [CrossRef]

16.   Stoter, J.; Salzmann, M. Towards a 3D cadastre: Where do cadastral needs and technical possibilities meet? *Comput. Environ. Urban Syst.* **2003**, *27*, 395–410. [CrossRef]

17.   Zevenbergen, J.A. *Systems of Land Registration: Aspects and Effects*; Netherlands Geodetic Commission (NCG): Delft, The Netherlands, 2002.

18.   Atazadeh, B.; Kalantari, M.; Rajabifard, A.; Ho, S.; Ngo, T. Building Information Modelling for High-rise Land Administration. *Trans. GIS* **2016**, *21*, 91–113. [CrossRef]

19.   Kadaster. 3D Deed. 2017. Available online: https://3d.bk.tudelft.nl/news/2016/03/21/3DKadaster.html (accessed on 30 March 2017).

20.   Thompson, R.; van Oosterom, P. Axiomatic Definition of Valid 3D Parcels, potentially in a Space Partition. In Proceedings of the 2nd Workshop on 3D cadastres, Delft, The Netherlands, 16–18 November 2011.

21.   Stoter, J.E.; van Oosterom, P.J.M. Technological aspects of a full 3D cadastral registration. *Int. J. Geogr. Inf. Sci.* **2005**, *19*, 669–696. [CrossRef]

22.   Döner, F.; Thompson, R.; Stoter, J.; Lemmen, C.; Ploeger, H.; van Oosterom, P.; Zlatanova, S. 4D cadastres: First analysis of Legal, organizational, and technical impact—With a case study on utility networks. *Land Use Policy* **2010**, *27*, 1068–1081. [CrossRef]

ISPRS Int. J. Geo-Inf. **2017**, *6*, 158

23. Döner, F.; Thompson, R.; Stoter, J.; Lemmen, C.; Ploeger, H.; van Oosterom, P.; Zlatanova, S. Solutions for 4D cadaster—With a case study on utility networks. *Int. J. Geogr. Inf. Sci.* **2011**, *25*, 1173–1189.

24. Van Oosterom, P.J.M.; Stoter, J.E.; Ploeger, H.D.; Thompson, R.; Karki, S. World-Wide Inventory of the Status of 3D Cadastres in 2010 and Expectations for 2014. In Proceedings of the FIG Working Week 2011 "Bridging the Gap between Cultures" & 6th National Congress of ONIGT, Marrakech, Morocco, 18–22 May 2011; Schennach, G., Ed.; Cadastre 2.0. pp. 117–122.

isprs International Journal of
Geo-Information

MDPI

Article

LandXML Encoding of Mixed 2D and 3D Survey Plans with Multi-Level Topology [†]

Rodney James Thompson [1,*], Peter van Oosterom [1] and Kean Huat Soon [2]

[1] Department OTB, GIS Technology Section, Delft University of Technology, P.O. Box 5030, 2600 GA Delft,
The Netherlands; P.J.M.vanOosterom@tudelft.nl
[2] Singapore Land Authority, 55 Newton Road, #12-01, Revenue House, Singapore 307987, Singapore;
soon_kean_huat@sla.gov.sg
[*] Correspondence: R.J.Thompson@tudelft.nl; Tel.: +31-15-2786950
[†] This paper is extended from the version presented at the 5th International FIG 3D Cadastre Workshop,
Athens, Greece, 18–20 October 2016.

Academic Editors: Efi Dimopoulou and Wolfgang Kainz
Received: 31 March 2017; Accepted: 5 June 2017; Published: 12 June 2017

Abstract: Cadastral spatial units around the world range from simple 2D parcels to complex 3D collections of spaces, defined at levels of sophistication from textural descriptions to complete, rigorous mathematical descriptions based on measurements and coordinates. The most common spatial unit in a cadastral database is the 2D land parcel—the basic unit subject to cadastral Rights, Restrictions and Responsibilities (RRR). Built on this is a varying complexity of 3D subdivisions and secondary interests. Spatial units may also be subdivided into smaller units, with the remainder being kept as common property for the owners/tenants of the individual units. This has led to the adoption of hierarchical multi-level schemes. In this paper, we explore the encoding of spatial units in a way that highlights their 2D extent and topology, while fully defining their extent in the third dimension. Obviously, topological encoding itself is not new. However, having mixed a 2D and 3D topological structure is rather challenging. Therefore, despite the potential benefits of mixed 2D and 3D topology, it is currently not used in LandXML, one of the main and best documented formats when representing survey data. This paper presents a multi-level topological encoding for the purposes of survey plan representation in LandXML that is simple and efficient in space requirements, including the question of curved surfaces, (partly) unbounded spatial units, and grouping and division of 2D and 3D spatial units. No "off the shelf" software is available for validating newly lodged surveys and we present our prototype for this. It is further suggested that the conceptual model behind this encoding approach can be extend to the database schema itself.

Keywords: 3D cadastres; survey plans; LandXML; InfraGML; multi-level topology; geographic information systems; rights; restrictions; responsibilities; spatial data infrastructure; real property

1. Introduction

In many jurisdictions, the cadastral survey plan is a critical instrument in the administration of property rights, being the starting point that defines the extent and location of the property. The secondary purpose of such a plan is as a data source for a database (and map) of cadastral information. With the growing trend towards digital submission of cadastral plans, there is a need to maintain the authoritative nature of the plan in the absence of a paper document. It is critical that the definitions of properties are correct and topologically sound, with adjoining properties identified in 2D and 3D.

2D land parcels (3D columns of space) or 3D spatial units may be subdivided into smaller spatial units, with the remainder being kept as common property for the owners/occupiers of the individual units—for example, townhouses within a 2D land parcel (spatial unit), with the driveways and gardens being held for common use, or a 3D building with both private spaces and common use spaces for the elevators, foyer etc. Multi-level schemes have been used to alleviate this complexity, with base 2D land parcels (3D columns of space) being subdivided into volumetric spatial units, which are in turn further subdivided into individual units.

In another example, a building may be placed on a base parcel, leaving property in common. It can be subdivided into volumetric spatial units for different classes of units (commercial, residential, etc.), leaving common property for entrance, elevators, etc. The volumetric spatial units can then be subdivided into individual units, with common property for the use of these unit owners/occupiers (but not for owners/occupiers of units in other volumes). It should be noted that when a volumetric unit is excised from a 2D parcel, the common property left will be partially unbounded: above and/or below.

The Land Administration Domain Model (LADM) provides for all of these levels of complexity [1,2], and it has been shown by Thompson et al. [3] that a mixed representation allows a relatively simple encoding of the full range of cadastral spatial units. The latter paper, however, does not address the issue of topological encoding of such a mixture of spatial units. The issues involved in encoding a survey plan (as distinct from a cadastral database) include some extra complexity.

This paper explores the practicality of topologically encoding spatial units, initially in LandXML [4], but with a view to also supporting InfraGML [5,6] when it is more mature; by demonstrating a topologically structured conceptual model for the purposes of survey plan representations, addressing the questions of curved surfaces, (partly) unbounded spatial units, and hierarchical grouping/division of 2D and 3D spatial units. The suggested method uses a form of mixed-dimensional topological structuring—sharing boundary definitions between spatial units that are simple and efficient in space requirements. It prevents problems of accidental overlap between spatial units in 3D, while providing a data source for a mixed 2D/3D digital cadastral database that minimizes redundancy and inconsistency. It is suggested that the conceptual model behind this approach can be extended to the cadastral database itself, including the requirement to maintain a historical record of the spatial unit structure (lineage).

Methodology

Several cases of plans of survey have been chosen, and "proof of concept" software written to (1) accept an encoding of the plan in a simplified form, (2) write the spatial units of the plan to a Postgres database, and (3) translate those spatial units into LandXML.

The path of the research has been: (1) the definition of axioms to ensure the validity of 3D spatial units [7], (2) the categorization of spatial units in terms of complexity and relative frequency of occurrence [8], (3) a representation of cadastral data enabling the mixing of 2D and 3D spatial units [3], and (4) the extension of the approach to a topological encoding scheme [9].

In what follows, Section 2 discusses the concept of survey plans and the research on topology for cadastral data. Section 3 describes a selection of 2D and 3D theoretical and real cases. Section 4 describes the conceptual encoding in these cases and demonstrates the actual encoding in LandXML and in future, InfraGML. In Section 5, the findings with respect to a proof-of-concept implementation are analyzed. Section 6 concludes the paper by summarizing the main results and indicating future work.

2. Background

This section first elaborates the purpose and role of the survey plans in the context of land administration (Section 2.1). Next, the importance of topology in a cadastral database is reviewed (Section 2.2).

2.1. Survey Plans

Typically, cadastral jurisdictions separate land administration into the act of defining the extent of a piece of land on a "Survey Plan" and the parties involved on a Title document (as in the case of the

Torrens Titling System) or Deed document. The "Rights Responsibilities and Restrictions" (RRR) recorded on a title or deed associates it with the land parcel as defined on the plan. Even though the term "Survey Plan" is used, in practice it may not always involve a survey in the conventional sense. Many types of technology are used, particularly where 3D spatial units are involved (e.g., laser scanner), to produce a combination of sketches, tabular data, and measurements that serve the purposes to which a survey plan is put. That is, the survey plan is the source document for the spatial unit.

Traditionally, a survey plan has been a paper document, which was reassuring in that it carried seals and signatures, and was suitable for long-term archive and storage. However, in this form, it was clumsy as a data source, especially when the preparation of these paper documents started to be a computer process. Recently, there has been an effort to switch to digital plans, containing, in structured and semantically enriched manner, the spatial and measurement data [10–13]. This has not changed the fundamental requirements of the survey plan as noted above. In addition, the move to 3D spatial units has led to a much greater complexity of the plans—needing to carry elevation diagrams and/or isometric views to make the geometry comprehensible. In Singapore, New Zealand, and several Australian States, the LandXML format [4] has been chosen to transport digital plans. The essential information carried on a survey plan is typically collected into a database to provide a multiple use cadastral database.

Currently, the Open Geospatial Consortium (OGC) is developing a standard named InfraGML [6], which is intended to replace LandXML [14]. This development is still in progress, but the standard has provision for survey data and land division. It is confidently expected that the techniques used to express this conceptual model in LandXML will carry across in large part to InfraGML, but not sufficient detail is present in the current draft to permit any detailed design at present.

2.2. Topology in Cadastral Data

The current way of using LandXML only defines geometry and not topology. For topology, it depends upon the software that reads it. For instance to encode a doughnut parcel, one not only has to define the two polygons but also to specify explicitly the relationship between these two polygons. When software reads in, it will only need to interpret the relationship that has syntactically been defined in the LandXML, even though the two polygons may follow the same parcel orientation.

The explicit encoding of the relationship will limit its use to a small group of software that can understand the relationship. By contrast, if a topology is encoded, say the outer ring of a parcel as counter-clockwise, and the inner ring as clockwise, a wider range of software can support the LandXML as this is the general rule of topology at least for ISO and OGC.

This becomes significantly more important in 3D modeling when perspective becomes significant. Let us imagine a void space (think it as a 3D empty box) contained within a 3D parcel. If one has to define explicitly every topological relationship between one surface to another, this is certainly not elegant and will be computing-power intensive. With topological encoding by specifying the parcel orientation into a certain direction for inner rings and outer rings, the proposed topological approach is a great improvement.

In 3D modeling for strata for instance, there exists typical floors or units within a building (typical floors refer to the exact same shape across different levels, while typical units refer to units of the same shape in different levels). The topological encoding approach will reduce tremendously the need to encode the common surfaces twice (or more), by simply adding an indicator of orientation to the encoding of a face or line.

The 2D spatial unit has a special place in a Cadastre. Often there is an identifiable "base layer" of 2D spatial units—interpreted as 3D prisms [15], which comprises a complete, non-overlapping coverage of the area administered by the jurisdiction. With the scarcity and value of land in modern cities, there is a strong trend to subdivision into explicit 3D spatial units. Typically, a 3D spatial unit which is to be associated via RRR with a party (person etc.) will be a closed volume, with a complete and well defined boundary (shell), but each time a closed volume is defined within a 2D spatial unit, it leaves a 3D "object" (a prism with a cavity). There is no volume to be determined for such a remainder spatial unit as it has an undefined top and bottom.

Alternatively, 3D spatial units may be defined, not by measurements but by the references to walls and floors/ceilings in a building that encloses them (or are planned to enclose them when the building is completed). There may be sketches on the plan of their location within the building, but the sketch is not the definition of the extents of the spatial unit (and may not have any measurements marked). In Queensland and Singapore, these do not have volumes defined, but are defined to have a certain floor area. These are known as Building Format Plans, and the spatial units defined by them as Building Format Units. This form of 3D spatial unit is the most common internationally, and it appears that all jurisdictions that recognize 3D subdivision use this form [11,16–18].

There has been considerable discussion on the subject of topological encoding of cadastral data in 2D over the years [19–21], One major advantage of the topological approach is the reduction in redundant storage of linework. There are different types of topology—from the simple single layer complete non-overlapping coverage, to the multi-layer with topological connections maintained between levels. In practice, a cadastral database needs to accommodate multiple levels of data—ranging from the simple property spatial unit, aggregated into administrative regions, and subdivided into 3D spatial units and into secondary interests (such as easements). This concept is addressed by the ISO19152 LA_Level class.

Figure 1 gives a rough schematic of the sub and super-sets of a basic spatial unit. Administrative areas may not consist of an integral number of whole spatial units, and secondary interests may span more than one base spatial unit.

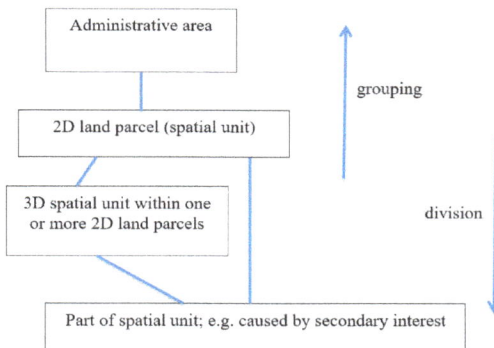

Figure 1. A rough "hierarchy" of spatial units.

It has been shown in [22] that such partial hierarchies can be accommodated in 2D, with the decision to use a number of single valued vector map (SVVM) layers, or alternatively more tightly as a multi-valued vector map (MVVM). Similarly, such SVVM or MVVM's could be defined in 3D. This is a question of balancing consistency against complexity. Integrating multiple levels in an MVVM enables the reuse of boundaries needed at two or more levels (good for consistency), but also causes some geometry fragmentation in other cases.

The question of extending the principle of topological encoding to 3D has been covered in detail, and in several ways, but always with the aim of a true 3D coverage—where all objects are volumetric [23,24]. By contrast, it has been shown that the vast majority of 3D cadastral parcels are relatively simple [8], and savings can be made by using that fact [3].

It is quite common that a complex 3D spatial unit can be comprised of a set of simpler units, either by the process of adding (forming the union of) smaller units, or subtracting inner excised areas. Regardless of what the process is, it raises a question of whether it is preferable to share common sub-unit definitions as far as the LandXML encoding is concerned. The following sections will address this question specifically.

3. Topological Representation of 2D and 3D Spatial Units

In this section, examples of three typical core spatial units is described: 2D Cadastral Spatial Units (Section 3.1), Simple 3D Spatial Units (Section 3.3), and Complex 3D Spatial Units (Section 3.4). Real-world cases of multi-level and hierarchical properties are introduced in Section 3.5 and 3.6. The scheme of categorization used here is from Thompson et al. [8], and is briefly:

- The 2D Land Parcel;
- The Building Format Unit (where the definition of the parcel is the actual building walls);
- The Polygonal Slice (where the units are described by vertical planes, with height limits);
- The Single-Valued Stepped Slice (where all surfaces are vertical or horizontal, and the unit does not have parts overlying other parts);
- The Multi-Valued Stepped Slice (where all surfaces are vertical or horizontal, but parts may overlie others); and
- General 3D Parcels (with few restrictions beyond validity).

3.1. 2D Cadastral Spatial Unit

2D topology can be modeled by representing cadastral boundaries as line strings, with encoding for the "left" and "right" base cadastral units [22,25,26]. For example, a "winged edge" structure can be extended to include non-base spatial units by including additional left/right encodings for non-base 2D spatial units (administrative areas, secondary interests, easements, etc.) [22]. This is the approach taken to develop a 2D topologically structured multi-layer Cadastral database such as the Netherlands Kadaster, with left and right references at multiple-levels: parcel, cadastral section, and municipalities [27].

For a mixed 2D/3D cadastral database, the 1D linestrings in 2D space that are used to define the cadastral boundaries are re-interpreted as 2D face strings in 3D space (as defined in the LADM) [1]. This does not change the database representation at all, adding nothing to the storage requirements because the storage of a LA_BoundaryFaceString is simply as a 1D linestring in 2D space. To illustrate this, let us consider Figure 2 as an example of a multi-valued-vector-map (MVVM) style encoding. In Figure 2,

(1)   Line segment *a* has Lot 25 and Easement C on **left**; Lot 26 on **right**;
(2)   Line segment *b* has Lot 25, Easement A and C on **left**; Lot 26 and Easement A on **right**;
(3)   Line segment *c* has Lot 25 and Easement A on **left**; Lot 26 and Easement A on **right**;
(4)   Line segment *d* has Lot 25 on **left**, Lot 26 on **right**;
(5)   Line segment *e* has Lot 25 on **left**;

- Therefore, Lot 25 is defined as being on the left of line segments *a,b,c,d,e,f,g,h,i,j* and both sides of *t,s,r,v,u*;
- Lot 26 is left of *k,l,m,n*, right of *a,b,c,d* and both sides of *q,p*;
- Easement A is left *i* and *m*, right of *t,p,q,r,s* and both sides of *b,c,v*.

Note that in some implementations, links where the same parcel is on both sides of a line are omitted. This will be the approach taken here.

The 2D spatial units are "converted" into 3D spatial units by replacing each line segment in their boundary by a face running from $-\infty$ to $+\infty$, and passing through the endpoints. The outside of any face is that side from which it appears to be anticlockwise (i.e., from the positive direction of the normal vector), so that the same definitions apply, with the words "on left" replaced by "positive side of" or

"+", and "on right" with "negative side of" or "−". In this way, Parcel 25 is now the positive side of face string *a,b,c,d,e*, and Parcel 26 is on the negative side of *a,b,c,d*.

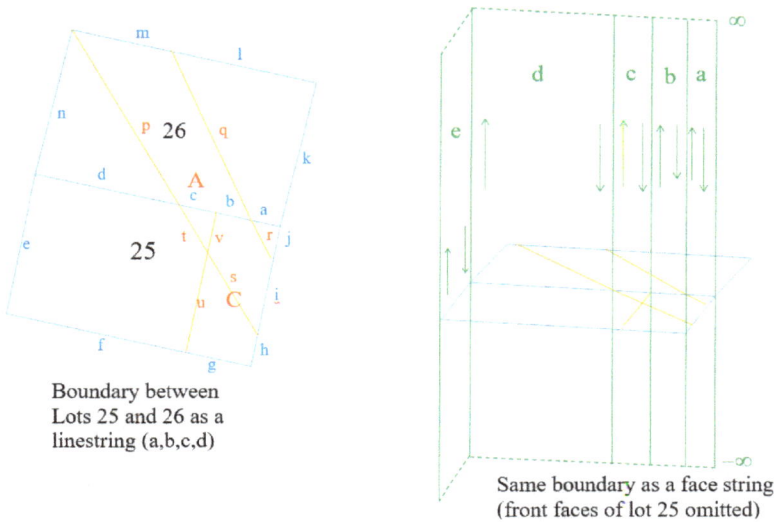

Boundary between
Lots 25 and 26 as a
linestring (a,b,c,d)

Same boundary as a face string
(front faces of lot 25 omitted)

Figure 2. Interpreting a line string topological boundary as a LA_BoundaryFaceString. Base Parcels 25 and 26 are burdened by two easements, A and C. A runs diagonally across both base parcels, while C overlays a rectangular area of Parcel 25 only.

3.2. Extension to 3D

The approach as suggested by [3] makes use of some of the specific features of cadastral data, and is suggested by the conventional form of survey plans. When defining a 3D subdivision, the survey is commonly first introduced by a "plan view," which defines the 2D shape of the subdivision. This is followed by various elevation and isometric views that define the faces that bound the volumes in question.

In brief, 2D spatial units (which correspond to the plan view) are viewed as a column of space, unbounded above and below. The "polygon" that conventionally defines a 2D parcel then is interpreted as the "footprint" of this column. Extending this, a 3D spatial unit is represented by a footprint, which is then restricted by faces above and below the actual parcel. That paper [3] did not address the topology either within individual parcels or between parcels. This paper and [9] use the mixed approach, extended to define a topological structure. Using the language of the LADM, the footprint is defined as a set of "face strings" (LA_BoundaryFaceString) [2], while the faces use LA_BoundaryFace definitions.

3.3. Simple 3D Spatial Unit

The simplest of the 3D spatial units are the "Polygonal Slice" and the "Above/Below Elevation" spatial units [3]. An example is shown in Figure 3. These are a prism of space with vertical faces, delimited above and/or below by surfaces—usually horizontal. Only slightly more complex, those with a well-defined top and bottom surface, which are not per se exactly planar (hence the triangulation of these surface in the example of Figure 3).

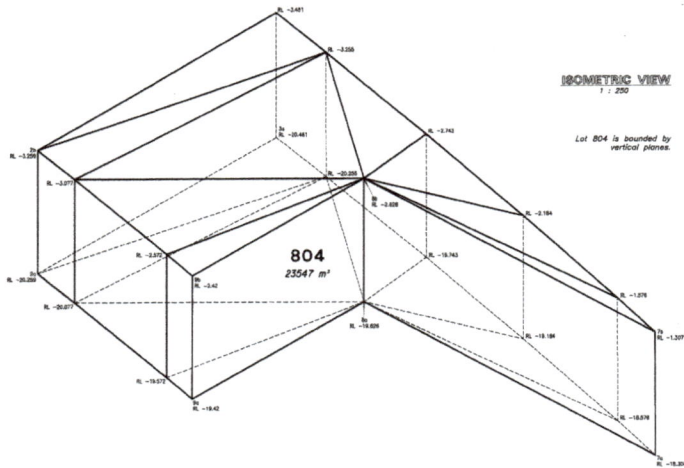

Figure 3. A simple 3D (underground) spatial unit with vertical walls corresponding to surface parcel boundaries, but with non-horizontal top and bottom reflecting the slope of the tunnel.

3.4. Complex 3D Spatial Units

This is the most generic situation, which does not necessarily define the 3D Spatial Unit as face strings, as the boundaries have arbitrary orientation and might even be curved. Figure 4 illustrates an artificial example of this kind of spatial unit. These Spatial Units are relatively rare, and in some cases, our proposed storage scheme may require more storage space for consistency than would the conventional 3D polyhedron stored as a set of faces. However, having the 1D line string in 2D space is still useful as it can be applied to depict the footprint of the 3D spatial unit on the traditional 2D cadastral map, so it is always included in this representation.

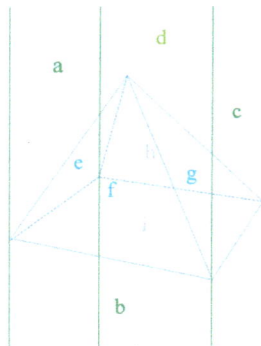

Figure 4. Complex spatial unit in the shape of a pyramid.

3.5. Multi-Level Case of a Tunnel Below a Building

This case consists of a partially unbounded spatial unit with development where a tunnel (3D spatial unit) has been created underneath. At and above ground level is a multi-unit building, while below ground is a road tunnel unit. Thus, it combines the issues of a building format plan, a volumetric excision from a 2D parcel, and a complex spatial unit unbounded above and below. This was described in some detail by [9], with the encoding addressed in Section 4.4.

The subdivision of surface parcels where they are affected by subterranean infrastructure, and the converse breaking of infrastructure at surface parcels is a complex question, which is described in Section 4.1.1 of Stoter and van Oosterom [15]. The case illustrated here (Figure 3) is defined as required by Queensland legislation, where new infrastructure must be legally "broken" at existing surface parcel boundaries, but newer surface boundary changes do not necessarily affect the subsurface parcels (Queensland Government Cadastral Survey Requirements, Section 10.2.4) [28].

3.6. Multi-Level Hierarchical Case—A High-Rise Building

This case is a highrise building (Figure 5). It is subdivided into sections, with different uses (residential, commercial etc.) e.g., Lot 3, and these sections are further subdivided into units—e.g., Unit 3301. As a result of this hierarchical subdivision, there are volumes of common property, which are available to all tenants of the building (Lot 5, defined as the remains of original 2D lot after the 3D excisions), but also more restricted volumes such as the remainder of Lot 3, which is only available to owner/occupiers of the units of Lot 3, but not those of Lots 2 or 4. This applies in particular to restricted lift wells. Note that the subdivision of lots into parts (Parts A to M) is for convenience of definition.

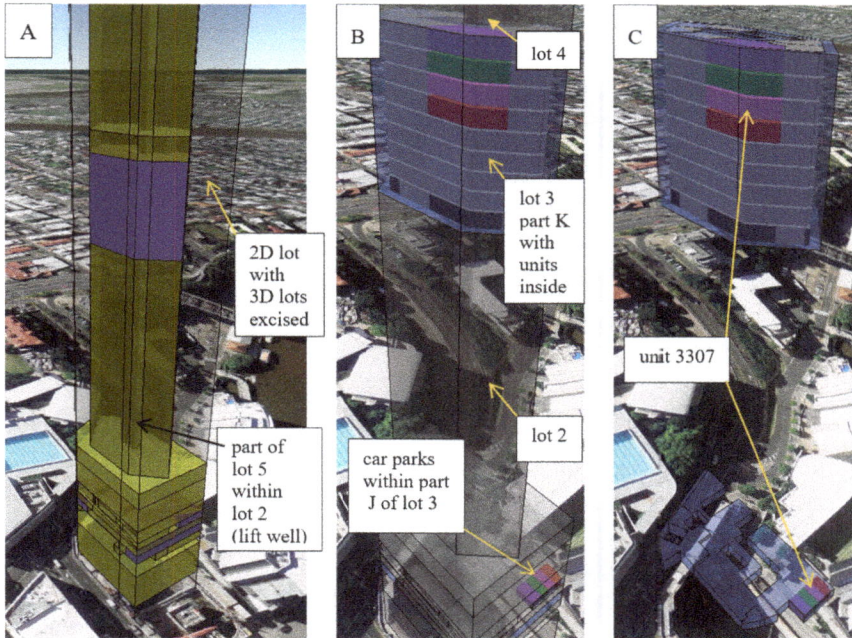

Figure 5. (A) The 2D lot shown in grey with the 3D lots (Lots 2, 3, and 4) shown in solid fill. The lot of interest here is Lot 3, shown in blue; **(B)** the 3D lots (2, 3, and 4) made transparent to allow the building format lots to be seen; **(C)** Lot 3 only is shown with some of its building format lots highlighted (Red Lot 3603, Purple Lot 3703, Green Lot 3803, Blue Lot 3903). It should be noted that this figure reflects quite a complex situation (together with the text in Section 3.6 and additional explanation of the structure in Section 4.5, the complex situation should become clear).

The subdivision began life as a simple 2D spatial unit, which was divided into four volumetric spatial units (defined by measurements before the building was constructed). Then three of these volumetric units were further subdivided into units, based on the walls of the building under construction. The remainder of the original 2D parcel is the grey prism in Figure 5A and is identified as Lot 5 on SP217742. On plan SP217742, it is subdivided in 3D into Lots 2, 3, and 4, and some road, with Lot 5 being the remainder of the prism of space after these are excised. In Figure 5A, the outer boundary of Lot 5 is made transparent, with Lots 2 and 4 being shown in yellow. Lot 3 is highlighted in blue.

Lot 3/SP217742 was then further subdivided on plan SP217743. Note that Lot 3 consists of two disconnected parts. These are joined by common property (a lift well, part of Lot 5 which runs up the interior and can be glimpsed in Figure 5A). The conceptual modeling and expression of the plan in LandXML is discussed in Section 4.5.

Lot 3 Part K

As an example of a relatively simple part of this complex building, consider Part K of Lot 3 (Figures 5B and 6). This is a simple slice, with a polygonal footprint, and three lift shafts running through it. These shafts are not the same cases—two of them (Lot 2 Part K, and Lot 4 Part K—with the lot number in dashed font) are sections of other volumetric spatial units, excised from Lot 3, but Lot 5 (with the lot number in solid font—indicating a base parcel) is a remnant of the original 2D spatial unit (and is connected to +∞ and to −∞). Encoding of this is discussed in Section 4.5.

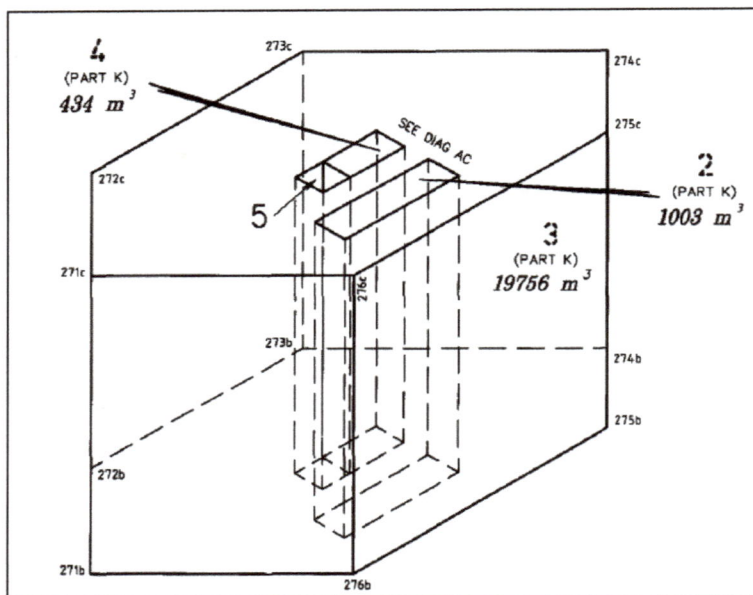

Figure 6. Excerpt from plan SP217742 showing part of Lot 3 (part K).

4. Representation in LandXML

The LandXML specification is quite complex, but for the purposes of this discussion, only three elements are required: **Parcel**, **Parcels**, and **CoordGeom**:

1. **Parcel:** in LandXML, the term "parcel" is overloaded to include volumes, faces, and face strings. The class attribute of the parcel element makes this explicit (Face, Lot, FaceString, etc.)
2. **Parcels:** Parcels are a collections of parcels that are used to define a higher-level parcel. For example, a set of Faces and FaceStrings that define a volume.
3. **CoordGeom:** Strictly speaking, this defines a chain of straight line segments, but traditionally in LandXML encoding, a closed CoordGeom is assumed to be a planar surface (polygon).

In this section, the topological modeling of the cases identified in Section 3 are addressed in terms of the conceptual structure and possible implementation in LandXML at varying levels of detail. The simpler cases are presented as a simple element model, while the more complex cases have been in part encoded in LandXML. Section 4.4 revisits the case introduced in Section 3.5, while Section 4.5 expands on the case from Section 3.6. The spatial logic of the encoding is addressed in Section 4.6, the encoding issues particular to curved surfaces in LandXML in Section 4.7, and alternate encodings in Section 4.8.

4.1. 2D Cadastral Spatial Units

Returning to the case of Section 3.1, Figure 2, a possible element structure, in part, would be as shown in Table 1. The sharing of face string elements should be noted, and compared with a non-topological approach, in which the elements for FaceString d and m would need to be duplicated.

Table 1. Excerpt from Face Strings Table.

Face String	+ Spatial Unit(s)	− Spatial Unit(s)
a	Lot 25	Lot 26
b	Lot 25	Lot 26
d	Lot 25	Lot 26
i	Easement A, Lot 25	
p		Easement A
q		Easement A

The encoding in LandXML could be as follows:

```
<Parcel class="FaceString" name="a">
   <CoordGeom><Line><Start pntRef="1"/><End pntRef="2"/></Line>
   </CoordGeom></Parcel>
<Parcel class="FaceString" name="b">
   <CoordGeom><Line><Start pntRef="2"/><End pntRef="3"/></Line>
   </CoordGeom></Parcel> (etc.)
<Parcel class="Lot" name="25" parcelFormat="Standard">
   <Parcels>
      <Parcel pclRef="a" />    <Parcel pclRef="b"/> (etc.)
      <Parcel pclRef="i"/> (etc.)
   </Parcels>
</Parcel>
<Parcel class="Lot" name="26" parcelFormat="Standard">
   <Parcels>
      <Parcel pclRef=" a" />    <Parcel pclRef=" b"/> (etc.)
      <Parcel pclRef="k"/> (etc.)
   </Parcels>
</Parcel>
<Parcel class="Easement" name="A" parcelFormat="Standard">
   <Parcels>
      <Parcel pclRef="i" />    <Parcel pclRef="r"/> (etc.)
   </Parcels>
</Parcel>   … etc.
```

Note that a negative reference to a face string is indicated by the character "¬" rather than a minus sign. This will be discussed later in the paper (Section 4.6).

4.2. Simple 3D Spatial Unit

The example in Figure 3, Section 3.3 (Reproduced in Figure 7), is clearly well catered for in this approach, where the top and bottom are not horizontal (as indicated by the reduced level values of the vertices—RL in the plan) and therefore have been triangulated (by the surveyor) to ensure the planarity of all faces.

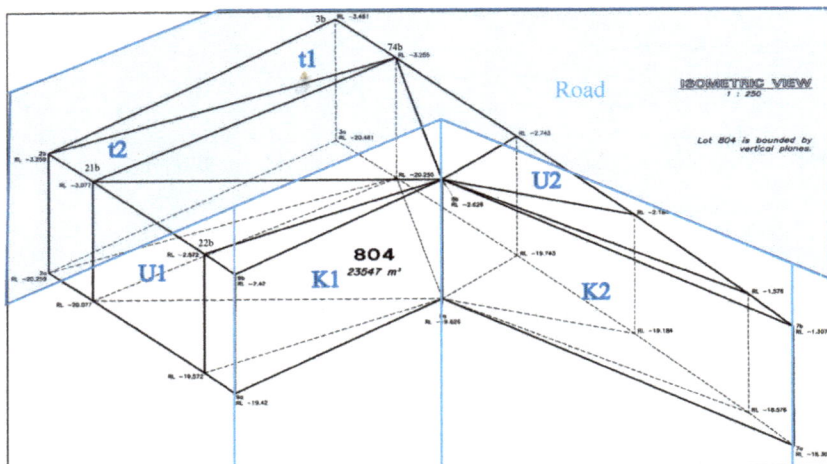

Figure 7. Figure 3, with the addition of the surface road parcel.

The encoding in LandXML could be as follows:

```
<Parcel class="FaceString" name="U1">
  <CoordGeom><Line><Start pntRef="2"/><End pntRef="21"/></Line>
<Line><Start pntRef="21"/><End pntRef="22"/></Line>
<Line><Start pntRef="22"/><End pntRef="9"/></Line>
  </CoordGeom></Parcel>
<Parcel class="FaceString" name="K1">
  <CoordGeom><Line><Start pntRef="9"/><End pntRef="8"/></Line>
  </CoordGeom></Parcel>
<Parcel class="FaceString" name="K2">
  <CoordGeom><Line><Start pntRef="8"/><End pntRef="7"/></Line>
  </CoordGeom></Parcel>
<Parcel class="Face" name="t1">
  <CoordGeom><Line><Start pntRef="2b"/><End pntRef="74b"/></Line>
    <Line><Start pntRef="74b"/><End pntRef="3b"/></Line>
      <Line><Start pntRef="3b"/><End pntRef="2b"/></Line>
</CoordGeom></Parcel>
(etc.)
<Parcel class="Lot" name="904" parcelFormat="Volumetric">
  <Parcels>
    <Parcel pclRef="U1"/>      <Parcel pclRef="K1"/>
    <Parcel pclRef="K2"/> (etc.)
    <Parcel pclRef="t1"/>      <Parcel pclRef="t2"/>
      (etc.)
  </Parcels>
</Parcel>
<Parcel class="Road" parcelFormat="Standard">
  <Parcels>
    <Parcel pclRef="K1" />      <Parcel pclRef="K2"/> (etc.)
    <Parcel pclRef="¬t1"/>      <Parcel pclRef="¬t2"/>
<Parcel pclRef="u1"/>  (etc.)
  </Parcels>
</Parcel>
etc.
```

Here, compared to the non-topological approach, face strings K1 and K2 are shared by the underground spatial unit (Lot 804) and the 2D road parcel. By contrast, the face string U1 is not a boundary of the road parcel, and would not be shared, but a face u1 would be created, as the part of

U1 that separates Lot 804 from the remainder of the road. The faces t1, t2, etc. are shared between Lot 804 and the remainder of the road parcel, but in this case, the sense is reversed.

4.3. Complex 3D Spatial Units

In the case shown in Figure 4, the face strings *a*, *b*, *c*, *d* of the outer face string do not add anything to the definition of the pyramid—which is fully defined by the tilted faces *e*. *f*, *g*, *h*, and horizontal face *i*, but do provide the consistency that permits the database to be viewed as a 2D repository, by simply accessing the boundary face strings as linestrings [3]. This is a small cost for a significant advantage. The encoding in LandXML could be as follows:

```
<Parcel class="FaceString" name="a">
    <CoordGeom> detail </CoordGeom></Parcel>
<Parcel class="FaceString" name="b">
    <CoordGeom> detail </CoordGeom></Parcel>
<Parcel class="FaceString" name="c">
    <CoordGeom> detail </CoordGeom></Parcel>
<Parcel class="FaceString" name="d">
    <CoordGeom> detail </CoordGeom></Parcel>
<Parcel class="Face" name="e">
    <CoordGeom> detail </CoordGeom></Parcel>
<Parcel class="Face" name="f">
    <CoordGeom> detail </CoordGeom></Parcel>
<Parcel class="Face" name="g">
    <CoordGeom> detail </CoordGeom></Parcel>
<Parcel class="Face" name="h">
    <CoordGeom> detail </CoordGeom></Parcel>
<Parcel class="Face" name="i">
    <CoordGeom> detail </CoordGeom></Parcel>
<Parcel class="Lot" name="L1" parcelFormat="Volumetric">
    <Parcels>
        <Parcel pclRef="a"/>    <Parcel pclRef="b"/>
        <Parcel pclRef="c"/>    <Parcel pclRef="d"/>
        <Parcel pclRef="e"/>    <Parcel pclRef="f"/>
        <Parcel pclRef="g"/>    <Parcel pclRef="h"/>
        <Parcel pclRef=" i"/>
    </Parcels>
</Parcel>
<Parcel class="Lot" name="Common Property" parcelFormat="Standard">
    <Parcels>
        <Parcel pclRef="a"/>    <Parcel pclRef="b"/>
        <Parcel pclRef="c"/>    <Parcel pclRef="d"/>
        <Parcel pclRef=" e"/>    <Parcel pclRef=" f"/>
        <Parcel pclRef=" g"/>    <Parcel pclRef=" h"/>
        <Parcel pclRef="i"/>
    </Parcels>
</Parcel>
etc.
```

This storage schema does not depend on the concept of a top surface and a bottom surface—it is only described in these words for clarity. Where there is no clear separation, the set of faces is sufficient as long as the correct orientation of the faces is maintained. Further, if in the survey plan there are multiple 3D spatial units defined, then faces of adjoining spatial units are also shared—one unit on the positive side and one on the negative. Again, this is realized by using signed references to the shared elements, i.e., the shared faces.

4.4. Case of a Tunnel Below a Building (from Section 3.5)

In the case shown in Figures 8 and 9, initially the spatial unit was a simple 2D lot with 4 sides (Lot 10), adjoining other lots on two sides, and road on the others. A tunnel was constructed below, the volume being excised from it below ground, leaving a remainder lot unbounded above and below.

At a later date, a five-story building was constructed, an easement (labeled as "EMT A" in Figure 8) was put through the lot, and the road corner was truncated. This resulted in the creation of the building format plan but had no effect on the tunnel lot (Lot 210). Although Lot 210 is noted on the new plan, its definition is still provided by the previous plan, and the truncation of the corner does not apply to it (This is in line with the Queensland regulations as mentioned in Section 3.4). Lot 10 then becomes the base lot from which the building format lots are excised. Lot 10 is known from this time on as Lot CP (for Common Property) on the new plan. The tunnel parcel (Lot 210) is unchanged except for the splitting of the face strings and top and bottom faces (Figure 9). The resultant face strings and faces tables are illustrated in Tables 1 and 2.

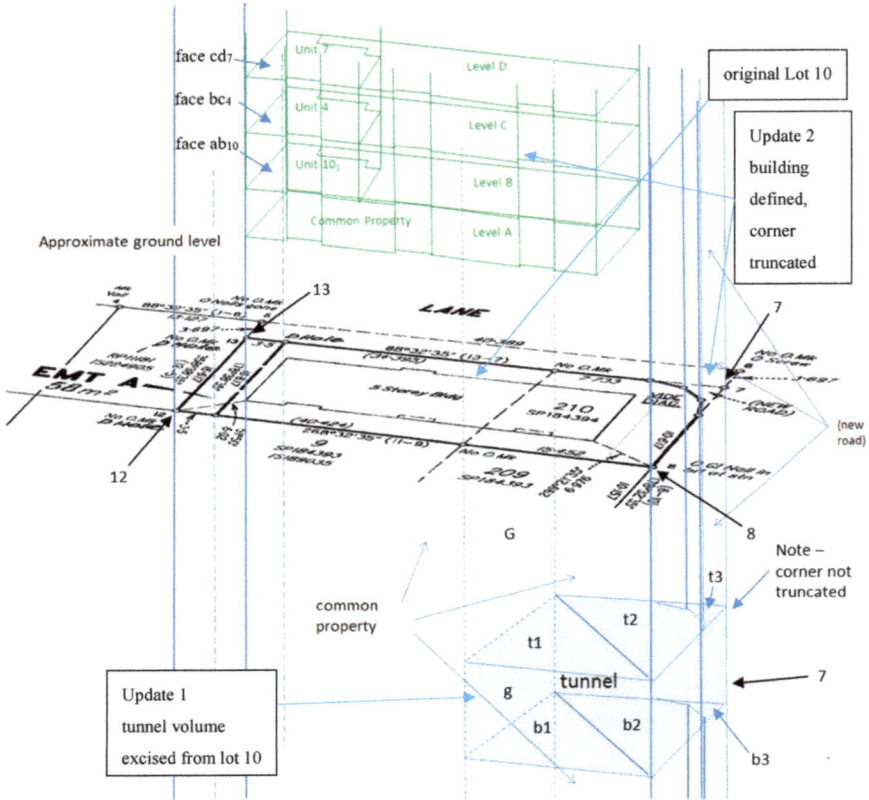

Figure 8. Side view, showing original 2D lot (Lot 10 bounded by corners 12, 13, 7, and 8), the volume of the part of the tunnel below this lot, the corner truncation (removing corner 7 from the surface lot), and four floors of the building near and above ground level. It should be noted that this figure reflects quite a complex situation with 3D parcels above and below the surface at the same location (together with the earlier text in Section 3.5 and the explanation in the current Section 4.4, the complex situation should become clear).

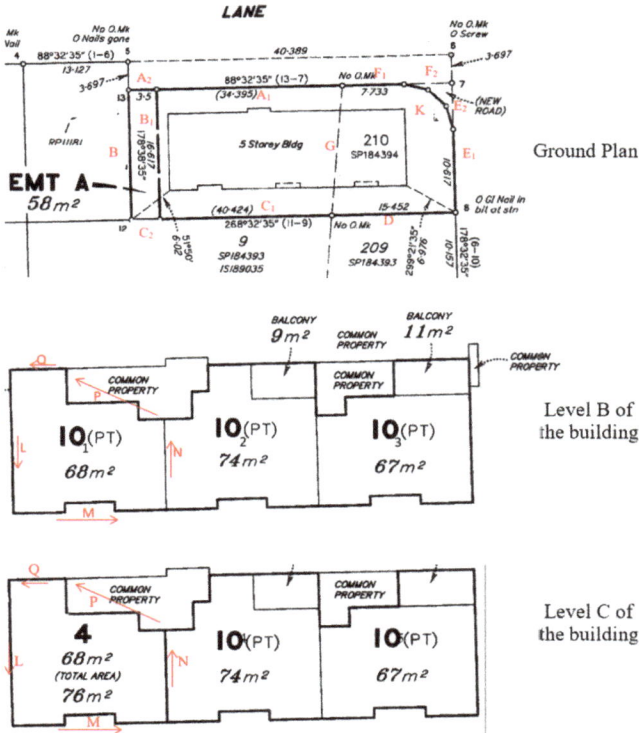

Figure 9. Detail of the building subdivision showing labels used in Table 2. Note that the face string definitions L, M, N, P, and Q are used in the definition of units on multiple floors.

Using the labels from Figures 8 and 9, where face strings are shown as capital letters, and faces as lowercase, the following encoding is possible (Tables 2 and 3).

Table 2. Face strings table.

Line	+ Spatial Unit(s)	− Spatial Unit(s)
A_1	Lot CP	Mark Lane (road)
A_2	Lot CP, Easement A	Mark Lane (road)
B	Lot CP, Easement A	Lot 1/RP11181
B_1	Easement A	
C_2	Lot CP, Easement A	Lot 9/SP184393
C_1	Lot CP	Lot 9/SP184393
D	Lot CP, Lot 210	Lot 9 and Lot 209/SP184393
E_1	Lot CP, Lot 210	Main Street (road)
E_2	Lot 210, New Road	Main Street (road)
F_2	Lot 210, New Road	Mark Lane (road)
F_1	Lot CP, Lot 210	Mark Lane (road)
G	Lot 210	
K	Lot CP	New Road
L	Unit 10 part 1, Unit 4	
M	Unit 10 part 1, Unit 4	
N	Unit 10 part 1, Unit 4	Unit 10 part 2, Unit 10 part 4
P	Unit 10 part 1, Unit 4	
Q	Unit 10 part 1, Unit 4	

Table 3. Faces table.

Face	+ Spatial Unit(s)	− Spatial Unit(s)
t_1	Lot 210	Lot CP
t_2	Lot 210	Lot CP
t_3	Lot 210	New Road
b_1	Lot CP	Lot 210
b_2	Lot CP	Lot 210
b_3	New Road	Lot 210
g	Lot 210	Lot CP
ab_{10}	Unit 10 part 1	Common Property
bc_4	Unit 4	Unit 10 part 1
cd_7	Unit 7 (not in diagram)	Unit 4

Thompson et al. (2016b page 148) presents excerpts from an encoding of this case, which is not repeated here.

4.5. Multi-Level Hierarchical Case—High-Rise Building from Section 3.6

This was pictured in Figure 5 as a 2D parcel subdivided into volumetric parcels, some of which are further subdivided by building format plans. The hierarchy of this decomposition is shown in Figure 10.

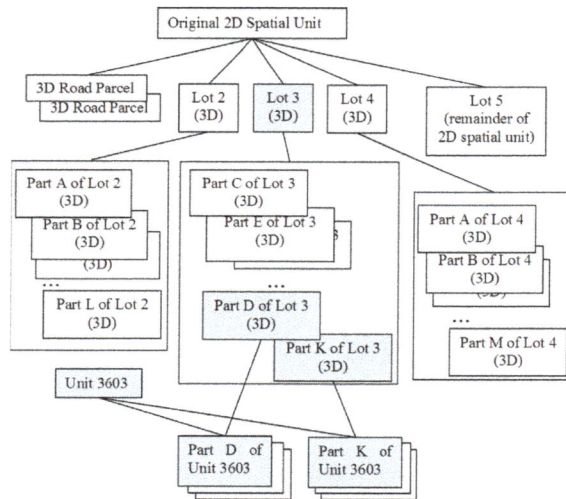

Figure 10. The hierarchy of spatial units and part spatial units, and their use in the definition of basic allocation units.

4.5.1. Encoding of Lot 3 Part K

Please see Figures 5, 6 and 10. The encoding proceeds by defining the outer and inner polygonal boundary of the slice as follows:

```
<Parcel class="FaceString" name="TowerOuter">
    <CoordGeom><Line><Start pntRef="271"/><End pntRef="276"/></Line>
     <Line><Start pntRef="276"/><End pntRef="275"/></Line>
     <Line><Start pntRef="275"/><End pntRef="274"/></Line>
     <Line><Start pntRef="274"/><End pntRef="273"/></Line>
     <Line><Start pntRef="273"/><End pntRef="272"/></Line>
     <Line><Start pntRef="272"/><End pntRef="271"/></Line>
    </CoordGeom></Parcel>
<Parcel class="FaceString" name="LiftWell5Inner">
    <CoordGeom><Line><Start pntRef="23"/><End pntRef="20"/></Line>
     <Line><Start pntRef="20"/><End pntRef="21"/></Line>
     <Line><Start pntRef="21"/><End pntRef="22"/></Line>
     <Line><Start pntRef="22"/><End pntRef="23"/></Line>
    </CoordGeom></Parcel>
```

Noting that both face strings are encoded in strict order—outer is anticlockwise, inner clockwise. Note also that the repetitive form of the line definition, repeating the node identifiers, is the only form permitted by LandXML at this time.

Then the top and bottom planes are defined. In this case, as described in earlier papers [3,9], the horizontal faces that define the tops and bottoms of the polygonal slice parcels can be expressed as an infinite plane and do not need to repeat the polygon definitions—so that:

```
<Parcel class="Face" name="KTop">
    <CoordGeom><Line><Start pntRef="271b"/><End pntRef="276b"/></Line>
     <Line><Start pntRef="276b"/><End pntRef="275b";/></Line>
     <Line><Start pntRef="275b"/><End pntRef="274b";/></Line>
     <Line><Start pntRef="274b"/><End pntRef="273b";/></Line>
     <Line><Start pntRef="273b"/><End pntRef="272b";/></Line>
     <Line><Start pntRef="272b"/><End pntRef="271b";/></Line>
    </CoordGeom></Parcel>
```

and, with a minor extension of the LandXML schema, this could be replaced by the following:

```
    <Parcel class="Face" name="KTop" a="0" b="0" c="-1" d="137.625"/>
```

(which is equivalent to defining the boundary face as "Z < 137.625", or equivalently $aX + bY + cZ + d \geq 0$ with $a = b = 0$, $c = -1$, $d = 137.625$).

Note that this simplification does require a change to the LandXML specification. If this is not possible, the original encoding can be used, but at the cost of more redundancy in encoding.

Next, the two other lift well face strings are defined as follows:

```
      <Parcel class="FaceString" name="LiftWell2Outer">
      <CoordGeom><Line><Start pntRef="25"/><End pntRef="24"/></Line>
       <Line><Start pntRef="24"/><End pntRef="28"/></Line>
       <Line><Start pntRef="28"/><End pntRef="29"/></Line>
       <Line><Start pntRef="29"/><End pntRef="25"/></Line>
      </CoordGeom></Parcel>
<Parcel class="FaceString" name="LiftWell4Outer">
      <CoordGeom><Line><Start pntRef="23"/><End pntRef="132"/></Line>
       <Line><Start pntRef="132"/><End pntRef="131"/></Line>
       <Line><Start pntRef="131"/><End pntRef="22"/></Line>
       <Line><Start pntRef="22"/><End pntRef="23"/></Line>
      </CoordGeom></Parcel>
```

Naturally, the face strings LiftWell2Outer and LiftWell4Outer will be used in the definitions of Lot 2 and Lot 4. Finally, Part K of Lot 3 can be defined as follows:

```
<Parcel class="LOT" name="3K" parcelFormat="Volumetric">
   <Parcels>
      <Parcel pclRef="TowerOuter"/>  <Parcel pclRef="LiftWell5Inner"/>
      <Parcel pclRef="KTop"/>             <Parcel pclRef="¬JTop" />
      <Parcel pclRef="¬LiftWell2Outer"/>
         <Parcel pclRef="¬LiftWell4Outer"/>
   </Parcels>
```

where the "¬" (not sign) preceding the pclRef of a 2D "parcel" indicates that the "minus" side of the face string or face is used in the definition.

LiftWell5 defines part of Lot 5, which is a 3D spatial unit, but is open above and below. It can be quite simply defined by taking the original 2D lot, excising the volumetric Lots 2, 3, and 4 and the 3D road parcels. So Lot 5, (the remainder) would be defined as follows:

```
<Parcel class="FaceString" name="Lot5Outer">
   <CoordGeom> (the definition of the original 2D lot as a footprint
   face string)
   </CoordGeom></Parcel>
<Parcel class="LOT" name="5" parcelFormat="Volumetric">
   <Parcels>
      <Parcel pclRef="Lot5Outer" />     <Parcel pclRef="-4" />
      <Parcel pclRef="-3"/>             <Parcel pclRef="-2" />
      <Parcel pclRef="-Road" />  ...
   </Parcels>
</Parcel>
```

where the "−" preceding a pclRef for a 3D parcel indicates that it is subtracted from the main parcel. Since Lots 2, 3, and 4 all individually will have LiftWell5 excised from them, the space is included in Lot 5.

4.5.2. Encoding the Units within Lot 3

The next plan further subdivides Lot 3 into building unit lots, most of which are in two parts—the actual apartment (shown in Figure 11) and the associated car park. The apartments are defined purely by the walls and floor/ceiling of the structure—hence no measurements have been shown on the plan. The car parks are defined by bearing and distance in plan (there being no walls separating car parks in this case), but by the building floor and ceiling.

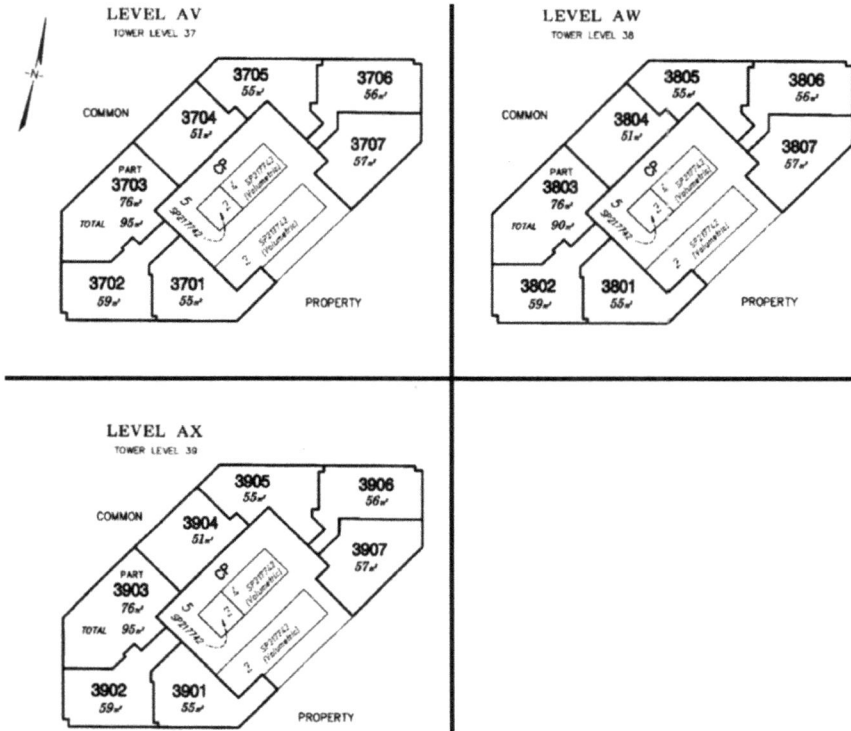

Figure 11. Subdivision of Lot 3 part K into 63 units and common property.

Floors 31 to 39 are subdivided in the same basic format, so the face string definition of the unit footprints can be shared between floors, thus allowing the residential parts of the units to be encoded quite simply as follows:

```
<Parcel class="LOT" name="3701U" parcelFormat="Building">
   <Parcels>
    <Parcel pclRef="Unit3101-3102" /> <Parcel
    pclRef="Unit3101Outer"/>
    etc.
    <Parcel pclRef="¬36Top"/> <Parcel pclRef="37Top"/>
   </Parcels>
</Parcel>
<Parcel class="LOT" name="3801U" parcelFormat="Building">
   <Parcels>
    <Parcel pclRef="Unit3101-3102" /> <Parcel
    pclRef="Unit3101Outer"/>
    etc.
    <Parcel pclRef="¬37Top"/> <Parcel pclRef="38Top"/>
   </Parcels>
</Parcel>
```

The car parks can be defined in a similar fashion as 3201P etc . . . , allowing the units to be defined as follows:

```
<Parcel class="LOT" name="3201" parcelFormat="Building">
  <Parcels>
    <Parcel pclRef="+3201U" />      <Parcel pclRef="+3201P"/>
  </Parcels>
</Parcel>
```

Note that the face 38Top could have be defined as 39Bottom, but with c = "1" and d = "−116.5". In that case, the reference in the definition of lot 3801U would have read pclRef = "¬39Bottom".

4.6. Spatial Logic Issues

The interpretation of the LandXML is critical, especially since XML requires that the physical order of elements within the file should never be assumed to be significant. Thus in this encoding:

```
<Parcel class="LOT" name="L" parcelFormat="Volumetric">
  <Parcels>
    <Parcel pclRef="OuterFootprint"/>
    <Parcel pclRef="CourtyardFootprint"/>
    <Parcel pclRef="Top1"/>         <Parcel pclRef="¬Top2" />
    <Parcel pclRef="-LiftWell"/>  <Parcel pclRef="+RoofShed"/>
  </Parcels>
</Parcel>
```

The references should be interpreted as follows: a pclRef to a 3D parcel (such as RoofShed and LiftWell) must have a "+" or "−", with "+" interpreted as the union of the referenced parcel with the current parcel, "−" indicates a subtraction (current parcel "and not" the referenced parcel). Two-dimensional referenced parcels (faces such as Top1 or face strings such as OuterFootprint) either have a not symbol "¬" to indicate that the face is to be reversed, or no special character. The priority should be that the 2D face strings and fully defined faces should be processed first, followed by the faces defined using the "a,b,c,d" notation, followed by the 3D referenced parcels, subtractions before unions. In this example, this would lead to the same order as the XML implies (if LiftWell and RoofShed are pre-defined 3D spatial units) (see Figure 12).

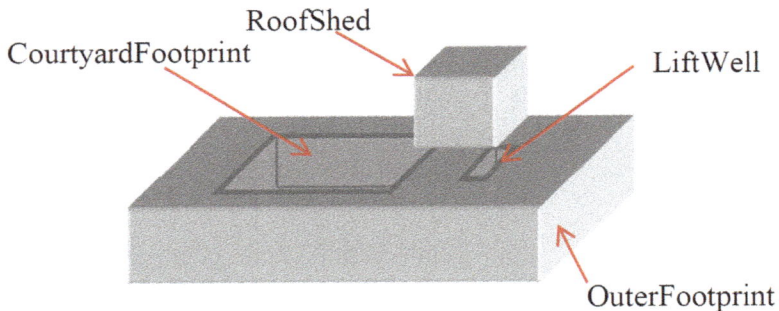

Figure 12. Note that the roof shed overlaps and partially blocks the CourtyardFootprint and LiftWell. This is because the "+" operation is done after the boundaries are assembled, and after the lift well is subtracted.

4.7. Curved Surfaces

Several jurisdictions, as in Queensland or the Netherlands, permit curved boundary lines, and curved boundary surfaces in the definition of cadastral spatial units. As mentioned before, where a spatial unit boundary is defined as curved, the plan must record the details of that curve. While approximating this with short lines and facets is acceptable and often necessary for presentation, the definition in a survey plan is the legal situation. As an example, LandXML is used in a number of jurisdictions [29] and does have some options to define curved lines and surfaces. In practice, the commonly used curves can be accommodated in LandXML (circular arcs, cylindrical faces, conic faces), but not spherical/ellipsoidal surfaces or elliptical lines, which would need to be notated as text on a paper survey plan, and could not therefore be processed automatically.

LandXML, having been defined originally for different purposes has an unusual combination of primitives: It has 0D primitives (Centre), 1D primitives (Line, Irregular line, etc.), and 3D primitives (VolumeGeom), but no explicit 2D primitive that can be used in the definition of a "Parcel" (spatial unit). The only geometric elements available within the Parcel element are Center (a point), CoordGeom (collection of line, curve, and/or spiral elements) (spiral curves are used in road and rail design), and VolumeGeom. An assumption is made that a CoordGeom that closes defines a surface and that, if any of the CoordGeom elements are curved, then the simplest curved surface that passes through those lines is intended. For example, in Figure 13, the definition of the two volumes would be identical in LandXML, but the simplest, A, would be inferred. This cannot be said to provide a truly unambiguous definition of the spatial unit. There is no agreed LandXML encoding for a negatively curved surface such as Figure 13B at present.

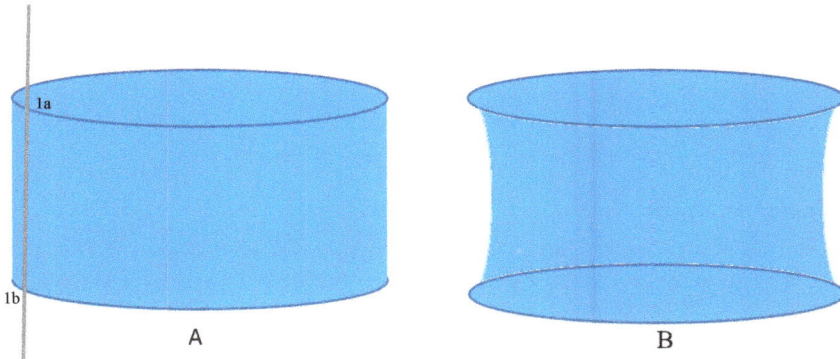

Figure 13. (**A**) A cylindrical volume which could be encoded in LandXML using the "simplest surface" assumption; (**B**) A volume with curved surfaces that would be problematic in LandXML.

While there is no unambiguous and agreed encoding for parcels with general curved surfaces in LandXML (as used for Cadastral plans), it should be noted that the volume of Figure 13A can be encoded in the mixed format as follows:

```
<Parcel class="FaceString" name="Outer">
  <CoordGeom><Curve radius="25" rot="acw">
    <Start pntRef="1"/><End pntRef="1"/></Curve>
  </CoordGeom></Parcel>
<Parcel class="Face" name="Top" a="0" b="0" c="-1" d="130"/> <Parcel
class="Face" name="Bottom" a="0" b="0" c="1" d="-120"/>
```

The storage of spatial units with curved boundaries is identical to that for conventional planar-faced or linear faced spatial units, but the processing is more complex. It must be remembered that the curve of the intersection of two simple curved surfaces can be complex to determine [30]. Note that the footprint boundary face strings may or may not need to be curved to define a volume with a curved surface.

4.8. Alternate Encodings

It is clearly possible, for any moderately complex spatial unit, that different encodings could be used. For example, the space in Figure 14, can be readily decomposed into a single "footprint" of vertical walls, with a well-defined set of faces defining the top and bottom. Alternatively, it could be encoded in parts (A to E)—each of which would be a simple slice with horizontal tops and bottoms. Any combination of these encodings is also possible. Similarly, as discussed with respect to the lift wells in Figure 6, an inner face string can be used; a separate 3D parcel could define the well, to be subtracted from the target spatial unit. It is important that the end-result is functionally equivalent, irrespective of the encoding. To ensure this requires a computable test for equality between alternate representations of the same spatial unit—see Section 5.5.

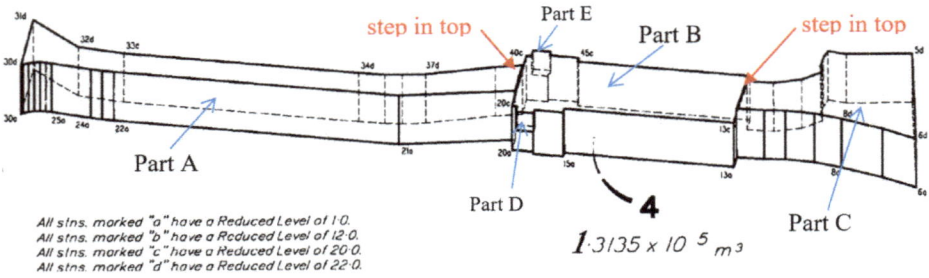

All stns. marked "a" have a Reduced Level of 1·0.
All stns. marked "b" have a Reduced Level of 12·0.
All stns. marked "c" have a Reduced Level of 20·0.
All stns. marked "d" have a Reduced Level of 22·0.

$1·3135 \times 10^5\ m^3$

Lot 4 is bounded by vertical & horizontal planes.

Figure 14. A Spatial unit that can be encoded in many ways.

5. Implementation Issues

A proof of concept implementation has been written to allow encoding of 2D and 3D into LandXML, and to read a limited subset of the LandXML as described here. The remainder of this section discusses various findings from or suggested by this work. Conversion to and from polyhedral form has been implemented, as a way of demonstrating that the necessary functionality is accommodated. Section 5.1 discusses the completeness of the proof of concept implementation, Sections 5.2–5.4 indicate the methods needed to convert to and from more conventional schemas. Section 5.5 addresses the critical issue of computability.

5.1. Completeness of Implementation

As part of the proof of concept, the encoding of face-string and face defined spatial units using this conceptual model, the conversion of them to or from conventional polyhedra (see Section 5.2), and the validation of these according to the axioms proposed by Thompson and van Oosterom [7,31] have been demonstrated. However, not all the required combinations of addition and subtraction of sub-units have yet been implemented. Enough has been implemented to allow the examples of Sections 4.1–4.5 to be encoded.

5.2. Conversion to Polyhedron Form

The proof of concept implementation converts plan data encoded using this schema to a set of polyhedra. Because some of the spatial units are not fully bounded (e.g., Lot 5 in the example of Section 4.5), special values of z to represent ±∞ have been chosen. As mentioned above, this implementation is not complete, but it is expected that no major obstacles will be encountered. The output polyhedra have been validated, using the proof of concept coding described by Thompson and van Oosterom [7].

5.3. Conversion from Polyhedra Form

In practice, the creation of plans using this schema would be the result of the calculations and data entry carried out in the survey process (in a "user-friendly" equivalent to the process described in Section 1, but conversion from polyhedra may be necessary where the 3D plan information is sourced from a CAD/CAM package, where the individual spatial units are already described as polyhedra. The secondary reason for this implementation is that it demonstrates that the storage method is complete.

Assuming planar faces, the procedure is as follows:

(1) The parameters (a,b,c,d) of each face are calculated. This is not trivial, involving a least-squares fitting of the function aX + bY + cZ + d to the set of vertices of the face, but this is a necessary action in any case to validate the face. Where c < 0, this is a top face, and it is replaced by a 2D polygon, with Z = 0 for each vertex. Where c $\geq$ 0, the face is discarded.

(2) These polygons are then dissolved to produce a (multi) polygon (with possible holes). This becomes the footprint boundary face string.

(3) The faces are then re-scanned, and any with c = 0 are tested against the boundary face string. If they fall within it, they are discarded. Otherwise, they are retained, along with all faces with c $\neq$ 0, as boundary faces.

(4) This produces a non-topological encoding of the spatial units of the plan. Normal 2D algorithms can be used to convert the face strings to a topological network that encodes the individual spatial units in 2D. In a similar way, shared faces can be detected and encoded.

A major issue in this, that must be handled very carefully is the test for c = 0. Rounding errors in this can lead to mis-calculation of the boundary face string, and overlaps and gaps in the final result. This will be discussed further in Section 5.5.

5.4. Viewing and Manipulating Data Using 2D Software

It is very simple to view a set of spatial units encoded in this form using 2D viewing software. If the face strings that define the spatial units are interpreted as line strings, and the faces are simply ignored, what is left is a 2D topologically encoded plan view of the data There is little software development needed to achieve this, and the resulting visualization can be useful for a first enquiry on a set of plans—including the production of thumbnail views.

It is even possible to allow updates to the data using 2D software. If the positions of the points are adjusted (in the X,Y position), the topological structure of the data will not be affected. In many cases, for example, where the spatial units are of the "stepped slice" form (or simpler), they will remain valid. For more complex spatial units, the adjustment can be checked by revalidating the data (in particular, checking planarity of the faces).

5.5. Calculations and Rounding Errors

As a result of the proof of concept implementation, the following observations are made on various issues—which are important to consider if a reliable and robust data transfer is to be achieved.

In calculations, it is important that the coordinates are comparable. Having X and Y coordinates in degrees of longitude and latitude while Z is in meters makes nonsense of the parameters (a,b,c,d) of a plane. It is, however, not necessary to convert to a rigorous rectangular coordinate system in general. The assumption is made that the Z direction is such that two points with the same X,Y coordinates are on a vertical line, and so a simple approximation is acceptable—such as choosing a local origin and multiplying the latitude and longitude displacements by a constant (about 111,000 for latitude and 111,000 $\times$ cos(*lat*) for the longitude)—so that all dimensions are approximately in meters.

This then allows the (a,b,c,d) parameters of a plane to be calculated such that $a^2 + b^2 + c^2 = 1$, and d is in meters (the distance from the plane to the local origin). The value of c indicates how nearly vertical the plane is, and $1/c$ is approximately the distance in the Z direction that a point needs to move on the plane to make a difference of 1 m in the X,Y position. Alternatively, if h is the maximum range of z values on the face (its "height"), then hc is the distance that the face deviates from the vertical. If this value for any face is smaller than the planarity tolerance (ε' in [7]), then it is worthwhile setting $c = 0$, adjusting a,b,d so that $a^2 + b^2 = 1$, and re-applying the tolerance test for the vertices of the face.

Ideally, an effort should be made to flatten a set of faces that were intended to be planar. This can be significant where a planar surface is common to a number of spatial units, but should be determined by calculating a set of plane parameters for all faces together, and checking that all vertices are within tolerance based on these parameters.

Because the same plan can be encoded in several different ways in general, the logic operations must be rigorous and repeatable. For example, there must be a method of testing whether two representations are equal. The point set definition ($A = B =_{def} p \in A \Leftrightarrow p \in B$) is not programmable, especially where a tolerance is permitted. As argued in [32], a more realistic approach is to provide a process that can detect an empty parcel (isEmpty(A) meaning that there can be no points within A), then use the subtraction operation to define $A \subseteq B =_{def}$ isEmpty(B–A), and then define $A = B =_{def} A \subseteq B$ and $B \subseteq A$.

There are broadly three approaches to fully rigorous logic within spatial database structures that are immune to rounding errors (where all faces are planar): 1. The infinite precision rational number approach [33], 2. The Dual Grid [34,35] (which also uses rational numbers, but restricts the sizes to which they can grow), and 3. The Regular Polytope [32,36], which uses a different form of storage of spatial objects.

An approach to rigor in the definition of spatial units in 3D is being adopted informally by some surveyors, and perhaps should be taken up as a standard or guideline. This is to ensure that all faces are either horizontal or vertical, or are fully triangulated. (Face strings are by definition vertical and thus fit this scheme very well. Spatial units defined in this way in the Queensland Cadastre can be found in Figure 3 and the underground portion of Figure 8. By contrast, it would be wrong to "correct" a non-planar surface by triangulating it post-registration of the plan, as this introduces an interpretation that could lead to disagreements.

The proof of concept uses an approach where two tolerances are imposed to ensure robustness [37], ε a larger value, which sets a limit to the closeness of points, flatness of faces, etc., and a much smaller value ι, which is assumed to be the accuracy to which the computations can be made. The difficulty here is that ε is a fixed distance—for example, 1 mm—while ι varies depending on the size of the arguments being applied to the calculation. It is not certain that the requirement $\iota \ll \varepsilon$ can necessarily be achieved with 8 byte floating point arithmetic and real-world sized spatial units.

6. Analysis of Results

6.1. Sharing of Data

The sharing of face and face string definitions in this approach had proved to operate in many cases at a very high level. The following variants occur in the plans investigated here:

1. Some definitions are not shared: Face strings at the edge of the plan, and faces that surround a secondary interest (not excised from the enclosing spatial unit) are examples.
2. Many definitions are shared by exactly two spatial units—where they form the boundary between them.
3. A significant number of face strings are shared a large number of times: For example, the face string "Unit3101–3102" defined in Section 4.5.2 would be used in the units 3101 to 3910 (9 units), and reversed in 3102 to 3902. A total of 18 usages.

Further, using the infinite plane form of the horizontal boundaries between floors (Section 4.5.2), which admittedly does require a change to the LandXML specification, the definition of "38Top" can be shared between 7 units and the common property above and the same below (16 uses in total).

7. Conclusions and Future Research

7.1. Major Results

This paper has presented a conceptual model applied to topology encoding of range of spatial units (2D, simple 3D, complex 3D). It supports elegant fusing of the 2D and the 3D representations with a key role for the 1D line string in 2D space, which becomes a 2D face string in 3D space. Additionally, this line string is used to depict footprints of the 3D spatial units in traditional 2D cadastral maps. The further benefits of the proposed conceptual model and the topology encoding are as follows:

1. It is a close fit to current cadastral survey practice (Section 3.2).
2. It permits efficient encoding, explicit semantics on neighbors, reduction of redundancy, and avoids inconsistencies. (Section 6.1)
3. It is capable of being viewed and even manipulated by software that is strictly 2D (not 3D-aware) (see Section 5.4).

Further, it achieves this with any combination of simple 2D and/or a mixture of 2D and 3D spatial units in an individual survey, combined with a hierarchical/multi-layered decomposition of spatial units.

The conceptual model for survey plans has been expressed in the language of the LADM. Two multi-step real world examples are given and encoded according to this conceptual model in LandXML for exchange purposes, including the initial registration.

7.2. Future Work

Future work includes investigating more types of 3D Cadastral parcel real world cases. It should be noted that, in the future not all survey plans may originate from surveys, but quite a number of them might result from design; e.g., BIM/IFC (Building Information Modeling/Industry Foundation Classes) [18]. Additionally, here too the role of topology is of equal importance, and topology encoding from this source needs further investigations. Currently, in much legislation, the survey plan (paper hard copy) is an official, legal document. The issue of the digital document being the official document needs further investigation (on aspects such as long-term preservation, digital signatures, approvals, etc.). This is quite similar to the treatment of administrative legal documents (Deeds) in the Netherlands. The digital deeds have become legal documents and are submitted via the ELAN system [38]. These digital deeds now form the large majority of submitted, approved, and recorded legal administrative documents.

It has been suggested in the introduction that the conceptual model as proposed for the survey plans, could also be applied to the cadastral database. Future work should explore how this can be realized: integrating the topology island from the survey plan into the topology structure of the complete cadastral database, and using the technique of "Versioned Objects" as used in ISO19152 [2]. The ultimate goal is to integrate the survey data with the cadastral map data in an integrated database [39]. Further investigation of the "two-tolerance" approach is needed to ensure that a rigorous logic can be assured.

The aim of the InfraGML project is to satisfy all of the functional requirements of LandXML, so that an expression of this conceptual schema in InfraGML should be relatively trouble-free. The appropriate sections of InfraGML are Section 6 (Survey) and Section 7 (Land Division), but the latter is not currently completed [6]. When the InfraGML standard is firmed up, an extension of this research to use that standard is strongly indicated.

Author Contributions: The research reported here is the work of all three authors. Rodney James Thompson wrote the proof of concept software, and all three authors contributed the text of the final paper.

Conflicts of Interest: The authors declare no conflict of interest.

References

1. Lemmen, C.; van Oosterom, P.; Thompson, R.J.; Hespanha, J.; Uitermark, H. The Modelling of Spatial Units (Parcels) in the Land Administration Domain Model (LADM). In Proceedings of the XXIV FIG International Congress, Sydney, Australia, 11–16 April 2010.

2. ISO TC211. ISO 19152:2012 Geographic Information—Land Administration Domain Model (LADM). Available online: https://www.iso.org/standard/51206.html (accessed on 6 May 2017).

3. Thompson, R.J.; van Oosterom, P.; Soon, K.H.; Priebbenow, R. A Conceptual Model Supporting a Range of 3D Parcel Representations through all Stages: Data Capture, Transfer and Storage. In Proceedings of the FIG Working Week 2016, Christchurch, New Zealand, 2–6 May 2016.

4. LandXML. Available online: http://www.landxml.org/ (accessed on 6 May 2017).

5. Scarponcini, P. InfraGML Proposal (13–121), OGC Land and Infrastructure Domain Working Group. Available online: https://portal.opengeospatial.org/files/?artifact_id=56299 (accessed on 6 May 2017).

6. OGC InfraGML 1.0. Available online: https://portal.opengeospatial.org/files/?artifact_id=72352 (accessed on 6 May 2017).

7. Thompson, R.; van Oosterom, P. Axiomatic Definition of Valid 3D Parcels, Potentially in a Space Partition. In Proceedings of the 2th International FIG 3D Cadastre Workshop, Delft, The Netherlands, 16–18 November 2011.

8. Thompson, R.; van Oosterom, P.; Karki, S.; Cowie, B. A Taxonomy of Spatial Units in a Mixed 2D and 3D Cadastral Database. In Proceedings of the FIG Working Week 2015, Sofia, Bulgaria, 17–21 May 2015.

9. Thompson, R.J.; van Oosterom, P.; Soon, K.H. Mixed 2D and 3D Survey Plans with Topological Encoding. In Proceedings of the 5th International FIG 3D Cadastre Workshop, Athens, Greece, 18–20 October 2016.

10. ICSM. ePlan Data Model Governance. Available online: http://www.icsm.gov.au/eplan/ePlan_Governance_v1.1.pdf (accessed on 6 May 2017).

11. Janečka, K.; Karki, S. 3D Data Management—Overview Report. In Proceedings of the 5th International FIG 3D Cadastre Workshop, Athens, Greece, 18–20 October 2016.

12. Dimipoulou, E.; Karki, S.; Miodrag, R.; de Almeida, J.-P.D.; Griffith-Charles, C.; Thompson, R.; Ying, S.; van Oosterom, P. Initial Registration of 3D Parcels: Overview Report. In Proceedings of the 5th International FIG 3D Cadastre Workshop, Athens, Greece, 18–20 October 2016.

13. Soon, K.H.; Tan, D.; Khoo, V. Initial Design to Develop a Cadastral System that Supports Digital Cadastre, 3D and Provenance for Singapore. In Proceedings of the 5th International FIG 3D Cadastre Workshop, Athens, Greece, 18–20 October 2016.

14. Zeiss, G. A Proposal to Replace LandXML with a New Standard: InfraGML. Directions Magazine 11th Dec. 2013. Available online: http://www.directionsmag.com/entry/a-proposal-to-replace-landxml-with-a-new-standard-infragml/371892 (accessed on 6 May 2017).

15. Stoter, J.E.; van Oosterom, P. *3D Cadastre in an International Context: Legal, Organizational, and Technological Aspects*; CRC Press: Boca Raton, FL, USA, 2006; p. 344.

16. Oldfield, J.; van Oosterom, P.; Quak, W.; van der Veen, J.; Beetz, J. Can Data from BIMs be Used as Input for a 3D Cadastre? In Proceedings of the 5th International FIG 3D Cadastre Workshop. Athens, Greece, 18–20 October 2016.

17. Paulsson, J. Sweedish 3D Property in an International Comparison. In Proceedings of the 3rd International FIG 3D Cadastre Workshop: Developments and Practices, Shenzhen, China, 25–26 October 2012.

18. Atazadeh, B.; Kalantari, M.; Rajabifard, A.; Ho, S.; Ngo, T. Building Information Modelling for High-rise Land Administration. *Trans. GIS* **2017**, *21*, 91–113. [CrossRef]

19. Van Oosterom, P.; Stoter, J.; Quak, W.; Zlatanova, S. The Balance between Geometry and Topology. In *Advances in Spatial Data Handling*; Richardson, D.E., van Oosterom, P., Eds.; Springer: Berlin/Heidelberg, Germany, 2002; pp. 209–224.

20. Hoel, E.; Menon, S.; Morehouse, S. Building a Robust Relational Implementation of Topology. In *Lecture Notes in Computer Science: Advances in Spatial and Temporal Databases. SSTD 2003*; Hadzilacos, T., Manolopoulos, Y., Roddick, J., Theodoridis, Y., Eds.; Springer: Berlin/Heidelberg, Germany, 1993; Volume 2750, pp. 508–524.

21. Louwsma, J. H. Topology Versus Non-Topology Storage Structures. Available online: http://www.gdmc.nl/publications/2003/Topology_storage_structures.pdf (accessed on 6 May 2017).

22. De Hoop, S.; van Oosterom, P.; Molenaar, M. Topological Querying of Multiple Map Layers. In *Lecture Notes in Computer Science: Spatial Information Theory A Theoretical Basis for GIS. COSIT 1993*; Frank, A.U., Campari, I., Eds.; Springer: Berlin/Heidelberg, Germany, 1993; Volume 716, pp. 139–157.

23. Ledoux, H.; Gold, C. Simultaneous Storage of Primal and Dual Three-Dimensional Subdivisions. *Comput. Environ. Urban Syst.* **2007**, *31*, 393–408. [CrossRef]

24. Boguslawski, P.; Gold, C. Construction Operators for Modelling 3D Objects and Dual Navigation Structures. In *Lecture Notes in Geoinformation and Cartography: 3D Geoinformation Sciences*; Zlatanova, S., Lee, J., Eds.; Springer: Berlin/Heidelberg, Germany, 2009; Part II; pp. 47–59.

25. Baumgart, B.G. Winged Edge Polyhedron Representation. Stanfort Artificial Intelligence Project, 1972, 44, Document STAN-CS-72-320 Memo AIM-179. Available online: http://www.dtic.mil/dtic/tr/fulltext/u2/755141.pdf (accessed on 6 May 2017).

26. Molenaar, M. Single Valued Vector Maps—A Concept in Geographic Information Systems. *GeoInform. Syst.* **1989**, *2*, 18–27.

27. Van Oosterom, P. *Research Issues in Integrated Querying of Geometric and Thematic Cadastral Information (2)*; Technical Report; Delft University of Technology: Delft, The Netherlands, 2000.

28. DNRM Cadastral Survey Requirements v7.1, Reprint 1. Department of Natural Resources and Mines, 2016. Available online: https://www.dnrm.qld.gov.au/?a=105601 (accessed on 6 May 2017).

29. ICSM. ePlan Protocol LandXML Mapping. Available online: https://icsm.govspace.gov.au/files/2011/09/ePlan-Protocol-LandXML-Mapping-v2.1.pdf (accessed on 6 May 2017).

30. Abdel_Malek, K.; Yeh, H.J. Determining Intersection Curves Between Surfaces of Two Solids. *Comput. Aided Des.* **1996**, *28*, 539–549. [CrossRef]

31. Thompson, R.; van Oosterom, P. Modelling and Validation of 3D Cadastral Objects. In *Urban and Regional Data Management: UDMS Annual 2011*; Zlatanova, S., Ledoux, H., Fendel, E., Rumor, M., Eds.; CRC Press: London, UK, 2012; pp. 7–23.

32. Thompson, R.J. Towards a Rigorous Logic for Spatial Data Representation. Ph.D. Thesis, Delft University of Technology, Delft, The Netherlands, 2007.

33. Weinrich, B.E.; Schneider, M. Use of Rational Numbers in the Design of Robust Geometric Primitives for Three Dimensional Spatial Database Systems. In Proceedings of the 13th Annual ACM Workshop on GIS, Bremen, Germany, 4–5 November 2005; pp. 163–172.

34. Lema, J.A.C. Dual Grid: A Closed Representation Space for Consistent Spatial Databases. Ph.D. Thesis, Universidade da Coruña, Coruña, Spain, 2012.

35. Lema, J.A.C.; Güting, R.H. Dual grid: A new Approach for Robust Spatial Algebra Implementation. *GeoInformatica* **2002**, *6*, 57–76. [CrossRef]

36. Thompson, R.J.; van Oosterom, P. Connectivity in the Regular Polytope Representation. *GeoInformatica* **2011**, *15*, 223–246. [CrossRef]

37. Belussi, A.; Migliorini, S.; Negri, M.; Pelagatti, G. Establishing Robustness of a Spatial Dataset in a Tolerance-Based Vector Model. *Trans. GIS* **2016**. [CrossRef]

38. Stolk, P.; Lemmen, C.H.J. Technical aspects of electronics conveyancing. In Proceedings of the 2nd FIG Regional Conference, Marrakech, Morocco, 2–5 December 2003.

39. Van Oosterom, P.; Lemmen, C.; Uitermark, H.; Boekelo, G.; Verkuijl, G. Land Administration Standardization with focus on Surveying and Spatial Representations. In Proceedings of the ACMS Annual Conference Survey Summit 2011, San Diego, CA, USA, 7–13 July 2011.

International Journal of
Geo-Information

MDPI

Article

Addressing Public Law Restrictions within a 3D Cadastral Context

Dimitrios Kitsakis * and Efi Dimopoulou

School of Rural and Surveying Engineering, National Technical University of Athens (NTUA), Athens 15780, Greece; efi@survey.ntua.gr
* Correspondence: dimskit@yahoo.gr; Tel.: +30-694-972-5897

Academic Editors: Peter van Oosterom and Wolfgang Kainz
Received: 5 April 2017; Accepted: 18 June 2017; Published: 22 June 2017

Abstract: Public law affects contemporary life by imposing various regulations that apply in 3D space. However, such restrictions are either literally described in legal documents or presented on a horizontal plane, resulting in ambiguities, especially in the case of vertically overlapping restrictions with a significant impact on land management. This paper investigates public law restrictions (PLR) applying to 3D space and their management within a 3D cadastral context. Within this framework, a case study is examined in Greece concerning the establishment of a subway station, focusing on public utilities, archaeological legislation, and building regulations. Relative legal documentation is compiled and mapped in a 3D PLR model, presenting inefficiencies and malfunctions that can be resolved if PLRs are addressed within a 3D cadastral context. Stipulations implying restrictions in 3D space within current legislation are presented, along with the restrictions deriving from the absolute character of ownership right, thus highlighting the significance of 3D definition, modeling and recording of PLRs.

Keywords: PLR; RRR; 3D Cadastre; legal framework

1. Introduction

The rapid expansion in urban areas over the years forced the vertical exploitation of real property, both for accommodation purposes and to respond to emerging public needs related to the amenities required by modern societies [1,2]. Technological advances in the field of construction address such needs, through the development of complex, overlapping, and interlocking structures. Furthermore, the development of surveying techniques as well as of information and computer technology provides the means for high-accuracy 3D data acquisition, modeling, and management. Research has extended to 3D/4D and even 5D representations, which is also available in the literature [3,4], taking into account aspects such as time and scale.

The exploitation of real property is regulated by national legal frameworks, based on Roman principles of real property, specifically regulated to address overlapping private and public rights. Public law restrictions (PLR) significantly impact contemporary life [5]. The 3D aspects of PLRs related to mineral activities, archaeology, environment, civil aviation, urban planning, building regulations, and utilities are presented in [6]. Despite the steadily growing effect of legal regulations applying to 3D space, the interest of legal professionals in 3D Cadastre issues remains limited [7,8]. Legal professionals' lack of interest mainly derives from: (a) the existence, until recently, of limited cases that would require a 3D cadastral approach; (b) the use of limited real rights and condominium concepts or; (c) the establishment of specific legislation to regulate large-scale underground infrastructures. This results in considerable complexities regarding land management and structural activities, thereby leading to costly and time-consuming judicial procedures. It can be concluded that the legislative regulation of subsoil is required in order to address emerging conflicts [9].

Taking into account technological abilities in 3D modeling and management, the variety of PLRs on land can be modelled to benefit the public, professionals and land administrators. Combined with advances in the field of web geographic information systems (GIS), high-quality 3D objects' visualization and management in web-based platforms can be provided. Within the field of 3D Cadastre, [10] a web-based 3D visualization prototype for both legal and physical 3D real property objects' space has been developed. The documentation and management of PLRs is becoming a complex issue, as they are defined by various legal documents. which also need to apply to specific volumes of space, dividing the absolute character of ownership rights, which is not familiar to traditional legal thinking. Efficient modeling and visualizing of PLRs can only be achieved if operating in combination with cadastral systems in a multipurpose approach, contributing to land administration, decision-making, and development. To this aim, [11] proposed the 3D Cadastral Domain Model (3DCDM), providing "a framework to identify 3D Cadastre and clarify its scope".

Research interest in 3D PLRs mostly focuses on specific fields, e.g., utility networks. The legal, organizational, and technical implications of 3D and 4D Cadastres, focusing on utility networks in three countries, are examined in [12], while [13] presents a 4D Cadastre prototype implemented on a case study in the Netherlands. In [14], the implementation of the first 3D cadastral registration of multi-surface real property rights in the Netherlands to a multi-surface complex, including a railway station, municipal offices, underground parking and commercial uses, based on existing legal and cadastral framework, is presented. Considering cases of underground infrastructure, [15] propose a 3D underground cadastral data model, specialized in Korean legal framework, based on the Land Administration Domain Model (LADM). Themed cadastres recording specific types of PLRs are established in many countries, particularly in Europe [16]. However, such repositories are 2D based, not interrelated with centralized cadastral systems, nor with other relative databases. Therefore, the full extent of applying restrictions, even on surface parcels, cannot be depicted. In Switzerland, the establishment of a PLR Cadastre, recording 17 types of restrictions classified in 8 fields, has been in progress since 2014, while already operating in 8 pilot cantons [17]. The possibility of introducing a 3D PLR Cadastre in the canton of Geneva in Switzerland, featuring emerging legal and technical obstacles required to be accommodated for the implementation of such a transition, is investigated in [18].

Emerging Building Information Modeling (BIM) technology provides for semantically rich and detailed 3D building models. Combining this with 3D GIS that enable the establishment of 3D spatial relations and querying, can contribute to 3D Cadastre purposes, especially in the field of PLR management as presented in [19].

This paper continues the research work that was presented at the 5th International Workshop on 3D Cadastres [6], further examining the range of PLRs with 3D aspects that have an impact on real property. Within this context, the 3D nature of specific PLRs is presented, as well as their interrelation to legal and cadastral framework, highlighting resulting deficiencies on land management. This paper explores overlapping PLRs deriving from construction restrictions, archaeology, and public utilities legislation that apply to an area of the city of Thessaloniki in Greece. The paper is structured as follows: Section 2 sets the legal background of PLRs in Greece, starting from general stipulations by the Greek Constitution, archaeological legislation, as well as legislation on public utilities. Analysis focuses on the legal framework regulating the development of Thessaloniki's subway in order to create the 3D model of associated PLRs, based on their descriptive legal documentation. Section 3 presents the results of the implementation of 3D related PLRs on the examined case study. In Section 4, different types of PLRs and their association with 3D cadastral concepts are discussed and analyzed, in the light of case study's results. Issues of further research are also discussed in this section. The paper ends with some concluding remarks on the deficiencies on land administration due to the lack of 3D PLRs and the benefits of implementing such a concept (Section 5).

2. Materials and Methods

In this section, 3D PLRs derived from archaeological regulations, public utilities' regulations, and building regulations that apply in Greece are presented and analyzed, depicting the need of stratification of rights, restrictions and responsibilities (RRRs). The above-mentioned 3D PLRs are modelled in 3D to a case study regarding the establishment of Thessaloniki's city subway line. In order to model in 3D these PLRs, Autodesk AutoCAD software version N.52.0.0 [20] has been used. The 3D model generated depicts 3D real property stratification of the examined PLRs, which is further compared to the models of existing repositories, in order to present deficiencies and malfunctions that need to be addressed by 3D Cadastre legislation.

2.1. Legal Framework

This subsection sets the legislative background in Greece, by presenting the basic characteristics pertaining to Greek legislation on Archaeology, public utilities and infrastructures, as well as building regulations. Section 2.2 presents relative legal documentation and "transforms" descriptive legal stipulations to 3D volumes of rights and restrictions.

2.1.1. Constitutional Regulations

The Greek Constitution explicitly provides for the protection of property (Art. 17, par. 1), while in the case of duly proven public benefit, property can be deprived from its owner under compensation (par. 2). However, works of evident public utility serving public benefit, e.g., "digging of underground tunnels at appropriate depth", may be allowed by law without compensation "on condition that normal exploitation of the property situated above shall not be hindered" (par. 7). Ownership and disposal of, inter alia, mines, quarries, archaeological sites and underground resources is stipulated to be regulated by special laws (Art. 18, par. 1). Par. 4 also allows for other "necessary deprivation of free use and enjoyment of property", under special circumstances, which will be provided by law.

On the other hand, Art. 24 (par. 6) sets monuments, historic areas and elements under protection of the State stipulating that "a law shall provide for measures restrictive of private ownership . . . as well as the manner and the kind of compensation payable to the owners".

2.1.2. Archaeological Legislation

The protection of cultural heritage is a field of public law that applies to 3D space, as it is related to the protection of underground and maritime antiquities. Thus, it is strongly related to 3D Cadastre, as stratification of real property or the application of 3D restrictions on real property facilitate archaeological research, as well as underground antiquities' protection and management. 3D aspects of cultural heritage protection have been considered, yet since 1956, UNESCO's Recommendations on International Principles Applicable to Archaeological Excavations has recommended that each member state should "Define the legal status of the archaeological sub-soil and, where State ownership of the said sub-soil is recognized, specifically mention the fact in its legislation".

In Greece, a dense legal fabric has been woven over the years for the protection of archaeological sites and monuments. Archaeological legislation in Greece does not provide for 3D restrictions or responsibilities on real property. However, even in the case that archaeological restrictions apply in height or depth, real property stratification does not take place, as involved parcels are expropriated. In this subsection, characteristic archaeological legislation implying 3D aspects is presented.

Law 3028/2002 constitutes the cornerstone of Greek archaeological legislation, combining regulations established by legislation dating from 1893 and adjusting it to the socioeconomic and cultural environment of modern Greece [21]. It stipulates restrictions on the exploitation of mines and quarries, restrictions on activities that may destroy or damage monuments (Art. 10), and imposes regulations on agricultural, farming and building activities (Art. 13) under relative Ministerial Orders. Article 14 forbids structural activities in settlements within archaeological sites without a license issued under Ministerial Order. Furthermore, it stipulates special regulations regarding ownership restrictions, building and land use, building regulations and permitted activities (par. 6). It also provides for temporary deprivation, use restrictions or even expropriation of real property under compensation to the landowners (Art. 19).

Law 2833/2000 allows for the expropriation or establishment of rights in rem by purpose of public benefit on necessary real property in order to facilitate constructions aiming to unify Athens archaeological sites (Art. 4). Although this law focuses on the city of Athens, considering city's dense population along with its rich archaeological background, it greatly impacts Athens's vertical expansion and intensifies the need of real property stratification and management.

The Decree of 29 July 1999 (Official Gazette D′/580) provides for special restrictions on historic buildings or on their neighboring buildings to ensure their protection and promotion. No reference on the type of these restrictions is made. The exception is the provision of transfer of built surface ratio in the case that existing restrictions inhibit its complete exploitation.

Law 5351/1932 sets a zone of 500 m around ancient monuments where mining is forbidden without license issued by the Ministry of Culture and Sports. Establishment of this zone does not directly constitute 3D restriction on real property; however, it is based both on monuments' protection from mining activities and on the protection of other ancient antiquities that may be traced within this region's subsoil.

2.1.3. Legislation on Public Utilities

General stipulations on the establishment of public utilities can be traced in the Greek Civil Code (Art. 1031), providing for the establishment of aerial or underground rights of way. Articles 1126 and 1127 define, respectively, the servitude of preserving structures situated on servient parcel and the servitude of preserving structures on another structure that is situated on a servient parcel. The performance of specialized construction works of significant importance is regulated by specific legislation to facilitate projects' implementation, to deviate from special clauses and to secure concessionaires' established rights [22]. An adaptation of Law 3389/2005 addresses issues requiring ratification of concession agreements by law, therefore such ratification is no longer necessary.

The establishment of utility servitudes for underground networks such as cables and pipelines is separate from utilities, as utilities are installed deeper and require more extensive construction works, e.g., subway lines. The former are regulated through Joint Ministerial Decision 725/23 (Official Gazette 5/B′/05 January2012), while the latter are established under individual laws, as mentioned above, also regulating procedures related to the discovery of archaeological findings during construction works. To facilitate the construction of utilities while also ensuring monuments' preservation, Law 4072/2012 provides for agreement between the Ministry of Culture and Sports and bodies undertaking construction works to a memorandum of understanding on archaeological research and excavations, while an individual interministerial committee has been established, addressing conflicts during the construction of infrastructure.

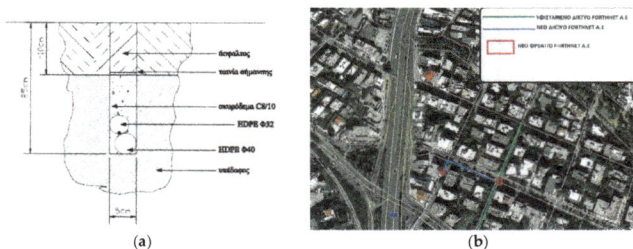

(a)   (b)

Figure 1. Documentation submitted for granting rights of way for public utilities. (**a**) Typical cross section; (**b**) location of network lines on drawing [23].

Figure 1 presents extracts from municipal granting of right of way for telecommunications network. The issuing of such a grant requires that agencies responsible for each network submit a drawing or extract of city/master plan, indicating the network's planned location (defining the position of ditches, drains or aerial networks by their street address and network's height or depth), along with typical cross sections (Joint Ministerial Decision 725/23).

2.1.4. Building Regulations

Building regulations stipulate allowed building height, depending on land parcel's location and area, as well as the building's intended use, thus defining the permitted building volume. Given the need to reduce energy consumption, building codes also define regulations regarding buildings' lighting, ventilation, and solar exposure pertaining to 3D aspects, and are influenced by constructions on 3D space, e.g., the shadow cast by a building onto neighboring buildings. Such regulations, combined with energy requirements of already built constructions, can also be used to export buildings' energy demands in urban areas, which can be further exploited within urban planning regulations.

Additionally, where urban landscape protection is required, specific building regulations apply to protect traditional architectural characteristics of buildings, e.g., facades or interior design [6].

2.2. Study Area

The above-mentioned legislative framework of PLRs is examined through a case study in the city of Thessaloniki in Northern Greece. Thessaloniki bears a longstanding history, established by 315 B.C. Due to the significance of its Paleochristian and Byzantine monuments it has been placed on UNESCO's World Heritage list since 1988 [24]. Thessaloniki's archaeological and historical significance, along with its leading character in the region of Macedonia in Northern Greece, result in the need for stratified land exploitation, especially in the field of construction projects.

To protect archaeological antiquities, subway tunnels were designed to be constructed at a depth of 14–31 m, while according to the archaeological report that was compiled, three stations have been characterized as being of "high archaeological risk" and other three of "medium archaeological risk" [25], as presented in Figure 2. However, Thessaloniki's long-standing history does not ensure that no other archaeological findings would be found within the rest parts of designated subway line. Construction works started in 2006 and are expected to be completed by 2020, while the discovery of significant archaeological findings has resulted in delays to the project's completion due to administrative and judicial procedures by the government, municipal authorities, archaeologists and the public regarding archaeological findings' protection and management.

Figure 2. Thessaloniki's subway line. Stations of archaeological risk are marked in blue (high archaeological risk) and green (medium archaeological risk) [26].

2.2.1. Legislation on the Establishment of the Subway Line

The establishment of Thessaloniki's subway line is regulated by Law 2714/1999, which constitutes the concession contract between the Greek State and the concessionaire company. The aforementioned concession contract was not put into force and it was decided that the whole project be implemented in the form of public infrastructure. However, its regulations are exploited within this work as the project's implementation followed similar requirements.

Article 5 of this concession defines real property rights during the concession period. Specifically, the Greek state grants "the right of exclusive use and possession of the totality of surface and underground spaces with adequate access thereto ... ", while it is also obliged not to encumber such real property by rights of any nature that may affect, limit or infringe the concessionaire's rights on the subway's construction, access or maintenance. In case of private real estate properties that are necessary for the carrying out of the subway's construction, the law provides for their expropriation at the expense of the Greek state (Art. 5, par. 2, indent 2). Indent 4 regulates rights relating to the boring of underground tunnels. It allows for executing related works without compensation of surface properties, "provided that the current use of subject properties is not affected by such boring", while the state guarantees that all parties that have rights on surface parcels are obliged to permit and withstand the boring of underground tunnels and of "any work or annoyance necessary for their construction, use, repair and maintenance".

Article 11 of the Law 2714/1999 also provides for antiquities and the protection of cultural heritage. However, regulations mainly focus on compensation issues, as well as on changes in workflow to facilitate archaeological research and not on imposing, stratified or even planar, restrictions or responsibilities on affected real property. During construction works, the ruins of an old Christian basilica were traced, Figure 3a, which has halted construction in order for the necessary archaeological excavations to be performed. Finally the station has been redesigned in order to preserve the basilica after the decision for its preservation in situ was made by the Central Archaeological Council.

The existence of public utilities hindering subway construction works is regulated in Article 19, which stipulates that the service responsible for pipes or network of pipes that hinder construction works is obliged to assist the concessionaire in their relocation by issuing the necessary permits.

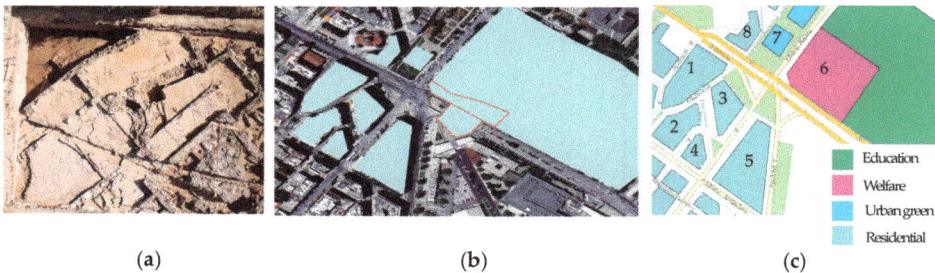

(a)   (b)   (c)

Figure 3. (a) Old Christian basilica revealed during subway excavations [27]; (b) building blocks within case study area (in cyan) and the area above the subway station (marked in red); (c) land use of case study blocks [28].

2.2.2. Building Regulations

There has been a significant amount of legal documentation regulating constructions over the years. This complicates access to building regulations, as their availability depends on the level of each municipality's spatial data infrastructure.

Within this work, access to building regulations was facilitated by Thessaloniki's municipal geoportal that provides detailed inventory of building regulations applying to city's building parcels, also referring to relative legal documentation and amendments. Related legal documents include the Presidential Decree defining built-surface ratio, the building codes of 1973 and 1985, and the new building code of 2012, defining building maximum height, and more than 10 town planning decrees regulating land use, land parcels' geometry and size, declaration of historic buildings, etc. In order not to disclose personal data, land parcellation, as well as existing buildings' footprints, were retrieved as presented in the Geographical Information portal of the Thessaloniki municipality.

2.2.3. Drawings and Basemaps

Apart from literal legal stipulations, the following drawings and basemaps of the case study area were used to model relative PLRs in 3D:

- Longitudinal section drawing of subway station, provided by Attiko Metro S.A. (Athens, Greece) (Figure 4).
- Base maps of Thessaloniki presenting land use and building regulations on each building parcel in the Hellenic Geodetic Reference System (HGRS87), available at the repository of Greek government agencies' databases [29] and the geoportal of the Thessaloniki municipality [28] (Figure 3b).

Apart from the above, it needs to be noted that the Hellenic Cadastre only records 2D land parcellation, buildings' footprints and servitude zones of passage as projections in the horizontal plane, while there is no provision of either underground or aboveground infrastructures, nor of public law restrictions (PLRs). Therefore, there can be no geographical or topological interrelation between cadastral land parcels, (legal/non-materialized) PLR 3D volumes, and physical 3D infrastructure and constructions' volumes.

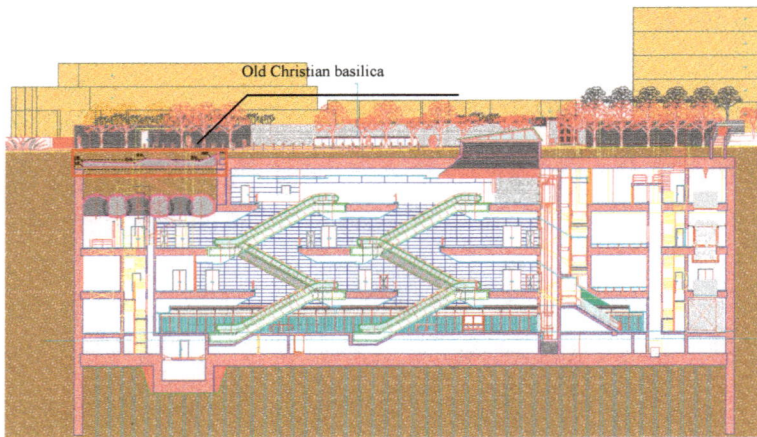

Figure 4. Subway station section.

3. Results

Based on the above documentation the 3D model of PLR volumes has been generated, as presented in Figure 5. 3D modeling of the subway station, the subway line, and legal building volumes not only clarifies the relations between different objects and of the rights establishing each object, but also allows for a concise, clear and complete presentation of their internal relations with the rest of the restrictions imposed by public law in the examined area. Findings of the preceding analysis are the following:

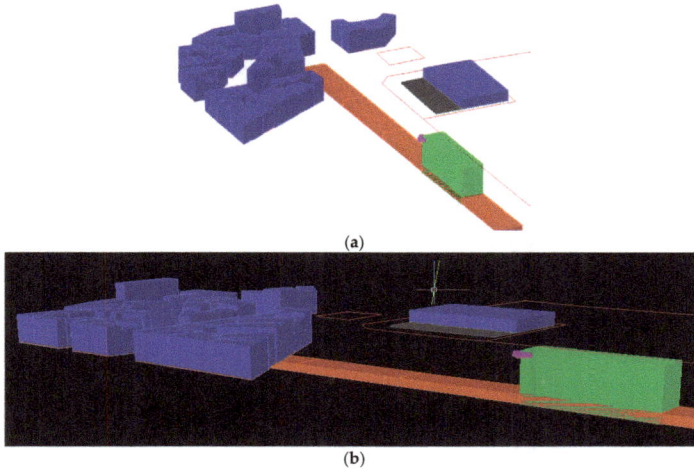

(a)

(b)

Figure 5. Implementation of 3D PLRs on case study's basemap in two different views (**a**,**b**) on Autodesk AutoCAD interface (green: space occupied by subway station; purple: remaining of old Christian basilica; orange: subway line; blue: permitted building volumes; grey: permitted land coverage; red: land parcels/building blocks).

- As PLRs constitute mainly legal (not materialized, invisible) spaces, their presentation in 3D maps, combined with general stipulations on the vertical extent of ownership, makes it possible to address ambiguities of RRRs' application in 3D space. Furthermore, 3D presentation of PLRs allows interested parties to understand the interrelation between different legal spaces or legal and physical spaces, as well as each one's impact on the other. This can also be derived from the examined case study, comparing the exported 3D model with existing cadastral and spatial data recordings that are presented in Figure 6. Cadastral documentation only presents land parcels on the examined area, along with the real property rights that apply to them in 2D, while no reference is made to the archaeological restrictions imposed. In case of real property expropriation for archaeological purposes, registration of the related administrative acts is required. Similarly, the municipal geoportal is limited in presenting the different building types, also including land parcels and building blocks. Places of archaeological interest are only presented as points.

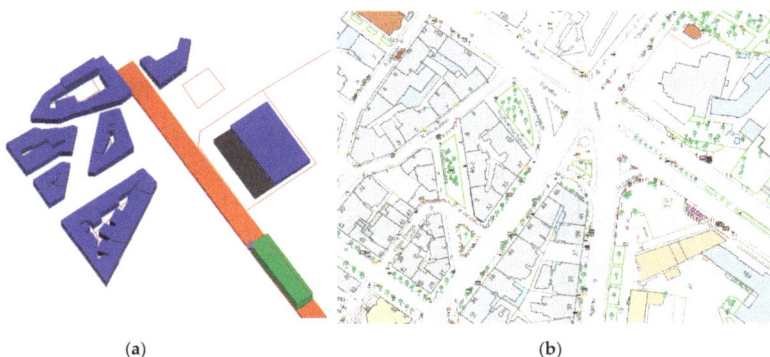

(a)   (b)

Figure 6. (**a**) Exported 3D PLR model; (**b**) current cartographic documentation of the case study area as presented in the Thessaloniki municipality geoportal [28].

- As mentioned above, current legislation cannot address complex situations of overlapping RRRs, given both the absolute character of the right of ownership and legal concentration on land parcels as 2D entities. This is also clear within the examined case study, as protection of the archaeological findings traced within the course of the subway line resulted in a long juridical dispute which has also led to delay in the project's completion.

- Current recording of archaeological PLRs mostly focuses on monuments or places of archaeological interest and not on the restrictions that are imposed on land. Even when such provision exists, it is limited to restrictions in the horizontal plane. In the examined case study, the combination of the archaeological reports on the designated subway route with establishment of 3D archaeological protection zones would accelerate the resolving of differences regarding preservation in situ, burial, or the relocation of monuments.

- 3D modeling of PLRs required the compilation of a variety of legal and cartographic documentation not available from a single source. Furthermore, given the significant number of legal amendments, tracing all legal documents related to a specific PLR or to a specific PLR applying to a specific region can become a difficult task.

- The implementation of 3D PLR models allows for the detailed representation of the legal space where each PLR applies, which contributes to non-ambiguous presentation of rights on land. Results of the 3D model of the examined case study, in contrast to existing cartographic documentation (Figure 6), clearly present differences in the delimitation of each real property, occupied 3D space by physical objects or 3D space encumbered by legal regulations. It needs to be noted that Greek subways are mainly established below state, municipal, or common use space, otherwise the involved, privately owned land parcels are expropriated by the state. Therefore, the whole process is facilitated as it is not affected by complications deriving from co-ownership of land parcels or limited real rights that may be imposed on land. It is evident that efficiency of 3D models would be significantly increased if combined with 3D subdivision or the encumbering of real property.

4. Discussion

In this section, legal areas previously examined are discussed. Deficiencies in land management that relate to PLRs with vertical aspects, as derived by the current legal and cadastral framework, are presented, along with options that 3D PLRs may provide.

4.1. Constitutional Regulations

- Constitutional stipulations on the protection of ownership need to be brought in line with stipulations concerning the exploitation of mines, quarries, underground resources, archaeological sites and monuments. The Greek Constitution sets the criterion of public benefit to either deprive privately owned land, or to oblige parcel owners to withstand boring activities that do not inhibit normal exploitation of surface parcel real property. Legislation on establishment of infrastructures, as well as Greek Civil Code, are in accord with constitutional stipulations. However, there is no specific definition of "normal exploitation" of real property. When the exploitation of a real property over a public utility needs to expand at a new, greater depth, this results either in the cancelling of planned exploitation, or in the expropriation of the surface parcel.

- In combination with the above, although the agencies responsible for the establishment and management of underground utilities benefit from their exploitation, surface parcel owners face restrictions on exercising to the fullest extent their ownership rights with no relative benefit.

- On the other hand, constitutional stipulations, both in article 18 and 24, as mentioned in Section 2.1.1, set the scene for stratification of real property, as they refer to "other necessary deprivation of free use and enjoyment of real property", as well as to provisions of "restrictive measures of private ownership necessary". Delays in infrastructure projects derive from the lack of expropriation funds, as well as from the reluctance of (surface) parcel owners to be deprived of their property (or objections to compensation values). This can be more easily addressed with restrictions that apply in 3D space providing for volume expropriation.

4.2. Archaeological Legislation

Archaeological legislation poses restrictions based on horizontal plane. Restrictions applying to 3D space can be traced to regulations regarding underwater antiquities, where protection zones can be established in their vicinity, forbidding or regulating specific types of activities. Archaeological regulations applying vertically are implied by relative legislation under the stipulation of "special regulations" or temporary deprivation of real property. However, these stipulations are not explicitly defined and described, while even such cases are addressed through land expropriation.

With the establishment of archaeological 3D PLRs on the other hand, registration of "restricted" volumes due to underlying antiquities would allow for exploitation of land parcel's parts that are not lying above archaeological antiquities. This would reduce land expropriation cost and facilitate interrelation with other 3D PLRs, e.g., the establishment of utility networks.

Given the complexities related to archaeological research, the stratification of real property and the imposition of specific RRRs on different land volumes may not resolve all types of emerging legal or technical conflicts on land exploitation. However, it can contribute, particularly in cases where archaeological findings need to be preserved within infrastructures or urban environment.

4.3. Public Utilities

The establishment of public utilities constitutes a field that is strongly related to 3D Cadastre. Regardless of their extent, height or depth, utilities are installed in 3D space, affecting nearby structures in the form of restrictions for installation, repairing, maintenance, protection, and securing of public health. The 3D space affected by public utilities is also related to other PLRs applying to 3D space, e.g., drone flights within a region where rights of way have been granted for passage of above ground powerlines.

The installation of cross-boundary public utilities is in most cases established through utility servitudes or servitudes of passage [2]. In case of pipelines or powerlines, building and agriculture restrictions apply to a specific distance from pipeline's or powerline's center-line. Such restrictions apply to building depth and height, cultivation restrictions, as well as the establishment of utility servitudes to allow access and maintenance of the utility. In such cases, compensation of the surface parcel owner [6] is provided. However, there is no provision of recording the content of relative restrictions or their delimitation in 3D space.

Apart from the lack of 3D data on public utilities' location and volume, their incorporation to cadastral infrastructure is partial: Servitudes of passage are registered by their projection on surface parcels, while the servitude's type is recorded in the descriptive cadastral database. Therefore, the exact location and extent of servitudes in 3D space, as well as the restrictions imposed on servitudes' zone, are not available. Internationally, similar issues on registering public utilities are presented by [12,13,30], noting that either servitudes on privately owned land require to be recorded, or that only land parcels encumbered by servitudes are recorded. This results in a lack of information on utilities' location and extent, even in the horizontal plane.

Finally, the establishment of passage servitudes for utilities entails that occupied space are also regulated under Civil Code stipulations on servitudes, which do not provide for use of such space as collateral. Therefore, 3D space covered by utilities cannot be acquired or further exploited by utility agencies in a proportionate way to other extensive linear networks, e.g., railways. The operation of 3D PLRs on utilities, apart from defining the exact location and extent of each network, also contributes to facilitation of compensation processes due to the establishment of rights of way and even allows utility operators to acquire ownership of relative land volumes, which can then be used as collateral.

4.4. Building Regulations

Building regulations, within 3D Cadastre context, define in detail the permitted legal volume of each construction, according to the geometrical characteristics and the area of the land parcel to which it pertains. Regulations on the protection of architectural heritage and physical environment are also stipulated within building regulations.

On the other hand, legal volumes defined through building regulations do not directly correspond to the "as-built" construction (physical space). Deviation between physical and legal space along with methods of their interrelation and modeling has been of significant scientific interest [10,31–35]. Given that PLRs mostly extend to multiple land parcels and, in most cases, only present the permitted or non-permitted 3D space of specific uses or activities within a region, such restrictions can be presented through the concept of legal space, as they are not directly related or presented through a physical structure. On the other hand, physical space can prove more useful on larger scale projects, e.g., land parcel or a building, where the real situation is required to clarify ambiguities and present the real situation concerning a structure and its constituent parts [35].

Administrative organization and cadastral infrastructure also affect 3D recording and presenting of building regulations. Depending on each country, building regulations are defined in national, regional or municipal level, while cadastral systems are, in most cases, centrally maintained and updated at national level. Consequently, efficient cooperation among different agencies at different levels of administration may be required.

4.5. Further Research

Investigating integration of PLRs to 3D Cadastre constitutes a challenging research task due to the complexities deriving both from its extended field of application and from technical issues in 3D data management.

The establishment and maintenance of a PLR Cadastre, even in 2D, involves significant cost, which also increases reluctance in introducing 3D systems. However, the benefits of resolving complexities in land management related to overlapping PLRs and real property rights, as well as advances in the field of 3D geoinformation, can provide sustainable, cost-effective 3D PLR recordings. Additionally, it needs to be noted that not all PLRs apply to 3D space or need to be recorded in 3D. Therefore, research on the fields that would allow the implementation of a sustainable PLR Cadastre would be required.

Variations between legal and physical space introduce complications in land management. Legal space efficiently accommodates the needs of PLRs as, in most cases, they neither apply to land parcel scale nor are implemented through a physical structure. On the other hand, physical space is the space that is feasibly perceived by professionals and public for their activities. These two types of spaces need to be compromised and interrelated in order to achieve efficient presentation and recording of 3D PLRs and, consequently, land management. The semantic enrichment of 3D models can also be applied to PLRs, allowing for the association of individual legal spaces defined by each field of public law, as well as their interrelation with overlapping physical space. The exploitation and integration of BIM and GIS technologies would significantly contribute to this, combining BIM's rich semantic information and GIS's spatial analysis capabilities [11,19].

Technical aspects of 3D PLRs' modeling, update and maintenance need to be investigated for the implementation of a 3D PLR Cadastre. Following the establishment of legal framework supporting stratification of real property, technical specifications defining (a) the type of 3D models that will be used to create 3D volumes along with their geometrical characteristics so that topological relations can be established; (b) semantic enrichment of 3D models in order to support the management of complex 3D object cases through queries and interoperability with other systems.

The introduction of 3D PLRs also entails the stratification of real property. Imposing 3D PLRs without the capability of vertically subdivision and encumbering real property would not have any effect. Therefore, concepts of real property stratification, both for land exploitation and establishment of 3D PLR management, within existing legal framework or through the introduction of 3D cadastral legislation, need to be investigated.

Finally, PLRs are related to different scientific fields, inter alia water and environment protection, archaeology or aviation. Consequently, imposed restrictions are based on various components, depending on each scientific field; therefore, investigation towards quantification of qualitative components or "translating" of physical attributes to legal restrictions and 3D volumes constitutes a great challenge for 3D PLR Cadastres, especially in the fields of environmental and water protection.

5. Conclusions

In this paper, the 3D nature of PLRs within a 2D legal framework has been highlighted, along with the impact of such 2D legal framework on real property management. Thereby, public law imposes restrictions on real property that apply to the vertical direction concerning archaeological legislation, regulations on the establishment of infrastructures, and building regulations. Despite their 3D nature, these regulations cannot be supported by 2D based cadastral systems, since the lack of legislation supporting real property stratification allows only for the partial implementation of 3D PLRs. This paper traces the malfunctions of current legal framework that involve (a) imposing restrictions or expropriating surface parcels in whole, instead of specific 3D spaces (thus leading to long lasting juridical and administrative procedure) and (b) introducing ambiguities in defining the 3D space that can be exploited by each surface parcel owner. Finally, the benefits of 3D cadastral legal framework in the field of PLRs are presented. These include the definition of the exact 3D space where a restriction is imposed (that allows for unobstructed exploitation of the rest of the parcel's volume), along with overcoming ambiguities in the management and presentation of complex, overlapping PLRs and real property rights.

Acknowledgments: The authors would like to thank Attiko Metro S.A for providing drawings of the examined subway station.

Author Contributions: Dimitrios Kitsakis carried out the main part of this paper, based on his PhD thesis research in School of Rural and Surveying Engineering of National Technical University of Athens. The paper was supervised by Efi Dimopoulou, who has provided important guidance concerning the methodological steps taken, as well as by providing her critical review, improvements and expertise on this research field, making this paper more reliable and original.

Conflicts of Interest: The authors declare no conflict of interest.

References

1. Stoter, J. 3D Cadastre. Ph.D. Thesis, Technical University of Delft, Delft, The Netherlands, 2004.
2. Kitsakis, D.; Dimopoulou, E. 3D Cadastres: Legal Approaches and Necessary Reforms. *Surv. Rev.* **2014**, *46*, 322–332. [CrossRef]
3. Van Oosterom, P.; Stoter, J. Principles of 5D modeling, full integration of 3D space, time and scale. In Proceedings of the Geospatial World Forum, Amsterdam, The Netherlands, 23–27 April 2012.
4. Ohori, K.A.; Biljecki, F.; Stoter, J.; Ledoux, H. Manipulating higher dimensional spatial information. In Proceedings of 16th AGILE Conference on Geographic Information Science, Leuven, Belgium, 14–17 May 2013; pp. 1–7.
5. Käser, C. PLRs in Switzerland—First Experiences with Eight Pilot Cantons. In Proceedings of the CLRKEN Workshop on the Documentation of Public Law Restrictions, Brussels, Belgium, 11–12 November 2015.
6. Kitsakis, D.; Dimopoulou, E. Investigating Integration of Public Law Restrictions to 3D Cadastre. In Proceedings of the 5th International FIG 3D Cadastre Workshop, Athens, Greece, 18–20 October 2016; pp. 25–46.
7. Paasch, J.M.; Paulsson, J. Legal Framework 3D Cadastres—Position Paper 1. In Proceedings of the 4th International FIG 3D Cadastre Workshop, Dubai, UAE, 9–11 November 2014; pp. 411–416.
8. Paasch, J.; Paulsson, J.; Navratil, G.; Vučić, N.; Kitsakis, D.; Karabin, M.; El-Mekawy, M. Building a Modern Cadastre: Legal issues in Describing Real Property in 3D. *Geodetski Vestnik* **2016**, *60*, 256–268. [CrossRef]

9. Käser, C.; Boss, H.Å. 3D-Kataster: Wohin geht's? Cadastre 3D: où va t-on? In Proceedings of the Swisstopo Colloquium, Wabern, Switzerland, 13 January 2017.

10. Shojaei, D.; Rajabifard, A.; Kalantari, M.; Bishop, I.D.; Aien, A. Design and development of a web-based 3D cadastral visualisation prototype. *Int. J. Digit. Earth* **2015**, *8*, 538–557. [CrossRef]

11. Aien, A.; Kalantari, M.; Rajabifard, A.; Williamson, I.; Bennett, R. Utilising data modelling to understand the structure of 3D cadastres. *J. Spat. Sci.* **2013**, *58*, 215–234. [CrossRef]

12. Döner, F.; Thompson, R.; Stoter, J.; Lemmen, C.; Ploeger, H.; Van Oosterom, P.; Zlatanova, S. 4D cadastres: First analysis of legal, organizational, and technical impact—With a case study on utility networks. *Land Use Policy* **2010**, *27*, 1068–1081. [CrossRef]

13. Döner, F.; Thompson, R.; Stoter, J.; Lemmen, C.; Ploeger, H.; Van Oosterom, P.; Zlatanova, S. Solutions for 4D cadaster—With a case study on utility networks. *Int. J. Geogr. Inf. Sci.* **2011**, *25*, 1173–1189. [CrossRef]

14. Stoter, J.; Ploeger, H.; Roes, R.; van der Riet, E.; Biljecki, F.; Ledoux, H. First 3D Cadastral Registration of Multi-level Ownerships Rights in the Netherlands. In Proceedings of the 5th International FIG 3D Cadastre Workshop, Athens, Greece, 18–20 October 2016; pp. 491–504.

15. Kim, S.; Heo, J. Development of 3D underground cadastral data model in Korea: Based on land administration domain model. *Land Use Policy* **2017**, *60*, 123–138. [CrossRef]

16. Cadastre and Land Registry Knowledge Exchange Network (CLRKEN). Documentation of "Public Law Restrictions"—Results of the Questionnaire in Preparation for the CLRKEN Workshop in November 2015. Available online: http://www.eurogeographics.org/sites/default/files/151209-ResultsOfQuestionnaireForPLRCadastre.pdf (accessed on 20 February 2017).

17. Federal Office of Topography Swisstopo. The Cadastre of Public-law Restrictions on Landownership (PLR-cadastre). Available online: https://www.cadastre.ch/en/home/meta/contact0/contact-1.detail.publication.html/cadastre-internet/en/publications/Broschuere-OEREB-Kataster-en.pdf.html (accessed on 12 February 2017).

18. Givord, G. Cadastre 3D des Restrictions de Droit Public à la Propriété Foncière. Master's Thesis, Conservatoire National des Arts et Métiers École Supérieure des Géomètres et Topographes, Le Mans, France, 2012.

19. Liu, X.; Wang, X.; Wright, G.; Cheng, J.C.; Li, X.; Liu, R. A State-of-the-Art Review on the Integration of Building Information Modeling (BIM) and Geographic Information System (GIS). *ISPRS Int. J. Geo-Inf.* **2017**, *6*, 53. [CrossRef]

20. Autodesk AutoCAD [Computer Software]. 2017. Available online: https://www.autodesk.com/education/free-software/autocad (accessed on 12 December 2016).

21. Akritidou, M. Archaeological Antiquities and Public Works in Greece: Egnatia Motorway Athens Subway. Master's Thesis, Faculty of Engineering Aristotle University of Thessaloniki, Thessaloniki, Greece, 2010.

22. Public & Private Partnerships. Available online: http://www.sdit.mnec.gr/ (accessed on 24 March 2017).

23. Municipality of Chalandri. Available online: http://www.halandri.gr/ (accessed on 15 March 2017).

24. Unesco World Heritage Convention. Available online: http://whc.unesco.org/ (accessed on 12 March 2017).

25. Attiko Metro, S.A. Available online: www.ametro.gr (accessed on 15 March 2017).

26. Metro Thessalonikis. Map of Thessaloniki subway line. Available online: http://www.ametro.gr/wp-content/uploads/2016/05/AM_Thess_Metro_map_Sept16_en.pdf (accessed on 19 June 2017).

27. Makropoulou, D. Findings of Byzantine Thessaloniki discovered during construction works of city's metropolitan subway line. In Proceedings of the 34th Symposium of Byzantine and post-Byzantine Archaeology and Art, Athens, Greece, 9–11 May 2014. Available online: http://www.chae.gr/en/other/dloads/announcements/The%20Byzantine-era%20finds%20brought%20to%20light%20by%20the%20excavations%20for%20the%20construction.pdf (accessed on 10 February 2017).

28. Thessaloniki Municipality Geoportal. Available online: http://gis.thessaloniki.gr/CityGuideThes/gis2014/index_en.html (accessed on 20 March 2017).

29. Repository of Greek Government's Agencies Databases. Available online: www.data.gov.gr (accessed on 5 March 2017).

30. Stoter, J.; Van Oosterom, P. *3D Cadastre in an International Context Legal, Organizational and Technological Aspects*; CRC Press: Boca Raton, FL, USA, 2006; p. 344.

31. Aien, A.; Kalantari, M.; Rajabifard, A.; Williamson, I.; Wallace, J. Towards Integration of 3D Legal and Physical Objects in Cadastral Data Models. *Land Use Policy* **2013**, *35*, 140–154. [CrossRef]

32. Ying, S.; Guo, R.; Li, L.; He, B. Application of 3D GIS to 3D cadastre in urban environment. In Proceedings of the 3rd International FIG 3D Cadastre Workshop, Shenzhen, China, 25–26 October 2012; pp. 253–272.

33. Karki, S.; McDougall, K.; Thompson, R.J. An overview of 3D Cadastre from a physical land parcel and a legal property object perspective. In Proceedings of the XXIV FIG International Congress, Sydney, Australia, 11–16 April 2010.

34. Dimopoulou, E.; Elia, E. Legal aspects of 3D property rights, restrictions and responsibilities in Greece and Cyprus. In Proceedings of the 3rd International FIG 3D Cadastre Workshop, Shenzhen, China, 25–26 October 2012; pp. 41–60.

35. Dimopoulou, E.; Kitsakis, D.; Tsiliakou, E. Investigating correlation between legal and physical property: Possibilities and constraints. In Proceedings of the SPIE 9535, Third International Conference on Remote Sensing and Geoinformation of the Environment (RSCy2015), 95350A, Paphos, Cyprus, 16–19 March 2015.

isprs International Journal of
Geo-Information

MDPI

Article

Toward the Development of a Marine Administration System Based on International Standards

Katerina Athanasiou [1],*, **Michael Sutherland** [2], **Christos Kastrisios** [1], **Lysandros Tsoulos** [1], **Charisse Griffith-Charles** [2], **Dexter Davis** [2] and **Efi Dimopoulou** [1]

[1] National Technical University of Athens, Athens 15780, Greece; christoskas@hotmail.com (C.K.);
   lysandro@central.ntua.gr (L.T.); efi@survey.ntua.gr (E.D.)
[2] University of the West Indies, St. Augustine, Trinidad and Tobago; michael.sutherland@sta.uwi.edu (M.S.);
   Charisse.Griffith-Charles@sta.uwi.edu (C.G.-C.); dexter.davis@sta.uwi.edu (D.D.)
* Correspondence: catherineathanasiou@gmail.com; Tel.: +30-694-887-9545

Received: 14 April 2017; Accepted: 17 June 2017; Published: 26 June 2017

Abstract: The interests, responsibilities and opportunities of states to provide infrastructure and resource management are not limited to their land territory but extend to marine areas as well. So far, although the theoretical structure of a Marine Administration System (MAS) is based on the management needs of the various countries, the marine terms have not been clearly defined. In order to define an MAS that meets the spatial marine requirements, the specific characteristics of the marine environment have to be identified and integrated in a management system. Most publications that address the Marine Cadastre (MC) concept acknowledge the three-dimensional (3D) character of marine spaces and support the need for MC to function as a multipurpose instrument. The Land Administration Domain Model (LADM) conceptual standard ISO 19152 has been referenced in scholarly and professional works to have explicit relevance to 3D cadastres in exposed land and built environments. However, to date, very little has been done in any of those works to explicitly and comprehensively apply LADM to specific jurisdictional MAS or MC, although the standard purports to be applicable to those areas. Since so far the most comprehensive MC modeling approach is the S-121 Maritime Limits and Boundaries (MLB) Standard, which refers to LADM, this paper proposes several modifications including, among others, the introduction of class marine resources into the model, the integration of data on legal spaces and physical features through external classes, as well as the division of law and administrative sources. Within this context, this paper distinctly presents both appropriate modifications and applications of the IHO S-121 standard to the particular marine and maritime administrative needs of both Greece and the Republic of Trinidad and Tobago.

Keywords: marine administration system (MAS); marine cadastre (MC); marine information data model; land administration domain model (LADM); ISO 19152; S-121 maritime limits and boundaries (MLB); marine rights; restrictions and responsibilities (RRRs)

1. Introduction

Over the last two decades, several countries with extensive coastlines and defined marine spaces, where they exercise sovereignty and administrative powers, have shown interest in the concept of Marine Administration Domain Model (MAS). Among others, Australia, Canada, the Netherlands and the United States have developed systems for the administration of marine interests and the sustainable management of marine resources [1]. Their efforts are at development stages, based on practices adopted in the fields of Marine Cadastre (MC), Marine Spatial Data Infrastructure (MSDI) and Marine Spatial Planning (MSP).

Many definitions have been provided for MC, as extensively described in [2]. It can be broadly defined as "an information system that records, manages and visualizes the interests and the spatial (boundaries and limits) and non-spatial data (descriptive information about the legal status, stakeholders, natural resources) related to them" [2]. MSDI is fundamental to the way marine information is developed and shared for more holistic and competent marine administration. MSP is a planning frame for balancing the rival human activities and managing their effects in the marine environment [3–9].

Research has been carried out concerning the correlation among these concepts and the way they interrelate—MC and MSDI relationship: MC is defined as a management tool, which can be added as a data layer in a marine SDI, allowing them to be more effectively identified, administered and accessed [4–9]. According to [10] (p. 4), there is a two-way relationship between the two: "Both of them function independently. However, MSP is designed and implemented safely and at a lower cost if it utilizes data from MC and MC will register and control the different rights and licenses in marine areas based on ecological environment when defined zoning from MSP exist." According to [11] "A MC is also different to a MSP as referred to in the directive 2014/89; a MSP is intended to regulate the use of the marine area/areas it covers; a MC is intended to describe and delimit distinct MC parcels and to indicate all relevant public and private rights, restrictions (including inter alia the restrictions resulting from MSP) and charges on those parcels."

MASs are generally designed to address various jurisdiction specific situations, while the tools developed for the management of marine environment tend to reflect the increasing institutional and research interests in this topic. However, there is a lack of a common standard and accepted base model to specifically handle the management of Marine Rights, Restrictions and Responsibilities (RRRs) and their spatial extents. Some research to date has focused on the development of MC data models that would serve as the basis of MAS, taking into account various existing standards. For example: Ng'ang'a et al. [12] described a marine property rights data model; Duncan et al. [13] advocated the integration of marine blocks with land volumes; Griffith-Charles et al. and Sutherland et al. [14,15] examined the development of a 3D Land Administration Domain Model (LADM) compliant MC in Trinidad and Tobago; and Athanasiou et al. [1,16,17] dealt with the conceptual classification of the marine entities and relationships and explored the adaptation of LADM to the marine environment. Furthermore, in several countries, the management of marine cadastral units tends to be included in the LADM implementation.

The most comprehensive MC modeling approach to date, which is currently still under development, is the proposed S-121 Maritime Limits and Boundaries (MLB) Standard [18]. It deals with this problem by providing a detailed conceptual model description for marine administration based on common national requirements. The product specification for MLB is based directly on the International Hydrographic Organization (IHO) S-100 Universal Hydrographic Data Model. The purpose is to support the legal aspects of marine data, providing a legal structure of sourced and versioned objects that is derived from International Organization for Standardization (ISO) 19152. The LADM is the first standard and approved base model for the land administration domain, and it establishes a rigorous mechanism for managing legal RRRs, their spatial dimension and the associated stakeholders. The implementation of this standard to the marine domain seems feasible because the triplet Object-Right-Subject, which forms the basis of LADM, may also be applied as well to the marine environment. This integration structurally bridges both the land and maritime domains and provides to the S-100 series a standard, which effectively supports the description of marine and maritime legal objects.

This paper explores the potential application of IHO S-121 to the marine and maritime administrative needs of both Greece and the Republic of Trinidad and Tobago. General data modeling approaches applicable to marine environments are first discussed. The paper then summarizes marine cadastral data model and standard criteria, distilled from relevant publications to date. Reviews of international data model standards that relate to marine environments are then presented. The paper then discusses the IHO S-121 Standard currently under development and concludes by proposing S-121 Code Lists for LADM related classes that may be applicable to both Greece and the Republic of Trinidad and Tobago. Implementation of the S-121 at a country level requires the modification of code lists to meet specific needs, since the values are different between the various legislation systems. This paper is a combined and revised version of two 3D Cadastres 2016 Workshop's papers: 'Toward the Development of LADM-based Marine Cadastres: Is LADM Applicable to Marine Cadastres?' and 'Management of Marine Rights, Restrictions and Responsibilities according to International Standards'. Section 2 presents a general review of modeling approaches applied to marine environments. Section 3 summarizes marine cadastral data model and standard criteria, distilled from relevant publications to date. In Section 4, international data model standards that relate to marine environments are reviewed, with extensive reference to S-121 product specification. Section 5 explores the potential application and modification of IHO S-121 to the marine and maritime administrative needs of both Greece and the Republic of Trinidad and Tobago. Finally, in Section 6, the code lists for the S-121 classes that are related to LADM, are presented for both Greece and Trinidad and Tobago.

2. General Review of Modeling Approaches in the Marine Environment

In marine environments, despite the fact that a number of jurisdictions have shown interest in the development of MAS and the academic community has dealt with the MC concept focusing on various technical, institutional, legal and stakeholder issues, there are only a few examples of literature dealing with marine data models in terms of data objects and the relationships among them [14]. Some of these are discussed below.

A Marine Rights data model is described by [12] (p. 463), as one "which provides a standard way to capture the laws that facilitate the allocation, delimitation, registration, valuation and adjudication of marine property rights; the interests that are allocated; the resources that the interests refer to; and their 3D spatial extent" (Figure 1). This model attempts to represent environmental, legal, and institutional elements that are generally associated with marine parcels. Their design is based on the fact that rights, responsibilities and restrictions in marine spaces relate to explicitly 3D/4D space i.e., to the sea surface, water column, seabed, and seabed subsurface. The design also incorporates the fact that rights, responsibilities and restrictions can be held by various types of stakeholders, in formal or informal ways. However, Ng'ang'a et al. [12] did not incorporate into their design any representation of stakeholders except government entities. The authors argue that there exists a marine parcel object as a base for data collection, storage, and retrieval on marine interests. This perspective was shared by Lemmen [19], who states, in relation to the LADM that "With some imagination, the laws (formal or informal) can be seen as parties; in fact, the laws allow people to have interests in "marine objects". The interests are RRRs and the MarineObject corresponds with the SpatialUnit in LADM Version C. Thus, it is expected that LADM can be used to marine space." The application of the LADM to the marine environment highlights the differences that may arise in the application of conceptual data models to different jurisdictions. The main issues refer to the decisions taken regarding how basic administrative units and any derivative spatial units are modeled. There are also differences relating to the linking or interfacing of land-based cadastres (that are based on existing standards) with newly defined standards applicable to marine spaces [14]. Another issue is the size of the marine parcel. The marine parcel shall not be of fixed size as human activity is greater closer to the coasts, e.g., ports, high density vessels traffic, seaplane landing areas, dumping ground, fisheries, and wind farms. Hence, the closer to the coast the smaller the marine parcel.

A marine data model that is based, in part, on the LADM and marine cadastre concepts was presented by Canadian Hydrographic Service [20]. The authors propose the development of the S-121 IHO standard, "built upon the Object Oriented structure of S-100", that would be able to handle MLB. The IHO S-121 product specification for MLB has now been published as a Draft International Standard and, in the future, can be a basis for combining data from different MASs.

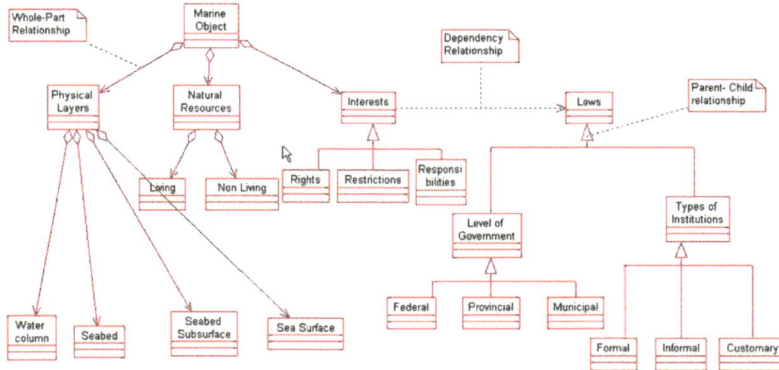

Figure 1. A marine parcel data model [12].

3. Marine Cadastral Data Model and Standard Criteria

The obvious 3D nature of marine spaces requires that specific criteria be met when developing marine cadastre logical and physical models. A review of literature relevant to the development of MC data models (and by implication data standards applicable to MC and MAS) imply that the following ought to be considered when developing MC or MAS [15]:

- The recording, updating and termination of various and overlapping types of formal and informal RRRs that affect multiple purposes and objectives in various jurisdictions [12,21];
- Diverse groups, individuals, and entities to whose rights, restrictions, interests, and responsibilities are ascribed [12,22];
- Various and overlapping types of 3D formal and informal boundaries and limits that relate to sea surfaces, water columns, the seabed and the seabed subsoil, some of which can change in space and time due to [12,23–26];

    1. Transfers of, or changes in, rights, restrictions, interests, and responsibilities;
    2. The ambulatory nature of some marine related boundaries;

- Various types of metadata, and other spatial information and surveying/mapping elements, such as datum, projections, data acquisition methodologies, etc., that are attributable to defined 3D spatial extents [12,27];
- Various types of non-spatial information (e.g., ecological, political, social, economic, etc.), attributable to defined 3D spatial extents [12,28,29];
- Complex levels of overlapping government laws, legislations, regulations, policies, treaties, conventions etc., that affect behaviors in defined 3D marine spatial extents [12,28–30];
- Possible integration of the marine cadastre datasets with other datasets in a spatial data infrastructure [12,16,20,31,32];
- The recording, updating and termination of various and overlapping types of formal and informal RRRs that affect multiple purposes and objectives in various jurisdictions [12,21].

According to [15], the LADM can handle the foregoing requirements, and, therefore, it is applicable for use in relation to marine environments. S-121, being in part built upon LADM classes and relationships, seem to be able to meet the aforementioned requirements. LADM and S-121 are discussed in the following sections.

4. International Standards Relating to Marine Environment

Standards are widely used, since they provide efficiency and support in communication between organizations and countries, as well as for system development and data exchange based on common terminology. Domain specific standardization is required to capture the semantics of marine administration. Such a standard will support marine registry and cadastral organizations utilizing a Geographic Information System (GIS) along with a spatially enabled Data Base Management System and applications, in order to implement and support marine policy implementation.

Current discussions and efforts in administration of marine spaces focus on the development and implementation of marine data modeling taking into account practices from Land Administration Standardization. Therefore, the registration of interests encountered in marine environments, along with their spatial dimensions, may be modeled in in the same manner as those modeled with land-based standardization techniques [18].

The following sub-sections discuss international standards that can be applied to marine environments. These include LADM, S-100, S-121, and Infrastructure for Spatial Information in Europe (INSPIRE).

4.1. LADM

The LADM was approved on the 6 November 2012 as an international standard, as ISO 19152, constituting the first adopted international standard in the land administration domain [33]. LADM provides a formal language for the description of existing systems, based on their similarities and differences. It is a descriptive standard, not a prescriptive one and can be expanded. LADM is organized into three packages and one sub package [19]. These packages are groups of classes with certain levels of cohesion. The three packages are: Party Package; Administrative Package; Spatial Unit Package and subpackage: Surveying and Spatial Representation Subpackage. The model contains thematic and spatial attributes. Furthermore, in several attributes, code lists are used rather than character strings in order to ensure consistency. The modification and adoption of code lists in national profiles is possible.

From the 3D perspective, LADM supports both 2D (LA_BoundaryFaceString) and 3D objects (LA_BoundaryFace) and distinguishes legal and physical objects by introducing external classes for BuildingUnit and UtilityNetwork. It covers the legal space, but the physical counterparts are not directly generated in LADM. At the semantic level, legal entities are not enriched by classifying data in relation to each other [34]. Furthermore, LADM through the VersionedObject class provides the attributes beginLifespanVersion and endLifespanVersion, allowing the recreation of a dataset at a previous point in time, which leads to a 4D visualization of the Cadastre [14].

The implementation of the model in marine environments is a user requirement in LADM version A. Furthermore, the scope of the standard explicitly addresses water-covered land when referring to land. The common relationships that can be observed in land administration systems may also be modeled in the LADM through a package of party/person/organization data, RRR/legal/administrative data, and spatial unit (parcel) data. The same pattern is also applicable in marine spaces.

4.2. IHO S-100 Universal Hydrographic Model

The IHO is an intergovernmental consultative and technical organization that was established in 1921 to support safety of navigation and the protection of the marine environment. IHO S-100 provides a contemporary hydrographic geospatial data standard that can support a wide variety of hydrographic-related digital data sources. Unlike its predecessor (S-57), which was primarily developed to meet the Electronic Navigational Charts (ENC) requirement for International Maritime Organization (IMO) compliant Electronic Chart Display and Information Systems (ECDIS) and which eventually became synonymous with ENC production and exchange standard specification, S-100 is inherently more flexible and aims to cover all aspects of hydrographic and marine information. S-100 is fully aligned with mainstream international geospatial standards, in particular the ISO 19100 series of geographic standards, thereby enabling the easier integration of hydrographic data and applications into geospatial solutions.

The primary goal of S-100 is to support a greater variety of hydrographic-related digital data sources, products, and customers. This includes the use of imagery and gridded data, enhanced metadata specifications, unlimited encoding formats and a more flexible maintenance regime. This enables the development of new applications that go beyond the scope of traditional hydrography—e.g., high-density bathymetry, seafloor classification, marine GIS and MLB. S-100 is designed to be extensible so that future requirements, e.g., 3D, time-varying data (x, y, z, and time), and Web-based services for acquiring, processing, analysing, accessing, and presenting hydrographic data, may be easily incorporated [35].

4.3. IHO S-121 Maritime Limits and Boundaries

Following the adoption of S-100, a number of product specifications are under development by the IHO S-100 specialized Working Groups (WGs), including S-101 for ENCs and S-121 for MLB. The intended purpose of S-121 is "to provide a suitable format for the exchange of digital vector data pertaining to maritime boundaries" and "for lodging digital maritime boundary information with the United Nations for purposes related to United Nations Convention on Law of the Sea (UNCLOS)" [36] (p. 1). Furthermore, as all ISO 19000 series product specifications, S-100 and the subordinate S-121 are intended to interwork with all similar products. In that sense, S-121 may serve as the bridge between the land and marine domains while the MLB of the S-121 standard may be used in an MAS.

S-121 is built upon the ISO 19152, which provides a rigorous mechanism for handling legal RRRs. Although the title of LADM refers to "land", the scope of LADM includes those elements "over water and land, [...] above and below the surface of the earth" [33], which makes it applicable to the marine environment as well. LADM has much in common with the management of MLB, while it provides a foundation to extend S-121 into the management of other regulated boundaries, such as marine reserves and fisheries. The alignment with the land domain model will facilitate consistent administration across the land and maritime domains for those states that adopt S-121 for use in relation to their marine spaces and ISO 19152 for use in relation to their exposed land jurisdictions [18]. Figure 2 illustrates the classes of S-121 that relate to ISO 19152 classes with realization relationships. Figure 2 illustrates the classes of S-121 that relate to ISO 19152 classes with realization relationships. Firstly, using this kind of relationship, any of the ISO classes could be introduced into the S-121 model if they are needed and secondly only the necessary attributes can be inherited from ISO 19152 to S-121 model.

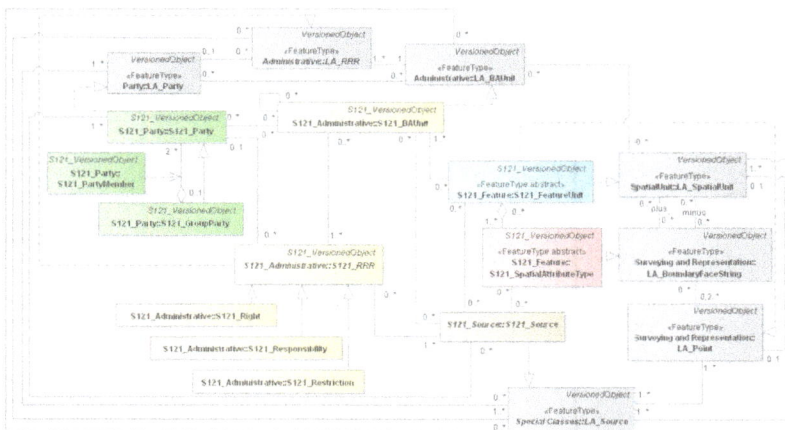

Figure 2. S-121 classes related to LADM.

A dataset of MLB consists of the entities pertaining to UNCLOS, i.e., Baselines, Zones, Limits, and Boundaries. More precisely, S-121 includes Shoreline and Coastline, Internal Waters, Archipelagic Waters, Territorial Sea, Contiguous Zone, Exclusive Economic Zone, Continental Shelf and International Boundaries. However, entities the states may declare (e.g., joint development areas) due national law or bilateral treaties are not included, and, therefore, for their use in a Marine Cadastre system, they need to be defined. In S-121, each real world feature is an object with properties represented as attributes (spatial and thematic) and associations which establish context for the feature. The four major components of each MLB object are [18]:

(1)   the party component which defines the different actors and their role associated with an object;
(2)   the geospatial component which defines the location and type of the object;
(3)   the legal component, which supports the description of the associated jurisdictions, and rights, associated with objects;
(4)   the administrative or spatial sources such as treaties, legal documents, charts.

The spatial attributes of the feature describe its geometric representation, whereas the thematic attributes describe its nature. The attributes associated with the geographic feature depend on the intrinsic type of the feature. A feature object may only have one intrinsic type that is the physical dimension of the feature in the real world based on the "truth on the ground" principle, i.e., point, curve, area, and volume. Subsequently, the feature is described in the dataset by a geometry property (point, curve, area-volume is area with elevation attribution), which is used for its cartographic representation that depends on the scale of the cartographic product. Finally, for the portrayal of each geometry property, which is separate from geometric representation, a variety of symbols may be used. For instance, the intrinsic type of an aquaculture installation is Zone. Depending on the scale of the cartographic product, the geometry of the farm may be area (large scale maps) or point (small scale maps). Finally, for the portrayal of, e.g., the point geometry, can be used a simple point, a star or a variety of other symbols. In S-121, portrayal is in compliance with the S-100 standards, whereas, for specific objects unique to MLBs, the use of new symbology may be required [18].

S-121 imports various primitives from LADM that are useful for an MLB standard and not supported by S-100, e.g., GM_Curve and GM_Surface from the ISO 19152 class LA_SpatialUnit. S-121 has also versioning capabilities for all objects within the model, meaning a historical tracking of the object's evolution. Versioning is another attribute that consists of begin and end dates, allowing attributes and features to be changed at the individual level, e.g., changing the right to a fishing zone among seasons. The Versioned Objects structure is clearly valuable in the MLB and MC application areas and also appropriate in an MSDI [18]. A fundamental difference between the LADM and the S-121 is the use of the Multi-primitive MultiSurface features in the land environment, whereas, in the marine environment, it is a requirement from S-100 to use complex features instead. In detail, when an object is crossed by another, in land, the crossed feature is defined as a multi-surface, whereas, in the marine environment, each spatial primitive is a simple rather than a complex one (Figure 3). Another issue is the use of 3D objects, which LADM addresses with the LA_BoundaryFace and LA_BoundaryString, whereas, due to the limitations of S-100, which does not address 3D objects, S-121 handles 3D objects as 2D objects with vertical extent (2.5 dimensions) [20].

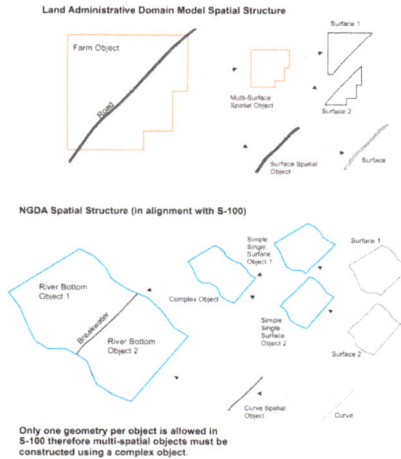

Figure 3. Terrestrial and maritime spatial structures according to LADM and S-100 [18].

4.4. INSPIRE

For cross-border access of geo-data, a European metadata profile based on ISO standards has been developed using rules of the implementation defined by the Infrastructure for Spatial Information in the European Community—INSPIRE [33]. INSPIRE (Directive 2007/2/EC) focuses strongly on environmental issues, while the LADM has a multipurpose character. One important difference is that INSPIRE does not include RRRs in the definition of cadastral parcels.

From the spatial perspective, according to [33], there is a compatibility between LADM and INSPIRE. More specifically, four classes are relevant in the INSPIRE context:

- LA_SpatialUnit (with LA_Parcel as alias) as basis for CadastralParcel;
- LA_BAUnit as basis for BasicPropertyUnit;
- LA_BoundaryFaceString as basis for CadastralBoundary; and
- LA_SpatialUnitGroup as basis for CadastralZoning.

Regarding the marine space, the expression and the definition of the above classes need to be examined.

INSPIRE data specifications are being developed across 34 themes. Many INSPIRE themes have a marine relevance, something that researchers have already pointed out (e.g., [37]). The marine related themes are presented in Figure 4. Two of the themes, i.e., Ocean Geographic Features (OF) and Sea Regions (SR), are related exclusively to the marine environment. According to [38], "INSPIRE is not marine nor land centric". Themes are considered independent of whether or not they refer to land or to sea, and therefore data can be brought together across land-sea boundaries. INSPIRE provides a level of interoperability to deliver integrated land-sea datasets. However, the data models (by design) will not solve the needs of all communities e.g., navigation. The marine themes on their own do not give all the information on the marine environment.

According to [19] (p. 169), "firstly, it is possible that a European country may be compliant both with INSPIRE and with LADM and secondly, it is made possible through the use of LADM to extend INSPIRE specifications in future, if there are requirements and consensus to do so". Given that IHO S-121 Product Specification for MLB is based on LADM, it is inferred that INSPIRE can cooperate as well with S-121 mainly in the spatial dimension.

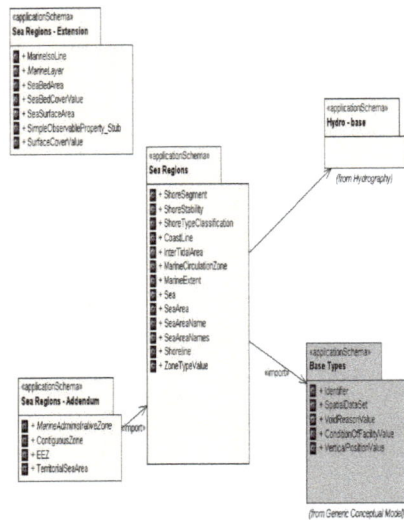

Figure 4. Marine related classes of INSPIRE [38].

5. IHO S-121 Maritime Limits and Boundaries and Marine Legal Objects

For the development of an MAS, the association of legal attributes with MLB information or marine parcels is necessary, in order to determine under whose authority or international treaty a particular limit or boundary is defined, and the restrictions around this specific marine parcel according to the legislation.

From an administrative modeling viewpoint where the focus is on abstracting the real world as a principle, sea is not a legal entity until an interest is attached to it. Therefore, the very close relationship between each interest and its spatial dimension in the real world should be identified and registered in information systems. These elements form a unique entity: the marine legal object.

Cadastres deal with entities consisting of interests on land that have three main components: spatial (spatial units), legal documents, and parties [34]. The same applies in the marine environment.

The S-121 Product Specification developed in order to support additional types of data that may be compatible with the eNavigation information, but they are not included in S-100 Universal Hydrographic Model. The theme behind these additional types of marine data is legal rights. Experience in this area does not derive from marine navigation, but it is largely based on the methods by which these legal rights issues are addressed in terrestrial mapping [18]. Therefore, the introduction of Parties, Legal Interests and Administrative Units is derived from ISO 19152. Since the LADM and IHO S-100 are both built on the ISO TC211 suite for Geographic Information standards, the elements of spatial units, spatial sources, spatial representations, basic administrative units and administrative sources are compatible and can be inherited into IHO S121. Appendix E of S-121 Product Specification describes how the LADM related classes are integrated into S-121.

The ISO suite of geographic information standards and the derivative IHO S-100 Universal Hydrographic Data Model are built upon an object structure [18]. The objects have two kinds of properties, spatial and thematic attributes. The IHO S-100 has added to the ISO General Feature Model the Feature Types (i.e., objects with existence in the real world) and Information Types (i.e., objects with no real geographic spatial position, only with thematic attributes). The whole structure of S-121 is built upon these objects.

Review of S-121 and adaptation of certain attributes and classes that are proposed by [1] are described below. The workflow of this research includes the identification of the unique characteristics of marine environment, the review of the marine related data models and finally the extension of the existing terrestrial methods and standards, through the enhancement of S-121 in order to address the special marine needs.

5.1. S-121 Party Unit Package

S-121 inherits the structure of ISO 19152 for the party package. The basic class is the S121_Party, which has a specialization of the S121_GroupParty. Between S121_Party and S121_GroupParty, there is an optional association class the S121_PartyMember (Figure 5). There is an aggregation relationship between S121_Party and S121_GroupParty, which means that an individual may be a member of a group, and a group as a whole can be treated as a party [18].

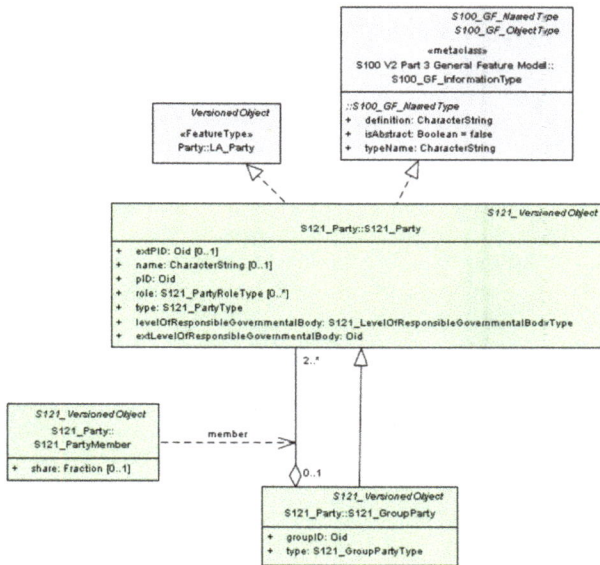

Figure 5. S-121 Party package ([18], p. 57, edited by the authors).

In a marine environment, the party can be a stakeholder or most likely a group of individuals or companies. The party can also be the state. According to [1,16], the additional attribute of the levelOfResponsibleGovernmentalBody will be useful in the marine environment. It specifies who has the supervisory control to exercise a RRR according to the relative law and its registration could be mandatory. Furthermore, the values of attribute "role" can be originated from a code list, instead of using character strings, as long as the roles are specific in the data update and maintenance processes. S-121 Party Unit package with the aforementioned additions in class S121_Party is presented in Figure 5.

5.2. S-121 Administrative Package

The operation of a Land Administration System (LAS) is based on the relationships between parties (stakeholders) and property units. Property is conceptualized as consisting of the rights, objects, and subjects. The suggestion of [39] is that property with its emphasis on 'rights' is a subset of land tenure, which is a much broader term with emphasis on RRRs. In the marine environment, property describes the resource, individual/s with an enforceable claim, and type of resource use claims [40]. Moreover, other than the state owned parcels in the territorial sea, parcel boundaries are determined according to usage only (e.g., minerals, aquaculture) and not as a property. There is not usually a market in ocean parcels where parcels are subdivided and consolidated and sold off, nor is the system designed to support this [41]. Therefore, in oceans, a different legal regime is shaped. The following types of RRRs can be found [17]:

- **State Interests:** The state RRRs are defined through the international Treaties (and bilateral agreements for states with maritime neighbors) and transposed into national legislation with laws. When referring to sovereign rights, we mean the power of the state and/or the sovereign entity (as regards the marine space, the sovereign entity is always the coastal state) to act as they deem appropriate for the benefit of their citizens. The legal term of the aforementioned power is "exclusivity of jurisdiction" that according to the international law implies that the state has complete control of its affairs within its territory, without being accountable for the means of exercising this control. The extent or the kind of the sovereign rights differ, according to the specific zone of the marine space it refers to. The full sovereignty or the sovereign rights of the coastal state means that, apart from the coastal state, private entities (natural or legal persons) can exercise an activity or use part of the marine space only by means of transferring of a right from the State for a specific activity under contract or licensing. This kind of rights are recorded by an MAS. Table 1 summarizes the rights of states in marine space as they are defined in UNCLOS, which may then be transferred to the private entities.

- **Public Rights:** Public rights refer mainly to the constitutional right of every citizen of the state having an unlimited/ without obstacles access statewide (terrestrial and marine space). These rights are not secured for an individual interest but for a public interest (e.g., public right to beach access). They may be described as protecting the public interest in the use and conservation of resources.

- **Environmental RRRs:** Refer to provisions that relate to the protection and conservation of water resources, places of preserved areas and cultural heritage. These places are pre-defined by law and the rights involved are of utmost importance and mandatory, in comparison to the following functional interests [1]. These RRRs include amongst others, the protection of archaeological and historical objects found at sea, the protection of Marine Protected Areas and the general MSP restrictions.

- **Usage and Exploitation Rights:** Progressively functional rights tend to acquire a private nature, associated with individual stakeholders that coexist with the state rights. In a wide sense, this term sets the limits of rights, which involve mainly the different ways of use and management. In other words, in the marine environment, the rights are limited in terms of space, duration and most importantly the extent, the content that refers only to the different kind of uses and management. The stakeholders are not owners but only beneficial "users" [1]. When private property rights are used as a basis of interpretation, these rights do not represent full ownership let alone absolute property rights; they can be classified into usage and exploitation rights. Usage rights are associated only with space, and exploitation rights are associated with the resources as well. Usage rights may be granted by a legal person that has been delegated the authority to provide usage rights. Rights granted in this manner are subject to restrictions in terms of the nature of the usage rights (e.g., type and temporal aspects of use) and the spatial extent linked to the usage rights (sometimes defined by boundaries). Functional rights are granted either by leasing contracts or through licensing. It has to be noted that the authority of granting remains national and no freehold ownership is involved. These rights are associated with specific stakeholders.

ISO 19152 defines an administrative package that associates parties with the Basic Administrative Unit. The same relationship applies in S-121. The S121_BAUnit is derived from the class LA_BAUnit and the definition remains the same. RRRs appear as attributes for an S121_BAUnit [18] and are subtypes of the abstract class S121_RRR. The class LA_Mortgage is not expressed in the model, since it is not normally applicable in the marine environment and as long as realization relationships are used between S121_RRR and LA_RRR classes. Therefore, any of the ISO classes could be included or excluded into the S-121 model. RRRs are implemented as information objects, and, accordingly, the RRRs can be shared.

In addition, there is one important aspect that is missing in S-121, the marine resources. When defining a RRR in marine space and the S121_BAUnit this RRR refers to, the registration of the natural resource that falls within its spatial extend is needed in an MAS, and could be occurred through the class S121_NaturalResource. The proposed attributes for this class are: the resourceID, the description, living (Boolean), resourceType and renewableType. The values of the last two attributes may be derived from predefined codelists. S-121 Administrative Unit package with the aforementioned additions is presented in Figure 6.

Table 1. Rights of states within maritime zones.

	Internal Waters	Territorial Sea	Contiguous Zone	Exclusive Economic Zone	Continental Shelf
Outer Limit	Baselines	12 NM	24 NM	200 NM	200 NM or Article 76
Vertical Domain	AS, WC, SB, SS	AS, WC, SB, SS	WC	WC, SB, SS	SB, SS
Rights of the Coastal State	Full sovereignty equal to that on the land. Exception: 8(2) UNCLOS	Full sovereignty with the exception of Innocent Passage	May put laws into in order to prevent and punish infringements of its customs, fiscal, immigration or sanitary laws committed within its land territory or territorial sea. Control of traffic of objects of an archaeological and historical nature found at sea. 303(2)	Exclusive sovereign rights for the purpose of exploring and exploiting, conserving and managing the natural resources, both living or non-living and the jurisdiction to establish artificial islands or installations, protection of the environment and to conduct scientific research.	Exclusive Sovereign rights for the exploration and exploitation of its natural resources and the sedentary species
Rights of the Other States	No rights, with the exception of right of innocent passage when Article 8(2) UNCLOS is applicable.	Right of Innocent Passage and transit passage where Articles 37 applies	All high seas freedoms	Freedoms of navigation, of overflight and of laying submarine cables and pipelines (subject to the rights of coastal and other states)	Freedom of laying submarine cables and pipelines (subject to the rights of coastal and other states)

AS: Air Space; WC: Water Column; SB: Seabed; SS: Subsoil.

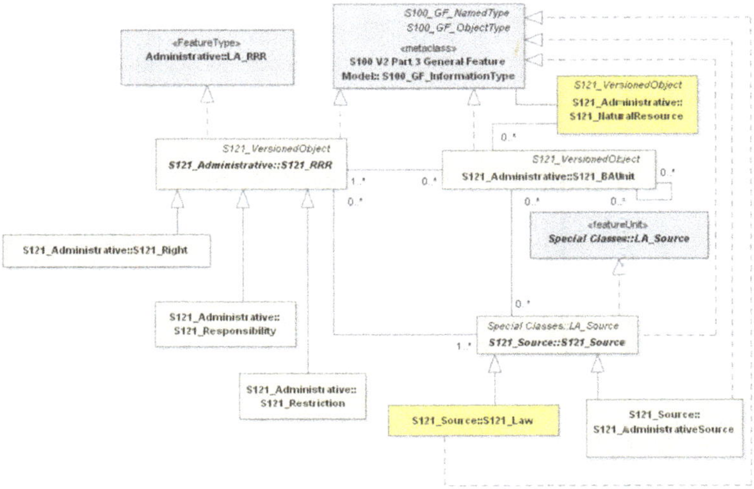

Figure 6. S-121 Administrative package ([18], p. 59, edited).

5.3. S-121 Spatial Unit Package

A plethora of research works and papers in literature deal with the definition of the marine parcel. Two alternative hypotheses are given by [12] about the marine parcel: "(1) that there either exists a multidimensional marine parcel that can be used as the basic reference unit in a MC, or (2) that there exists a series of (special purpose) marine parcels that can be used as basic reference units for gathering, storing and disseminating information. In either case, then, whereas the definition and spatial extent of a parcel is still not clarified, there still exists a parcel."

Another definition of marine parcel refers to: "A confined space having common specifications for its internal, mainly used as reference to locate a phenomenon. A marine parcel facilitates the distinction between contiguous territories and provides information concerning this phenomenon through appropriate codification" [3].

For the definition of the marine parcel, certain issues must be taken into account:

- *The third dimension:* The inherent volumetric 3D nature of marine space is apparent. Marine RRRs, such as aquaculture, mining, fishing, and mooring and even navigation, can coexist in the same latitude and longitude but in different depths. The question is if the 3D representation is necessary for an MAS. So far, the geomatics' community supports the idea of the 3D registration and visualization of marine interests. According to [12] (p. 446) " ... Clearly, the right to explore for minerals may have an impact on the surface of the land, but it will also affect a 3D cross-section of the parcel below the land's surface. Policy-makers would no doubt benefit from an understanding of the upper and lower bounds of the exploration rights, and how these may affect the environment or other property entitlements within the same parcel." Additionally, the registration of the restrictions that are defined by the laws and structure the marine legal object are related with the third dimension for most activities. They define in which vertical or horizontal distance is allowed to exercise other marine interests. Furthermore the multipurpose nature of the MAS demands access to additional types of information (geology, hydrology etc.), additional types information (geology, hydrology etc.), except of the RRRs, in relation to marine spatial extents. The use of the third dimension is considered important. However, the existing MAS have only used the third dimension for the representation of the seafloor.

- *The fourth dimension:* It is clear that time has always played an important role as the fourth dimension in cadastral systems. In the marine environment, most activities can co-exist in time and space and can move over time and space. Therefore, the registration of the fourth dimension will capture the temporary nature of many particular rights.

- *Spatial Identifiers:* Every land parcel or property recorded in a land registry or a cadastral information system must have an identifier. In fact, identifiers are the most important linking data elements in land administration databases. There are various ways for referencing land parcels and property [42]. In the Hellenic (Land) Cadastre for each individual property, a 12-digit code number is assigned, the *"KAEK"* , which is unique nationwide. It is proposed by Arvanitis et al., the use of a unique code to the marine parcels. "The 12-unit code will be based on the legislated zone, the Sea, the Greek Prefecture, the Head Office of the Port Authority Jurisdiction/Municipality, the use and number of the marine parcel". The code will be unique and will record the existence of multiple uses in the third dimension [17].

The Spatial Unit as defined in S121 is derived from the class LA_SpatialUnit defined in ISO 19152, but it also inherits from S121_GF_SpatialAttributeType [18]. However, the ISO 19152 has a very different structure for its Spatial Unit Package and Surveying and Representation SubPackage, since it is land oriented.

The S121_SpatialAttributeType describes the spatial properties of an S121_FeatureUnit. The subtypes of the S121_SpatialAttributeType are the S121_Point, S121_Curve, S121_Surface, S121_Volume and S121_Composite. The geometry types are inherited from S-100. Additionally, there are four feature types: Location, Limit, Zone and Space, which are connected with realization relationships to predefined feature types, named MLB Feature Types. However there are subtypes only for the Location, Limit, Zone ([18], Appendix F). According to IHO S-121 [18], there are no MLB Space objects defined. Any Space object needs to be generated from the generic object Space and registered in the Feature Catalogue for a particular product. It is proposed that Space Objects can be further specialized to LegalSpaceTunnel, LegalSpaceUtilityNetwork, LegalSpaceObject and LegalSpaceConsrunctionUnit, which can be externally linked to their physical counterparts using extID within these classes. In this way, the integration of physical and legal objects by virtue of international standard models is attained, as it occurs through the use of LA_LegalSpaceBuildingUnit and LA_LegalSpaceUtilityNetwork in ISO 19152. S-121 Spatial Unit package with the aforementioned additions is presented in Figure 7.

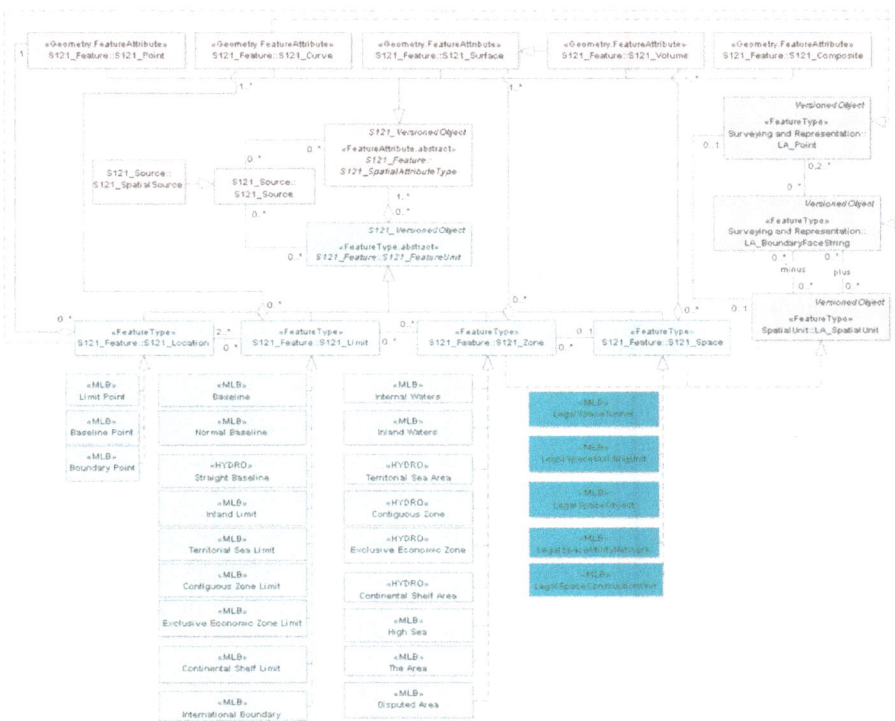

Figure 7. S-121 Spatial Unit package ([18], p. 68, edited).

Spatial Dimension and Associated Issues

The basic reference unit, could be spatially defined as: a multidimensional marine parcel or a series of (special purpose) volumetric marine parcels [12] or as sea surface objects, water volume objects, seabed objects, and sub seabed objects [43], as shown in Figure 8, or as a single part of marine space deriving from the determined and standard division of the maritime surface using a grid of specific dimensions and subdivisions if needed, as shown in Figure 9. It is specified by geodetic coordinates of the surrounding boundaries. This method is already in use for defining the blocks in the domain of minerals exploitation. The combination of these methods is feasible [1].

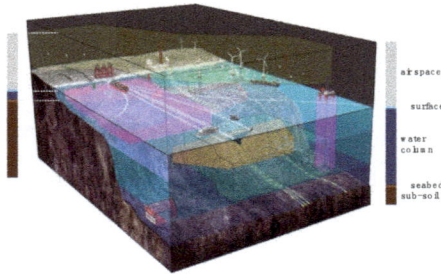

Figure 8. 3D nature of marine parcel ([44], edited).

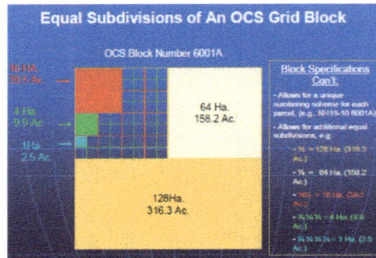

Figure 9. Grid System for oil and gas exploitation in USA [45].

The selection of the geodetic datum, on which the coordinates will be referred, is one of the issues that should be taken into account in the development of a MC. A geodetic datum specifies the reference ellipsoid and the point of origin from which the coordinates are derived. Different states, even different mapping authorities of the same state, use different geodetic datums.

Consequently, coordinates derived from one system do not agree with the coordinates from another datum, with their differences between adjacent states, as [46] points out, amounting to several hundred meters. In addition to the horizontal datums, the utilization of different vertical datums has a significant impact as well. Hydrographic Services, which are assigned with the task to map the marine environment, as their priority is the safety of navigation they depict depth soundings from a mean low water level, such as the Lowest Astronomical Tide (LAT), or the more conservative Lowest Low Water (LLW). On the other hand, the Land cadastral services usually use the Mean Sea Level (MSL). The difference between the two, needs to be precisely calculated. One of the factors affecting the calculation is the distance of the permanent tide gauges from the location. The different sea levels and the precise calculation of the sea level have also a significant impact to the development of a MC. More specifically, the delineation of the coastline may vary greatly depending on the vertical datum, which consequently has a significant impact on the outer limits of the maritime zones over which the states exercise their rights. For instance, as Leahy et al. [47] describes, for a foreshore of 0.5% gradient, a difference of 0.5 m in sea water level results 100 m error in the location of the coastline, a value that may exceed 200 m in some cases. In extreme cases and depending on the techniques followed for the delineation of the coastline (e.g., digitization from small scale charts), Leahy et al. [47] calculated that the horizontal displacement of the coastline may reach 3 NM when (the coastline) has been derived from topographic maps of scales 1:100,000.

In addition, here comes another issue: where does the data come from? Is it data acquired in situ using techniques according to specifications, or data derived from paper charts/maps compiled years ago with obsolete and error prone techniques? Another issue with the different sea levels is the potential reclassification of a sub-surface feature to a low-tide elevation, which may expand the maritime zones of the coastal state [36] (Article 13(1)).

The precise delineation of the baselines is one of the most important issues [17] as they are the reference to measure the maritime zones. However, it is not the only issue that affects the precise division of marine space. As nicely put by Carrera [48], "marine boundaries are delimited, not demarcated, and generally there is no physical evidence, only mathematical evidence left behind", hence the reference surface used for the delimitation of the outer limits and boundaries is another source of error. While technical publications, e.g., TALOS, state their preference towards the ellipsoidal earth, something of the kind is not stated in UNCLOS. The maximum relative error with approximating the earth as a sphere is 0.5%, but if projected plane was to be used for creating buffers of the baselines (e.g., Mercator projection) the produced error would be significantly greater.

Unfortunately, UNCLOS remains silent in many of its provisions regarding the technical aspects of the delimitation, including the horizontal and vertical datums, which the states need to consider and agree with neighboring states towards an effective MC.

5.4. S-121 Sources

The MLB data is primarily acquired from multiple external and internal sources, which can be registered in the model in the same way as in the ISO 19152. There are two subtypes of source S121_SpatialSource and S121_AdministrativeSource, which inherit the attributes from the abstract class S121_Source. The fact that all different rights find their base in some kind of transacting document is represented by the association between S121_RRR and S121_AdministrativeSource and this transacting document is recorded in the latter class. However, in marine environments, the existence of rights may be not created through the transaction, but from the implementation of law.

Therefore, two kinds of legal documents define the marine legal object [17]:

- **Laws:** The legislation which defines all RRRs of the marine space and constitutes the basis upon which the content of the administrative resources is developed. The term "law" leads to the main division between substantial and typical law (e.g., the legislation produced by the legislative power of the House of Representatives). Thus, the substantial law includes the principles of Common Law and equity, the administrative acts of the Administrative Authorities (Ministerial and Presidential Acts) as well as acts of legislative content. Needless to say, the European Law (Treaties, Regulations, Directives, Decisions) and the International law are main legal binding sources.
- **Administrative Sources:** The legal sources which include the administrative regime of the RRRs are defined. The administrative sources are: the legal contracts that relate to the disposal of the functional rights of the state to private entities (as defined by the legal framework). The functional rights of the state granted are either by means of an administrative contract (administrative long leases or public works contracts) or the right is conferred by an administrative act, most usually by a license agreement. The administrative sources that need to be recorded in an MAS are different depending on each activity. For several resources, the processes relating to registration of issue are standard. For example, in Greece, the registration associated with exploration for and extraction of gas and petroleum is highly refined. It is of high importance that all the activities that take place in the marine space need to be recorded accurately. This systematic recording could help to identify: the multiple licenses required for specific activities, regulated access rights, existing legal gaps.

5.5. External Classes

ISO 19152 provides stereotype classes for external datasets, which indicate what dataset elements the LADM expects from these external sources, if available. However, these datasets are outside the scope of LADM. They do not have the 'LA_' prefix, but they do give an exact definition of what the LADM is expecting of these external classes [33].

The product specification for MLB data provides attributes for the connection with external databases only for the Party Package and the Sources. More specifically: the exPID in S121_Party class, as an identifier of the party in an external registration and the extArchiveID in S121_Source class, as the metadata about an external archive. The use of the following classes is suggested in the proposed specializations of MLB Space Object: extPhysicalSpaceObject and extPhysicalSpaceUtilityNetwork, extPhysicalSpaceTunnel, and extPhysicalSpaceConstructionUnit.

Current considerations focus on the incorporation of 3D legal and physical objects. It is critical for data integration on legal spaces and on physical features to occur also in the marine environments, since the coincidence of boundaries of legal and physical spaces is very rare, if ever existent. Taking as reference the underwater tunnels, pipelines, utility networks, submerged archaeological cities or even underwater luxury hotels and facilities. Integration enables the reuse of geometrical data in different domains and the definition of legal spaces based on physical constructions [42].

Marine physical objects can be modelled by using appropriate 3D modeling techniques, constituting an area of research interest towards the development of an integrated land-marine administration system.

6. Code Lists of S121 Attributes Inherited from LADM for Greece and Trinidad and Tobago

Code lists are used to describe a more open and flexible enumeration. Code lists are useful for expressing a long list of potential values, aiming to allow the use of local, regional and/or national terminology [19].

Why use code lists instead of character strings? Code lists are used, rather than character strings in order to ensure consistency. Code lists of draft version of S-121 are currently generic and their content needs to be defined as part of the S-121 project development [18].

Therefore, implementation of the S-121 at a country level requires the modification of code lists to meet specific needs, since the values differ between the various legislation systems. In reference to the values of code lists provided by [18] and belong to the LADM related classes, those that are inherited from ISO 19152 are the following:

- S-121 Spatial Unit Package: LA_AreaValue, LA_AreaType, LA_SurfaceRelationType, LA_InterpolationType, LA_Transformation;
- S-121 Sources: LA_AvailabilityStatusType, DQ_EvaluationMethodTypeCode, Cl_OnlineFunctionCode, Cl_PresentationFormCode, Cl_RoleCode.

However, there are code lists inherited from ISO 19152, but different values are given in order to meet the marine environment special needs. These values can be further expanded to address each country's national jurisdiction.

In the following subsections, values of the code lists for the LADM related classes for Greece and Trinidad and Tobago are proposed, which either differ from the predefined values of S-121 or are not defined at all.

6.1. Code Lists for Greece

Greece is a state virtually surrounded by sea located in the Eastern part of the Mediterranean sea. Concerning the international law, Greece ratified the 1982 Convention on 21 July 1995 (Law 2321/1995) and since then has enacted a number of laws for the areas where sovereign rights are exercised (Figure 10). However, there is no comprehensive strategy to deal with the fractured and incomplete sets of data that are the legacy of the complex administrative and legal structures. A wide range of the existing marine RRRs is managed by the different governmental agencies. Figure 10 presents Greece's territorial sea and the potential Exclusive Economic Zone (EEZ), Continental Shelf.

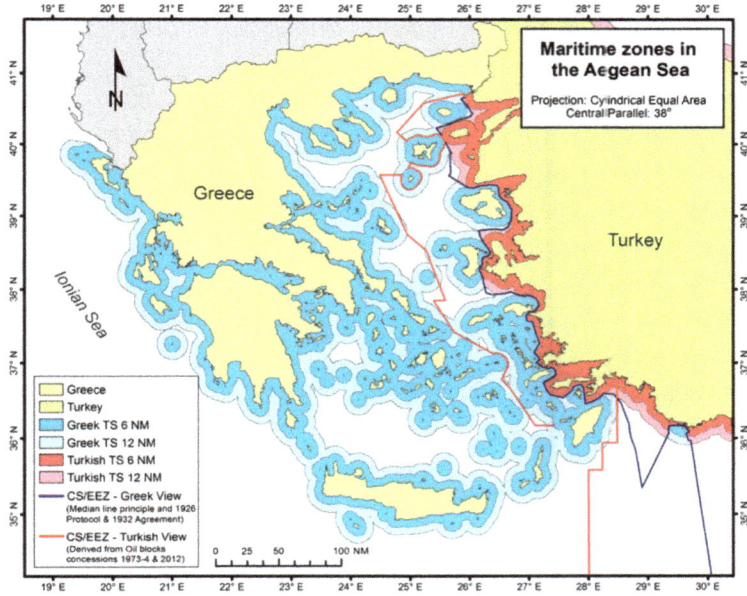

Figure 10. Greece's territorial sea and potential Exclusive Economic Zone (EEZ), Continental Shelf in the Aegean Sea [49].

The development and implementation of an MAS, based on the S-121 standard, seems to be applicable to the management of Greece's marine RRRs. In this section, the code lists relevant to the Greek case are presented (Figure 11), as derived from the current legal regime.

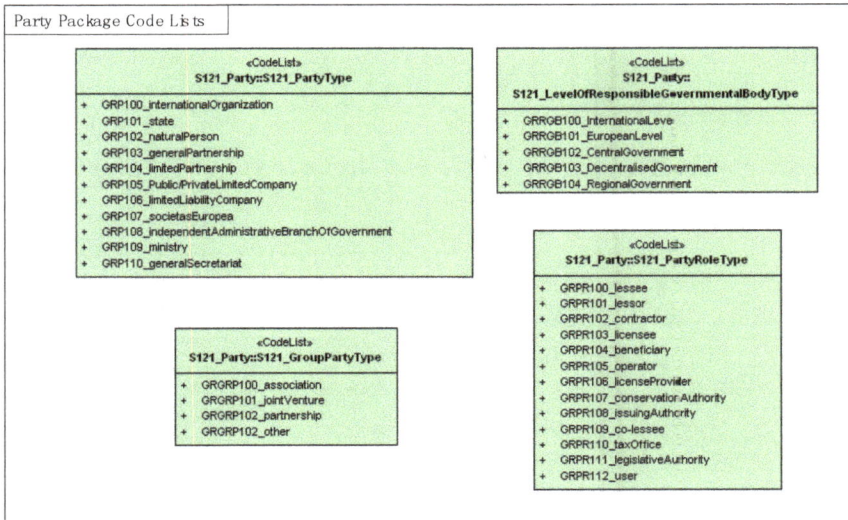

Figure 11. *Cont.*

Administrative Package Code Lists

«CodeList»
S121_Administrative::S121_RightType
+ GRR100_fullSovereignty
+ GRR101_sovereignRight
+ GRR102_rightOfEconomicExploitation
+ GRR103_jurisdiction
+ GRR104_militaryExercise
+ GRR105_access
+ GRR106_easement
+ GRR107_harvest
+ GRR108_scientificResearch
+ GRR109_seismicResearch
+ GRR110_exploration
+ GRR111_extraction
+ GRR112_pipelineInstallation
+ GRR113_establishmentOfAProductionPlant
+ GRR114_thirdStateRights
+ GRR115_construct

«CodeList»
**S121_Administrative::
S121_RestrictionType**
+ GRREST100_access
+ GRREST101_jurisdiction
+ GRREST102_passage
+ GRREST103_resource
+ GRREST104_timeBased
+ GRREST105_use
+ GRREST106_otherORR

«CodeList»
S121_Administrative::S121_ResponsibilityType
+ GRRESP100_maintenance
+ GRRESP101_leasePayment
+ GRRESP102_submissionOfCharges
+ GRRESP103_complianceWithPreventiveMeasures
+ GRRESP104_declarationOfASecurityZone
+ GRRESP105_submissionOfRequiredDocuments

«CodeList»
**S121_Administrative::
S121_RenewableType**
+ GRRNT100_renewable
+ GRRNT101_nonRenewable
+ GRRNT102_potentialRenewable

«CodeList»
S121_Administrative::S121_BAUnitType
+ GRBAU100_sovereigntyUnit
+ GRBAU101_sovereignRightUnit
+ GRBAU102_economicExploitationUnit
+ GRBAU103_stateJurisdictionUnit
+ GRBAU104_leasedUnit
+ GRBAU105_culturalHeritageUnit
+ GRBAU106_protectionUnit
+ GRBAU107_specialRightsUnit
+ GRBAU108_publicRightUnit

«CodeList»
**S121_Administrative::
S121_ResourceType**
+ GRRT100_hydrocarbons
+ GRRT101_gas
+ GRRT102_fish
+ GRRT103_sedentarySpecies
+ GRRT104_catadromousSpecies
+ GRRT105_windEnergy
+ GRRT106_waveEnergy

Spatial Package Code Lists

«CodeList»
S121_Feature::S121_Object
+ GROB100_archaelogicalHeritage
+ GROB101_archaeologicalSite
+ GROB102_shipwreck
+ GROB103_divingUnit
+ GROB104_natura2000

«CodeList»
S121_Feature::S121_UtilityNetworkType
+ GRUN100_oil
+ GRUN101_gas
+ GRUN102_water
+ GRUN103_renewableEnergySources
+ GRUN104_telecommunications

«CodeList»
S121_Feature::S121_ConstructionUnitType
+ GRCUT100_port
+ GRCUT101_platforms
+ GRCUT102_shipyard
+ GRCUT103_refinery
+ GRCUT104_desalinationUnit
+ GRCUT105_windPark
+ GRCUT106_windTurbine
+ GRCUT107_electricityEnergyStations
+ GRCUT108_drilling
+ GRCUT109_sewage
+ GRCUT110_aquaculture
+ GRCUT111_buildingUnit

«CodeList»
**S121_Feature::
S121_ConstructionUnitKind**
+ GRCUK100_floating
+ GRCUK101_permanent
+ GRCUK102_temporary

Source Classes Code Lists

«CodeList»
S121_Source::S121_LawType
+ GRL100_statuoryLaw
+ GRL101_precedentialDecree
+ GRL102_ministerialOrder
+ GRL103_jointMinisterialDecision
+ GRL104_europeanRegulation
+ GRL105_europeanDirective
+ GRL106_internationalConvention

«CodeList»
S121_Source::S121_SpatialSourceType
+ GRSS100_nauticalChart
+ GRSS101_bathymetricMap
+ GRSS102_hydrographicChart
+ GRSS103_bathymetricChart
+ GRSS104_orthophoto
+ GRSS105_topographicMap

«CodeList»
S121_Source::S121_AdministrativeSourceType
+ GRAS100_lease
+ GRAS101_publicWorkContract
+ GRAS102_contractForDistributionOfProduction
+ GRAS103_license
+ GRAS104_environmentalImpactStudy
+ GRAS105_waveConditionStudy
+ GRAS106_geotechnicalStudy
+ GRAS107_feasibilityStudy

«CodeList»
S121_Source::S121_ExistenceWayType
+ GREW100_law
+ GREW101_lease
+ GREW102_implementationOfSpecialFramework
+ GREW103_europeanDirectiveImplementation
+ GREW104_maritimeSpatialPlanningImplementation
+ GREW105_europeanRegulationImplementation
+ GREW106_actOfEstablishingLimits
+ GREW107_individualAdministrativeAct
+ GREW108_declaration
+ GREW109_applicationOfTheInterestedPartyForTheArea
+ GREW110_openCall

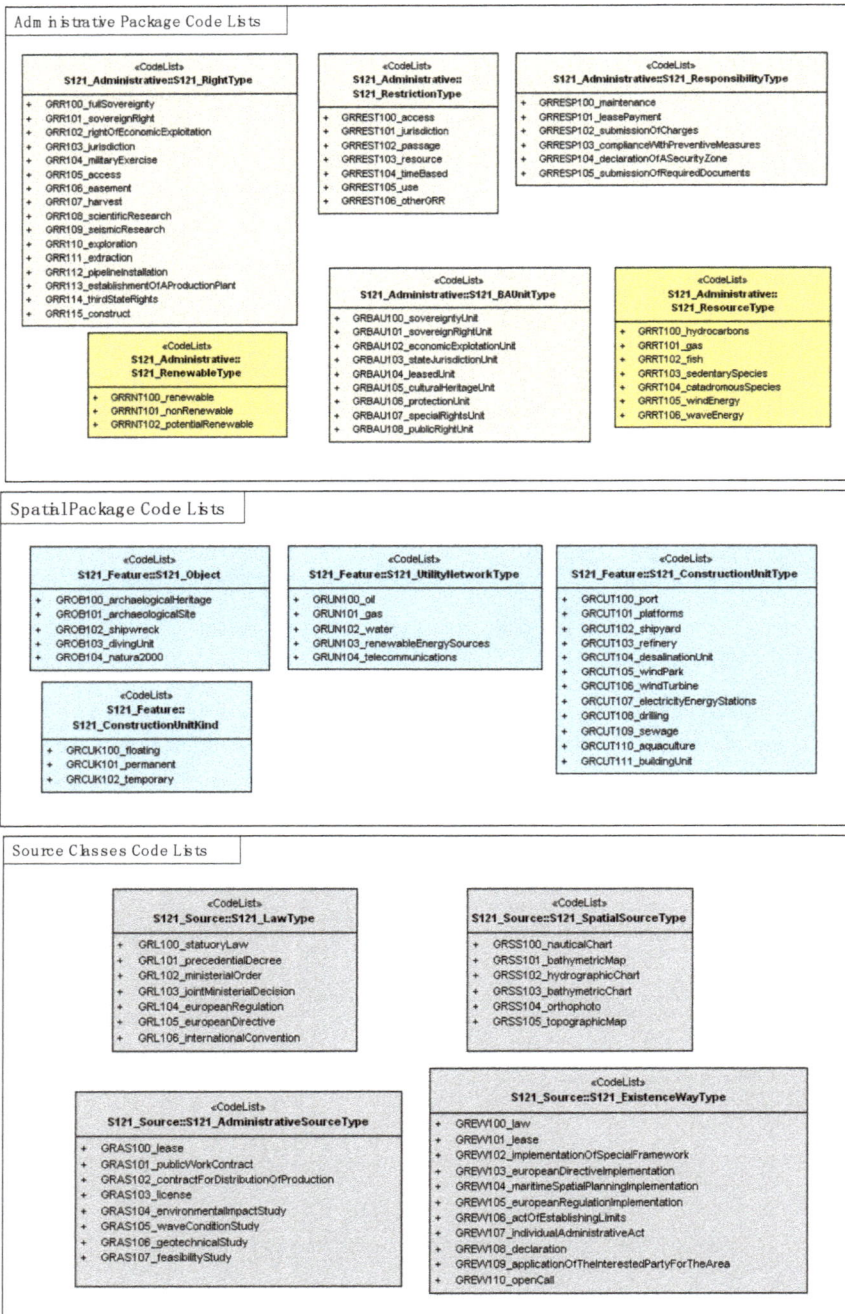

Figure 11. Code Lists for S121 Party, S121 Administrative, S121 Spatial Packages and Source Classes for Greece.

6.2. Code Lists for Trinidad and Tobago

The Republic of Trinidad and Tobago (Figure 12) is a Caribbean archipelagic twin-island country that is located approximately seven kilometers off the northeastern coast of Venezuela. It is the wealthiest country among Caribbean States, mostly because of oil and natural gas deposits in land and marine spaces under its sovereignty. Apart from oil and natural gas, the country's marine spaces are subject to a variety of other marine and maritime interests and rights relating to navigation and fishing, among others. Trinidad and Tobago ratified the United Nations Convention on the Law of the Sea on 25 April 1986. To manage all the marine RRRs in the country, the development of an MAS based on international standards such as S-121 would be ideal. This section presents code lists that would be applicable (Figure 13), should such an MAS be considered for development.

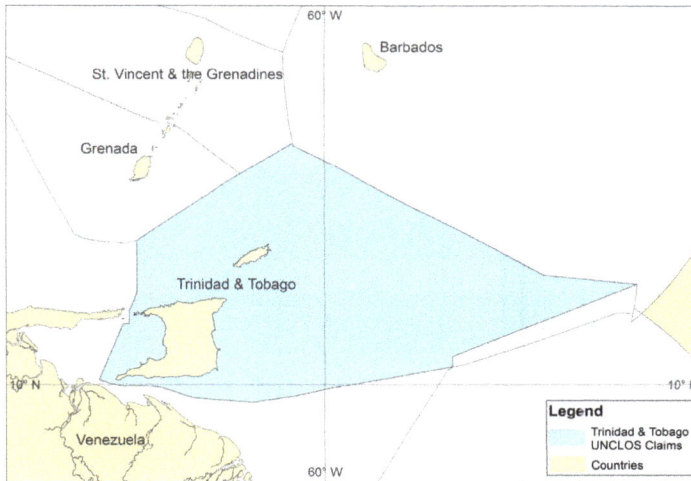

Figure 12. Trinidad and Tobago UNCLOS claim.

Figure 13. *Cont.*

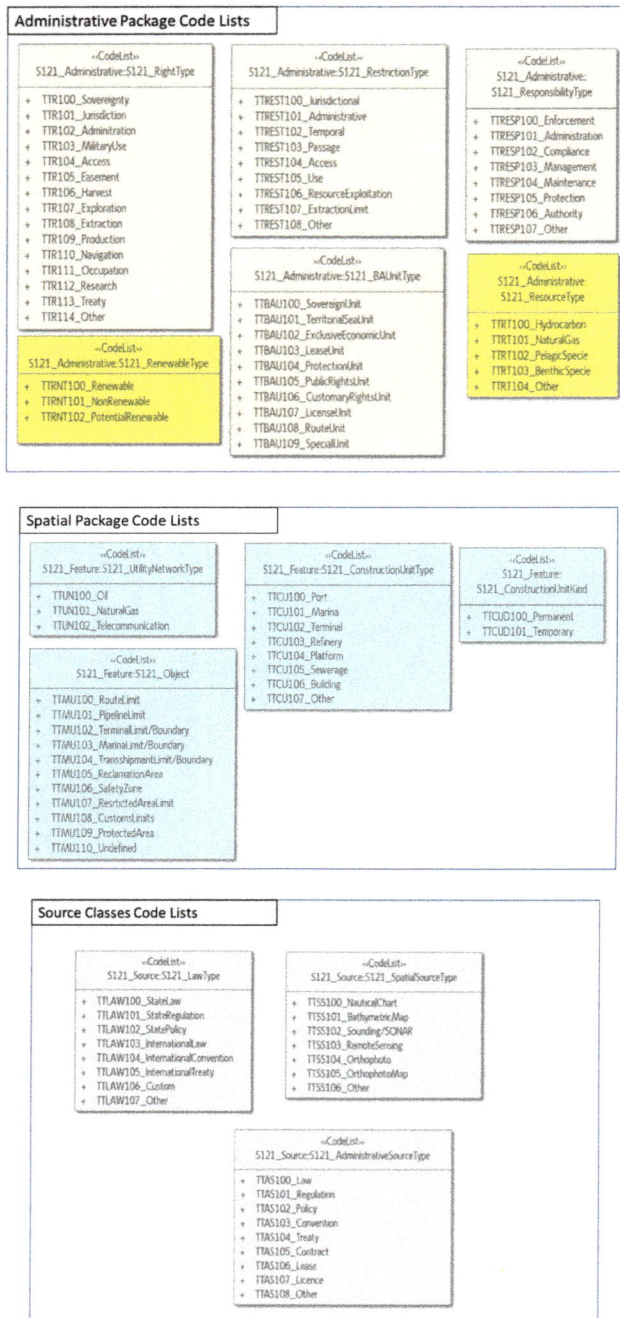

Figure 13. Code Lists for S121 Party, S121 Administrative, S121 Spatial Packages and Source Classes for Trinidad and Tobago.

7. Results

The emphasis of S-121 is on the legal description of marine objects. Different descriptions of legal rights and associated restrictions and responsibilities can be defined for multiple parties, even if these parties have potentially different and conflicting claims. The introduction of class marine resources is considered as critical since the ultimate goal in the administration and management of marine environment is sustainability and protection of the resources.

The definition of the code list values, according to each country legal regime, is considered as an important aspect in the implementation of S-121, while adjusting and extending the predefined values proposed by S-121 is also considered necessary. Regarding the comparison between the values of Greece and Trinidad and Tobago, it seems that, in their vast majority, they share a large common part. Some differences have been identified in the levels of responsible government, due to the different governmental structures and in the RRR types in regards to the different legislation systems.

8. Conclusions

Recent research focuses on regulating the establishment of basic principles, semantics, rules and procedures relating to the creation of an MAS. So far, standardization is a requirement to support the development of a National Land Information System. The same applies to the marine environment, since the term "land" encompasses the water element, as [33] states. S-100 provides the appropriate tools and framework to develop and maintain hydrography related data, products and registers. The extension of this standard to support the LADM, in order to include the registration of additional types of marine data, specified by the law is addressed through the development of S-121. S-121 may serve as the bridge between the land and marine domains while the MLB following the S-121 standard may be used in the marine administration domain. Part of the S-121 project development is the specialization of the generic code lists of the various attributes to marine environment for every state. Proposed values are given for the implementation of the model in Greece and Trinidad and Tobago.

Additionally, this paper addresses several issues that relate to the development of S-121 data model, as follows:

- The introduction of class marine resources into the model. The ultimate objective for the development and implementation of an MAS for the management of the legitimate interests is the conservation of the marine resources in the context of sustainable development.
- The integration of data on legal spaces and on physical features through the external classes into the specializations of MLB Space Object.
- The division of Sources into S121_Law and S121_AdministrativeSource. In marine environments, the existence of rights may be not created through transaction, but derive from the law.
- Regarding the Greek case, a conclusive approach is gradually an issue of growing priority, to support the state and European MSP initiatives. Determining maritime boundaries with its neighbors needs to be agreed, in order to define the area where the MAS applies. In addition, a national ocean's policy would be a first step towards the development of an MAS to manage the complex legal regime and overlapping jurisdictions, based on S-121 product specification.
- Organization of national marine legislation, considering EU orientations and directives in a structure similar of S-121. To this purpose, appropriate legislation should be introduced and maintain a database adapted to international standards. In this direction, any differences and conflicts in marine space need to be identified and resolved by specific regulations.
- Regarding the specialized code lists for Greece and Trinidad and Tobago as presented in Section 5, most of them have common values and the differences are to be found specifically in the RRR types and in the levels of responsible government. In the first place, the specialization of S-121 generic code lists is proposed, and, secondly, their adjustment, depending on the country needs.

Acknowledgments: The authors would sincerely like to thank Thijs Ligteringen, Hydrographic Service, Royal Netherlands Navy, Ministry of Defence, for providing the most recent documentation concerning S-121 Maritime Limits and Boundaries Standard.

Author Contributions: More specifically, Katerina Athanasiou carried out the main research for the potential application and modification of IHO S-121 to the marine and maritime administrative needs based on her diploma thesis, that took place at the School of Rural and Surveying Engineering of the National Technical University of Athens under the supervision of professor Efi Dimopoulou, and proposed the code lists for the Greek case. Similarly, Michael Sutherland, dealt with the review of modeling approaches applied to marine environment, developed a list of reasonable criteria that may apply to assess the LADM standard and proposed the code lists for Trinidad and Tobago, with supporting input from Charisse Griffith-Charles and Dexter Davis. Moreover, Lysandros Tsoulos and Christos Kastrisios supported the marine related international standards review and dealt with the spatial issues related to the marine environment. Finally, Efi Dimopoulou, provided important guidance concerning the methodological steps taken, by offering her critical review, improvements and expertise on this research field.

Conflicts of Interest: The authors declare no conflict of interest.

References

1. Athanasiou, A.; Pispidikis, I.; Dimopoulou, E. 3D Marine Admininstration System Based ON LADM. In Proceedings of the 10th 3D Geoinfo Conference, Kuala Lumpur, Malaysia, 28–30 October 2015; Abdul-Rahman, A., Ed.; Springer Nature: Berlin, Germany, 2017; pp. 385–408.

2. Nichols, S.; Ng'ang'a, S.M.; Sutherland, M.D.; Cockburn, S. Marine Cadast re Concept. In *Canada's Offshore: Jurisdiction, Rights and Management*, 3rd ed.; Calderbank, B., MacLeod, A.M., McDorman, T.L., Gray, D.L., Eds.; Trafford Publishing: Bloomington, IN, USA, 2006; Chapter 10.

3. Arvanitis, A. Development of an Integrated Geographical Information System for the Marine Space. Ph.D. Thesis, School of Rural and Surveying Engineering, Aristotle University of Thessaloniki, Thessaloniki, Greece, 2013.

4. Rajabifard, A.; Binns, A.; Williamson, I. Administering the Marine Environment. The Spatial Dimension. *J. Spat. Sci.* **2005**, *50*, 69–78. [CrossRef]

5. Rajabifard, A.; Williamson, I.; Binns, A. *Marine Administration Research Activities within Asia and Pacific Region—Towards a Seamless Land-Sea Interface*; Publication No. 36, International Issues; FIG, Administering Marine Spaces: Copenhagen, Denmark, 2006; pp. 21–36.

6. Strain, L.; Rajabifard, A.; Williamson, I.P. Marine Administration and Spatial Data Infrastructure. *Mar. Policy* **2006**, *30*, 431–441. [CrossRef]

7. Widodo, M.S. The Needs for Marine Cadastre and Supports of Spatial Data Infrastructures in Marine Environment—A Case Study. In Proceedings of the FIG Working Week, Paris, France, 13–17 April 2003.

8. Widodo, M.S. Relationship of Marine Cadastre and Marine Spatial Planning in Indonesia. In Proceedings of the 3rd FIG Regional Conference, Jakarta, Indonesia, 3–7 October 2004.

9. Widodo, S.; Leach, J.; Williamson, I.P. Marine Cadastre and Spatial Data Infrastructures in Marine Environment. In Proceedings of the Joint AURISA and Institution of Surveyors Conference, Adelaide, Australia, 25–30 November 2002.

10. Arvanitis, A.; Giannakopoulou, S.; Parri, I. Marine Cadastre to Support Marine Spatial Planning. In Proceedings of the Common Vision Conference 2016 Migration to a Smart World, EULIS, Amsterdam, The Netherlands, 5–7 June 2016.

11. De Latte, G. Marine Cadastre—Legal Framework UNCLOS & EU legislation. In Proceedings of the Common Vision Conference 2016 Migration to a Smart World, EULIS, Amsterdam, The Netherlands, 5–7 June 2016.

12. Ng'ang'a, S.M.; Sutherland, M.; Cockburn, S.; Nichols, S. Toward a 3D marine Cadastre in support of good ocean governance: A review of the technical framework requirements. *Comput. Environ. Urban Syst. J.* **2004**, *28*, 443–470. [CrossRef]

13. Duncan, E.E.; Rahman, A. A Multipurpose Cadastral Framework for Developing Countries—Concepts. *Electron. J. Inf. Syst. Dev. Ctries.* **2013**, *58*, 1–16.

14. Griffith-Charles, C.; Sutherland, M.D. Governance in 3D, LADM Compliant Marine Cadastres. In Proceedings of the 4th International Workshop on 3D Cadastres, Dubai, UAE, 9–11 November 2014; pp. 83–98.

15. Sutherland, M.D.; Griffith-Charles, C.; Davis, D. Toward the Development of LADM-based Marine Cadastres: Is LADM Applicable to Marine Cadastres? In Proceedings of the 5th International FIG 3D Cadastre Workshop, Athens, Greece, 18–20 October 2016; pp. 301–315.

16. Athanasiou, A. Marine Administration Model for Greece, based on LADM. Bachelor's Thesis, Department of Spatial Planning and Regional Development, School of Rural and Surveying Engineering, National and Technical University of Athens, Athens, Greece, 2014.

17. Athanasiou, A.; Dimopoulou, E.; Kastrisios, C.; Tsoulos, L. Management of Marine Rights, Restrictions and Responsibilities according to International Standards. In Proceedings of the 5th International FIG 3D Cadastre Workshop, Athens, Greece, 18–20 October 2016; pp. 81–104.

18. IHO S-121. Product Specification for Maritime Limits and Boundaries. Available online: https://www.google.com/url?sa=t&rct=j&q=&esrc=s&source=web&cd=1&ved= 0ahUKEwjhwZniv9DUAhUlCcAKHbrGBvEQFggiMAA&url=https%3A%2F%2Fwww.iho.int%2Fmtg_docs% 2Fcom_wg%2FS-100WG%2FS-121PT%2FS121%2520Draft%2520Product%2520Specification%2520Revised% 252001%2520Dec%2520v2.3.8.docx&usg=AFQjCNFv_PaTC96xHdMR-9tigRZ6s9ERLg&cad=rja (accessed on 1 December 2016).

19. Lemmen, C.H.J. A Domain Model for Land Administration. Ph.D. Thesis, Technical University of Delft, Delft, The Netherlands, 2012.

20. Canadian Hydrographic Service; Geoscience Australia. Supporting the ISO 19152 Land Administration Domain Model in a Marine Environment. Available online: https://www.iho.int/mtg_docs/ com_wg/S-100WG/S-100WG1/S100WG01-10.3A_IHOPaper_IntegrationOfLADM_Rev1.pdf (accessed on 26 February 2016).

21. Fowler, C.; Tremi, E. Building a marine cadastral information system for the United States—A case study. *Comput. Environ. Urban Syst.* **2001**, *6*, 493–507. [CrossRef]

22. Collier, P.; Leahy, F.; Williamson, I. Defining a Marine Cadastre for Australia. In Proceedings of the Institute of Australian Surveyors Annual Conference, Brisbane, Australia, 25–28 September 2001.

23. Hirst, B.; Robertson, D. Law of the Sea Boundaries in a Marine Cadastre. In Proceedings of the Institute of Australian Surveyors Annual Conference, Brisbane, Australia, 25–28 September 2001.

24. Todd, P. Marine Cadaster—Opportunities and Implications for Queensland. In Proceedings of the Institute of Australian Surveyors Annual Conference, Brisbane, Australia, 25–28 September 2001.

25. Sutherland, M.; Nichols, S.; Monahan, D. Marine Boundary Delimitation for Ocean Governance. In Proceedings of the Institute of Australian Surveyors Annual Conference, Brisbane, Australia, 25–28 September 2001.

26. Ng'ang'a, S.M.; Nichols, S.; Sutherland, M.; Cockburn, S. Toward a Multidimensional Marine Cadastre in Support of Good Ocean Governance; New Spatial Information Management Tools and their Role in Natural Resource Management. In Proceedings of the International Conference on Spatial Information for Sustainable Development, Nairobi, Kenya, 2–5 October 2001.

27. Ng'ang'a, S.M.; Sutherland, M.; Nichols, S. Data Integration and Visualisation Requirements for a Canadian Marine Cadastre: Lessons from the Proposed Musquash Marine Protected Area. In Proceedings of the ISPRS Commission IV, Ottawa, ON, Canada, 9–12 July 2002.

28. Binns, A.; Rajabifard, A.; Collier, P.A.; Williamson, I.P. Developing the concept of a marine cadastre: An Australian case study. *Trans-Tasman Surv.* **2004**, *6*, 19–27.

29. Binns, A. Building a National Marine Initiative through the Development of a Marine Cadastre for Australia. In Proceedings of the East Asian Seas Congress, Kuala Lumpur, Malaysia, 8–12 December 2003.

30. Peyton, D.; Kuwalek, E.; Fadaie, K. Managing Hydrographic and Oceanographic Information for Maritime Spatial Data Infrastructure: New Paths. New Approaches. Available online: http://slideplayer.com/slide/ 9352316/ (accessed on 28 August 2016).

31. Sutherland, M. *Marine Boundaries and the Governance of Marine Spaces*; Technical Report No. 232; University of New Brunswick: Fredericton, NB, Canada, 2005; p. 372.

32. Sutherland, M.; Nichols, S. Issues in the governance of marine spaces. In *Administering Marine Spaces*; International Issues, FIG Publication No. 36; Sutherland, M., Ed.; International Federation of Surveyors: Copenhagen, Denmark, 2006; pp. 6–20.

33. ISO 19152. Geographic Information—Land Administration Domain Model. Available online: https://www. iso.org/standard/51206.html (accessed on 25 October 2012).

34. Aien, A.; Kalantari, M.; Rajabifard, A.; Williamson, I.; Bennett, R. Utilizing Data Modeling to Understand the Structure of 3D Cadastres. *J. Spat. Sci.* **2013**, *58*, 215–234. [CrossRef]

35. IHO S-100. *Universal Hydrographic Data Model*; Publication S-100; 2.0.0; International Hydrographic Organization (IHO): Monaco, 2015.

36. McGregor, M. S-10X Maritime Boundary Product Specification—Explanatory Notes. Presented at the 26th IHO Transfer Standard Maintenance and Application Development Working Group (TSMAD) and the 5th Digital Information Portrayal Working Group (DIPWG), Silver Spring, MD, USA, 10–14 June 2013.

37. Longhorn, R. Assessing the Impact of INSPIRE on Related EU Marine Directives. In Proceedings of the Hydro12 Conference, Rotterdam, The Netherlands, 13 November 2012.

38. Millard, K. *Inspire 'Marine'—Bringing Land and Sea Together*; HR Wallingford: London, UK, 2007.

39. Nichols, S. Land Registration in an Information Management Environment. Ph.D. Thesis, Department of Surveying Engineering, University of New Brunswick, Fredericton, NB, Canada, 1992; p. 340.

40. Ng'ang'a, S. *Extending Land Management Approaches to Coastal and Oceans Management: A Framework for Evaluating the Role of Tenure Information in Canadian Marine Protected Areas*; ProQuest: Ann Arbor, MI, USA, 2006.

41. Barry, M.; Elema, I.; van der Molen, P. Ocean Governance in the Netherlands North Sea. New Professional Tasks. Marine Cadastres and Coastal Management. In Proceedings of the FIG Working Week, Paris, France, 13–17 April 2003.

42. Kalantari, M.; Rajabifard, A.; Wallace, J.; Williamson, I. Spatially referenced legal property objects. *J. Land Use Policy* **2008**, *25*, 173–181. [CrossRef]

43. Rahman, A.; van Oosterom, P.; Hua, T.H.; Sharkawi, K.H.; Duncan, E.E. 3D Modelling for Multipurpose Cadastre. In Proceedings of the 3th International Workshop on 3D Cadastres, Shenzhen, China, 25–26 October 2012; pp. 185–202.

44. NOAA. An Ocean of Information. Available online: http://marinecadastre.gov (accessed on 5 August 2016).

45. BOEMRE. Development of Marine Boundaries and Offshore Leases. Management of Marine Resources. Available online: http://www.mcatoolkit.org/pdf/ISLMC_11/Marine_Boundaries_Offshore_Lease_Areas_Management.pdf (accessed on 5 July 2014).

46. Beazley, P.B. *Technical Aspects of Maritime Boundary Delimitation*; International Research Unit, Durham University: Durham, UK, 1994; Volume 1.

47. Leahy, F.J.; Murphy, B.A.; Collier, P.A.; Mitchell, D.J. Uncertainty Issues in the Geodetic Delimitation of maritime Boundaries. In Proceedings of the International Conference on Accuracies and Uncertainties, Issues in Maritime Boundaries and Outer Limits, International Hydrographic Bureau, Monaco, 18–19 October 2001.

48. Carrera, G. *Lecture Notes on Maritime Boundary Delimitation*; University of Durham: Durham, UK, 1999.

49. Kastrisios, C.; Pilikou, M. Nautical Cartography Competences and their Effect to the Realisation of a Worldwide Official ENC Database, the Performance of ECDIS and the Fulfillment of IMO Chart Carriage Requirement. *Mar. Policy* **2017**, *75*, 29–40. [CrossRef]

International Journal of
Geo-Information

MDPI

Article

Assessing Performance of Three BIM-Based Views of Buildings for Communication and Management of Vertically Stratified Legal Interests

Behnam Atazadeh *, Abbas Rajabifard and Mohsen Kalantari

The Centre for Spatial Data Infrastructures and Land Administration, Department of Infrastructure Engineering, The University of Melbourne, Parkville, VIC 3010, Australia; abbas.r@unimelb.edu.au (A.R.); mohsen.kalantari@unimelb.edu.au (M.K.)
* Correspondence: behnam.atazadeh@unimelb.edu.au; Tel.: +61-383-444-431

Academic Editors: Peter van Oosterom, Efi Dimopoulou and Wolfgang Kainz
Received: 31 March 2017; Accepted: 29 June 2017; Published: 3 July 2017

Abstract: Multistorey buildings typically include stratified legal interests which provide entitlements to a community of owners to lawfully possess private properties and use communal and public properties. The spatial arrangements of these legal interests are often defined by multiplexing cognitively outlined spaces and physical elements of a building. In order to support 3D digital management and communication of legal arrangements of properties, a number of spatial data models have been recently developed in Geographic Information Systems (GIS) and Building Information Modelling (BIM) domains. While some data models, such as CityGML, IndoorGML or IFC, provide a merely physical representation of the built environment, others, e.g., LADM, mainly rely on legal data elements to support a purely legal view of multistorey buildings. More recently, spatial data models integrating legal and physical notions of multistorey buildings have been proposed to overcome issues associated with purely legal models and purely physical ones. In previous investigations, it has been found that the 3D digital data environment of BIM has the flexibility to utilize either only physical elements or only legal spaces, or an integrated view of both legal spaces and physical elements to represent spatial arrangements of stratified legal interests. In this article, the performance of these three distinct BIM-based representations of legal interests defined inside multistorey buildings is assessed in the context of the Victorian jurisdiction of Australia. The assessment metrics are a number of objects and geometry batches, visualization speed in terms of frame rate, query time, modelling legal boundaries, and visual communication of legal boundaries.

Keywords: building information modelling; stratified legal interests; legal view; physical view; multistorey buildings

1. Introduction

1.1. Background

The legal boundaries of properties in multistorey buildings are usually difficult to be discerned due to the existence of complex physical structures inside these buildings. Over the last decade, 3D digital data environments have emerged to provide a more communicable method of representing these boundaries for the inexpert community of owners residing in multistorey developments [1]. The basis of structuring, managing and representing 3D data in these environments is underpinned by three distinct types of 3D spatial data models, namely purely legal, purely physical and integrated models. Existing spatial models in land administration, such as Land Administration Domain Model (LADM), mainly focus on the purely legal conception of the built environment, taking less notice of physical aspects [2]. On the other side, purely physical notions of buildings are supported by 3D spatial models developed in Geographic Information Systems (GIS) and Building Information Modelling (BIM) domains. The widely known examples of physical models are CityGML [3], IndoorGML [4] and Industry Foundation Classes (IFC) standards [5]. Integrated models enable simultaneous management of both legal and physical aspects of buildings. These models have been developed based on either extending physical models with legal data elements (e.g., [6]) or defining linkages between legal models and physical models [7].

BIM-based approaches are increasingly being used in the Architecture, Engineering and Construction (AEC) sector to enable smart management and operation of buildings [8]. BIM is an intelligent, collaborative, 3D digital model-based process that equips AEC professionals with an integrated approach to more efficiently plan, design, construct, and manage built assets [9]. BIM models provide detailed spatial and semantic information about physical components of buildings, which are architectural, structural and utility elements. It is also possible to manage and represent cognitive spaces, such as a corridor or airspace, inside these models. Interoperability of BIM tools is considered to be a fundamental challenge in the AEC industry [10]. Use of proprietary formats to exchange BIM models typically results in inefficiencies, inconsistencies, repetitions, misinterpretations, and errors over the course of the building lifecycle [11]. To address these issues, the BuildingSMART organization defines and publishes the specifications for the IFC standard, which provides a universal data structure of buildings, facilitating data sharing among BIM tools [12]. However, each proprietary BIM tool implements their own import and export capabilities of IFC format using a different subset of IFC schema. This sometimes results in inconsistency issues in using IFC format to exchange BIM data.

1.2. Aim

The main focus of BIM is on modelling physical aspects of buildings. However, it is extendable to manage cognitive notions of built assets such as the legal arrangements of properties. The flexibility of BIM environments provide the ability to define either a purely physical model, or a purely legal model or an integrated legal and physical model of buildings [13]. These models provide three distinct alternatives to manage and communicate legal arrangements of properties in buildings. So far, however, there have been limited investigations about the performance of these three types of 3D digital BIM-based models of buildings. Therefore, the purpose of this article is to evaluate and compare the capabilities of these models using assessment metrics associated with representation, structure and management of legal arrangements of properties in multistorey buildings.

1.3. Methodology

The methodology consists of three main phases as explained below:

1. Understanding current practice: For studying the current practice of subdividing legal interests, the Victorian jurisdiction in Australia has been selected. The reason for choosing this jurisdiction is that a wide range of boundaries are used to define spatial extent of legal interests in multistorey building developments. In this phase, two main legal concepts were studied, namely arrangements and boundaries of legal interests. Arrangements refer to defining constituting parts of each legal interest. Boundaries delineate the spatial extent of a part of a legal interest. The identified legal boundaries and legal interests are modelled in BIM environment in the next phase.
2. Implementing models: In this phase, an appropriate multistorey development was selected as a case study. The selected development comprised broad types of legal interests and boundaries. For this development, three distinct BIM models were implemented. These include legal, physical and integrated models.
3. Assessing models: This last phase is dedicated to the assessment of the implemented BIM models using some metrics. The assessment metrics in this study include a number of objects and geometry batches, visualization speed regarding frame per second, query time, modelling legal boundaries, and visual communication of legal boundaries.

1.4. Structure

The next section provides a review of relevant literature on the spatial problems in legal partitioning of buildings as well as 3D spatial data models which are either defined specifically or used potentially for managing legal interests in multistorey buildings. This is followed by a summarized explanation of various boundaries and arrangements used for defining legal interests in current subdivision practices (Section 3). Section 4 provides the logical modelling of legal boundaries in purely legal, purely physical and integrated views of BIM environment. The "Assessment of BIM models" section describes implementation of the BIM models. Moreover, comparisons between these models using the assessment metrics are reported in this section. The last section comprises the main findings of this research and potential directions for future investigations.

2. Literature Review

2.1. Spatial Problems in the Legal Partitioning of Buildings

In reality, the legal partitioning of buildings is essentially 3D [14]. However, the medium of communication used to represent 3D reality of legal interests in buildings has been mainly predicated on 2D analogue plans. Various countries have proposed different approaches to manage the spatial complexity associated with legal interests in urban built environments [15–18]. This is due to the fact that each country has their own legislative framework, in which they should address problems associated with stratified legal interests. The common spatial problems in visualization and management of legal spaces in buildings include:

- Combining descriptive and 2D plans provided in the deed with 2D parcel boundaries to form a 3D image of a vertically located legal interest is a very difficult task [1].
- In most cadastral databases, there are no adequate geometric and solid models to store 3D spatial extent of vertically stratified legal objects [19].
- 3D legal objects are only delineated by coordinates and edges on paper plans and it is impossible to check the spatial integrity and validity of these objects [20]. Spatial integrity indicates that 3D legal objects should be volumetrically closed and there should be no gap and undesirable overlap between two legal spaces.
- 3D legal objects are linked to their corresponding land parcel in the cadastral maps in some jurisdictions. Others use a separate database to maintain spatial data about 3D properties [21]. Both approaches result in ambiguous and fragmented representation of 3D legal objects, which affects the authoritative nature of the cadastral maps.

In order to address the above problems, different kinds of 3D spatial data models have been developed. These data models underpin the basis of 3D digital environments. This article is focused on using BIM environment to compare three approaches (purely legal, purely physical and integrated models) in terms of modelling legal interests in buildings. Therefore, existing and potential models (such as CityGML and LADM or integration of these standards), which use similar methods but not essentially in BIM environment, for modelling vertically stratified legal interests will be reviewed in the next subsection. This will position the work presented in this article in a wider context of 3D cadastre field.

2.2. 3D Spatial Data Models

The purpose of managing and representing legal arrangements and boundaries are not specifically considered in purely physical models. However, legal information can be accommodated into these models using their built-in spatial entities and mechanisms for extending them. Purely physical models provide spatial and semantic entities to manage multistorey buildings in different levels of detail. IFC, CityGML and IndoorGML [4,22] standards provide a considerable level of detail to manage information about indoor physical structures and spaces within buildings [23]. Purely legal models are typically developed according to jurisdictional settings, differing from one jurisdiction to another one.

The concept of legal arrangements is usually communicated via 2D-based and planar representation schemes adopted in these models. Among purely legal models, LADM is one that supports the 3D digital representation of legal information [24]. Therefore, in this section, the relevant literature about IFC, CityGML, and IndoorGML standards (purely physical models), LADM (purely legal model) and integrated (legal and physical) models will be reviewed.

2.2.1. IFC

The IFC standard is an open and platform independent data model for storing and managing BIM models [25]. The purpose of this standard is to facilitate interoperability and exchange of BIM models among various BIM platforms. The EXPRESS data modelling language underpins the foundation of this open BIM data model [26]. A comprehensive number of spatial and semantic concepts are used in the IFC standard to enable the entire lifecycle modelling of buildings. From the perspective of spatial sciences, the subclasses of the "IfcProduct" entity describe the geometric and semantic aspects of building elements as well as spatial relationships between these elements [12]. The spatial structure and physical components of a building are hierarchically modelled within the IFC standard (see Figure 1). The order of the IFC hierarchy is "IfcProject", "IfcSite", "IfcBuilding", "IfcBuildingStorey", "IfcSpace" and "IfcBuildingElement" [27]. "IfcProject" provides contextual information about BIM projects such as measurement units, project phase, adopted geometric representations and world coordinate system. "IfcSite" represents the spatial extent of the surface of the land on which buildings are developed. In addition, geographical coordinates, datum elevations, land title number and the address of the construction site can be stored as attributes in this entity. "IfcBuilding" is used for modelling each building and recording information about the address of the building and its elevation from sea level. "IfcBuildingStorey" is used for modelling each building level and elevation of its base relative to the reference height (0.00) of the building. The "IfcSpace" entity describes various kinds of cognitively defined indoor spaces of buildings. "IfcBuildingElement" is a generalized concept for modelling any type of physical building elements. However, specialized entities, such as "IfcWall", "IfcWindow", and "IfcDoor", have been defined to model spatial and semantic aspects of distinct types of physical building elements. The "IfcSpace" entity is a suitable entity for modelling the spatial extent of legal spaces inside buildings. The space boundaries are defined by the "IfcRelSpaceBoundary" entity which not only defines the geometry and topology of a space boundary but also provides the semantic linkage between the physical elements (such as walls and ceilings) and spaces [28]. In other words, this entity provides the ability to query the type of physical structure defining the space boundary.

Besides the hierarchical decomposition of buildings, the IFC standard also supports non-hierarchical approaches to link or reference physical elements and cognitive spaces of buildings in spatial zones. In other words, the concept of spatial zones in IFC defines a non-hierarchical aggregation of building components and spaces for a peculiar purpose such as the thermal activity of the building [29]. This concept can be useful when various parts of a legal interest are in different levels of a building. For instance, a private property typically comprises an apartment unit, carpark, and storage area, which are located in various parts of a building. Therefore, the concept of the spatial zone can be used to establish the linkage between an apartment unit and its carpark or storage area.

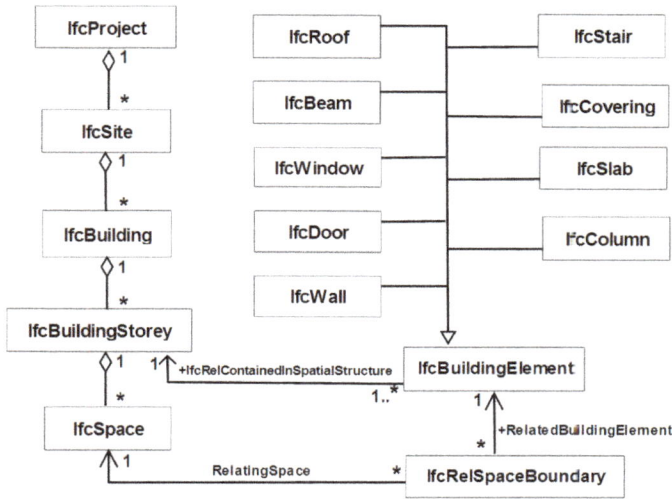

Figure 1. Hierarchical storage of building information in IFC standard.

2.2.2. CityGML

The Open Geospatial Consortium (OGC) has defined and promoted the open City Geography Markup Language (CityGML) standard to facilitate interoperability of urban information models [30]. CityGML provides a universal 3D data model for managing semantic, geometric, and topological aspects of urban built and natural environments in both outdoor and indoor scales. CityGML adopts a multi-granular approach to representing 3D urban objects in five Levels of Details (LoD0–LoD4). CityGML is conceptually divided into various modules. Each module comprises entities used for modeling a specific built (such as buildings, bridges, and roads) or natural (such as vegetation and water bodies) urban feature [22]. The building module in CityGML includes entities for modelling architectural structures and spaces of buildings. The fundamental abstract entity in the building module is "_AbstractBuilding", which is specialized into "Building" and "BuildingPart" subclasses (see Figure 2). These instantiable entities are respectively used to model individual buildings and parts of complex building developments. In CityGML, the cognitively defined and non-overlapping indoor spaces are modelled by "Room" entity. Therefore, this entity can potentially be utilized for representing legal spaces inside buildings. Since CityGML uses boundary representation (B-Rep) solid models for storing the geometry of 3D objects, the legal spaces defined by using the "Room" entity are topologically enclosed and valid for cadastral purposes [22]. The semantic boundaries of indoor spaces are also specified through "_BoundarySurface" entity and its instantiable subclasses, namely "WallSurface, InteriorWallSurface, CeilingSurface, FloorSurface, RoofSurface, GroundSurface, OuterCeilingSurface, OuterFloorSurface, and ClosureSurface" (see Figure 2).

These entities are used to define the boundary relationships between a physical structure, such as wall or ceiling, or a virtual element and an indoor space. These spatial relationships can be used to query the location of legal boundaries. However, the existing semantic entities for modelling boundary types in CityGML are limited to specifying only internal and external surfaces of architectural elements. The legal boundaries are sometimes defined in the median location of building elements. For example, a wall between two abutting townhouses is split into two equal parts, each of which belongs to the owner of each townhouse. In some complex situations, the legal boundary can be located in any position inside the physical structure, which is called "other" in the context of Victorian jurisdiction. Modelling the geometry of "median" and "other" is supported in CityGML. Nevertheless, the semantic entities associated with these types of physical boundaries need to be considered in CityGML. One may argue that these boundaries can be defined as virtual boundaries (ClosureSurface) because they are defined as imaginary surfaces inside the physical structure. However, we argue that such definition would result in ambiguities in semantic querying of the physical structure associated with the legal boundary. This indicates that the semantic query cannot determine whether the median boundary is defined by a wall or ceiling, although this can be visually observed in the 3D building model.

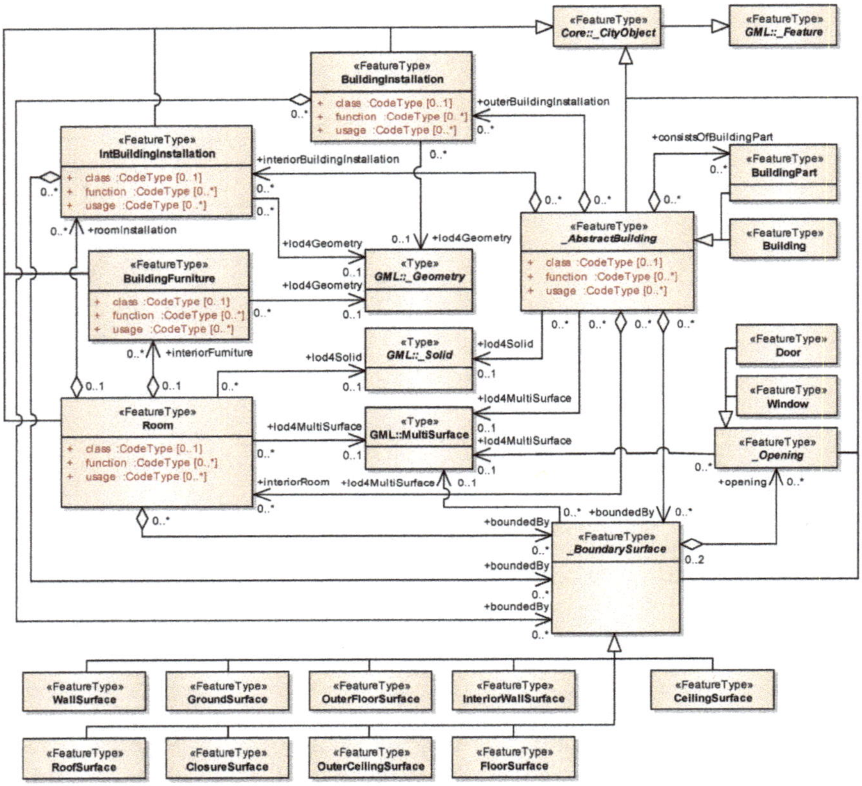

Figure 2. Building model of CityGML, adapted from [31] (p. 57).

Similar to the concept of spatial zones in IFC, CityGML uses the notion of grouping urban objects (CityObjectGroup entity) based on spatial or non-spatial constraints for a peculiar purpose. Various parts of a private or communal property located in different parts of a building can be arranged and grouped using "CityObjectGroup" entity [31]. For example, corridor spaces located in each building level can be grouped into an instance of "CityObjectGroup" to show the entire structure of a common property legal interest in a multistorey building.

2.2.3. IndoorGML

IndooGML is a relatively new 3D standard developed by OGC for managing and representing navigation network models of indoor environments [4]. This physical data model comprises fundamental topological and semantic entities required in indoor navigation tasks. Subdivision of indoor spaces and their connectivity and adjacency relationships are supported within IndooGML standard. In this purely physical model, a physical notion of indoor environment is considered to be one space layer. In order to provide indoor navigation services comprehensively, IndoorGML defines inter-layer connectivity relationships between space layers. "CellSpace" and "CellSpaceBoundary" are respectively used for modelling indoor spaces and their boundaries. Thus, these entities can be potentially adapted for modelling legal interests and legal boundaries in indoor spaces.

2.2.4. LADM

LADM is an ISO standard developed for defining a common ontological reference for communicating legal information in a consistent and effective way among various jurisdictions across the globe [24]. This legal data model lays the foundation for progressive development and enhancement of current land administration systems. The main concepts of LADM are parties, spatial units, basic administrative units, and spatial sources [32] (see Figure 3). Parties refer to people and organizations playing a role in transacting legal interests. Spatial units provide a form of representing legal interests associated with the land, buildings or properties. Common forms of spatial units include areal land parcels or volumetric legal spaces around buildings or utility networks. Basic administrative units are defined based on the arrangement of spatial units associated with the same legal interest. For instance, a private property in a multi-storey building can be considered as one basic administrative unit comprising three spatial units, namely legal spaces of an apartment unit, carpark and storage. Spatial sources refer to data sourcing methods used for delineating boundaries of spatial units. Common examples of spatial sources include field surveys, photogrammetric methods or point cloud data.

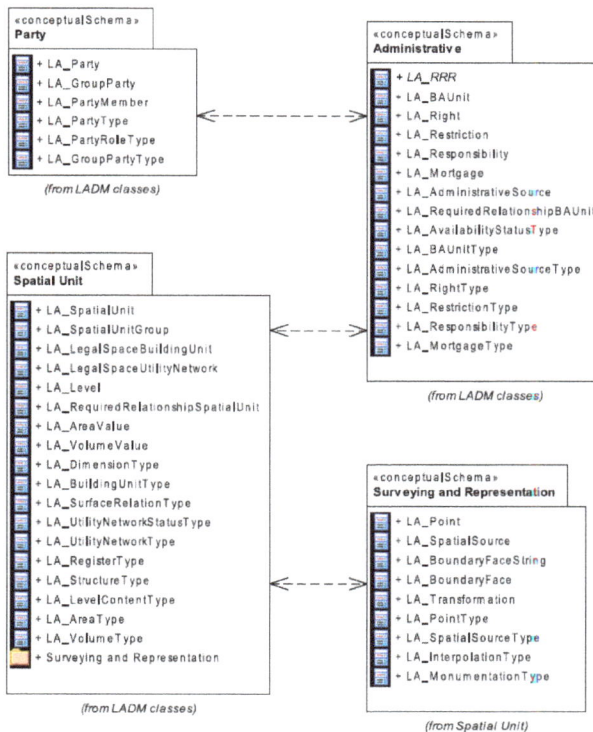

Figure 3. Packages of LADM standard, adapted from [24] (p. 8).

Two concepts are used to define boundaries of spatial units in LADM. The former one is *boundary face string* which is used to represent either 2D boundaries of areal spatial units or vertical boundary faces of unbounded volumetric spatial units [33]. The latter concept is *boundary face* which represents boundaries of bounded volumetric spatial units [24]. The location of boundary face strings are geometrically defined by at least two points which define the start and end of the boundary line. For boundary faces, a minimum of three points are required to define the geometry since the simplest form of boundary face is a triangle defined by three vertices. In some cases, the "locationByText" attribute is used in LADM to describe the location of boundary face strings and boundary faces in textual rather than geometric terms.

2.2.5. Integrated 3D Spatial Data Models

In some jurisdictions around the world (such as Australia), the legal and physical notions of multistorey buildings are linked to each other in design, management and operation phases of these buildings. A purely legal view sufficiently addresses the requirements of subdivision and registration of vertically stratified legal interests. However, when it comes to using legal information for broader purposes such as managing assets during the building lifecycle, a physical representation of buildings plays an important role in unlocking the value of legal information. Additionally, 3D legal interests (cadastral parcels) sometimes exist when there is no physical construction. A good example of this is airspace parcels. Airspaces cannot be spatially represented in the purely physical models. This means that a purely legal or an integrated model should be used in this case. Therefore, there have been some research projects investigating logical interdependencies between legal objects and their physical counterparts. These investigations have adopted two major approaches to defining an integrated 3D spatial data model. One method is to incorporate legal information into physical models using their built-in extension mechanisms, while another technique is to use external links provided in legal models and connect them to physical models.

Integrated models in the GIS domain mainly rely on enriching the CityGML standard using its Application Domain Extension (ADE) mechanism. One of the earliest investigations was done by Dsilva [34] who developed a rudimentary extension of CityGML for managing legal interests. The main incorporated legal data element in his model was "_KadasterApartment" used for managing legal information associated with apartments. Dsilva's proposed extension was more of a purely theoretical model and did not consider details of registration associated with each land parcel. In addition, it does not provide the capability to distinguish privately owned properties with communal ones in buildings. To address these shortcomings , a more advanced ADE of CityGML was developed by Çağdaş [6] to integrate legal and physical information in the context of Turkish jurisdiction. This ADE comprised three legal feature classes, namely "PropertyUnit", "CadastralParcel" and "CondominiumUnit". The "PropertyUnit" class is the abstract superclass for the "CadastralParcel" and "CondominiumUnit" classes. "CadastralParcel" represents legal land parcels which are defined as closed polygons on the Earth's surface. "CondominiumUnit" is used for modelling privately owned properties in buildings. This class has also composition relationships with "JointFacility" and "Annex" classes which represent communal properties used jointly by all and a specific group of owners, respectively. More recently, Li et al. [35] proposed an LADM-based ADE of CityGML aligned with the requirements of Chinese cadastral system. This extension of CityGML was able to model the ownership structures of condominium units inside buildings, reflecting the specifics of spatial relationships between legal objects and physical building elements.

Linking LADM with physical objects is another method to integrate legal and physical information. Soon et al. [7] adopted a semantic-based framework to link LADM elements to those of CityGML. In light of this, the "LA_LegalSpaceBuildingUnit" entity in LADM was externally connected to the "_AbstracBuilding" entity in CityGML.

In the BIM domain, the IFC standard was extended to model legal information using two approaches. In the first method, new subclasses are defined for appropriate IFC entities [19]. For instance, for modelling legal spaces, a new subclass of the "IfcSpace" entity was defined as the "IfcLegalPropertyObject" entity which includes attributes, such as lot liability and lot entitlement, for managing legal interests. The second approach relied upon the existing IFC extension capabilities such as property sets and user defined values [29]. In this case, instead of defining subclasses, legal attributes are defined through creating instances of the "IfcPropertySet" entity and attached to the proper IFC entity using "IfcRelDefinesByProperties" objectified relationship. For example, lot entitlement and lot liability attributes can be defined as the "Pset_Lot" property set which can be assigned to the "IfcSpace" entity using the mentioned relationship. All the previous studies published in the topic of using BIM for managing legal interests were mainly focused on development and implementation of the integrated legal and physical BIM models [2,19,29,36]. Those investigations have proposed extensions of IFC standard to incorporate legal information within the data schema of this standard. However, these investigations did not provide how well the integrated BIM model can perform in comparison with purely legal and purely physical models. Therefore, the major difference between this investigation and previously published studies is the objective assessment of the integrated BIM model and comparing it with purely legal models and purely physical models using some metrics. This study is a substantially expanded version of the article presented in the fifth 3D cadastre workshop [13].

3. Overview of Boundaries and Arrangements of Legal Interests in Victoria, Australia

In Victoria, Australia, two main types of legal interests are applied, namely primary legal interests and secondary legal interests. Primary legal interests are considered as the base level land parcels, which have spatial extent in either 2D or 3D, without any overlap or gap between them. Lots, common properties, roads, reserves, and crown parcels are the primary legal interest defined in the Victorian cadastral system. The secondary legal interests affect the primary ones by providing benefits or imposing restrictions on them. These include easements, restrictions, depth limitations and airspaces. Secondary legal interests can be spatially overlapped with primary legal interests.

Volumetric lots, common properties and easements are the commonly defined legal interests, which have 3D spatial extent inside multi-level building developments (Atazadeh, Kalantari, Rajabifard, Champion, and Ho, 2016). The main focus of this investigation is on those legal interests.

Volumetric lots usually comprise two components, main part and accessory parts. Examples of the main part are an apartment or an office. Accessory parts can be a carpark or a storage area. A typical instance of a volumetric lot is represented in Figure 4.

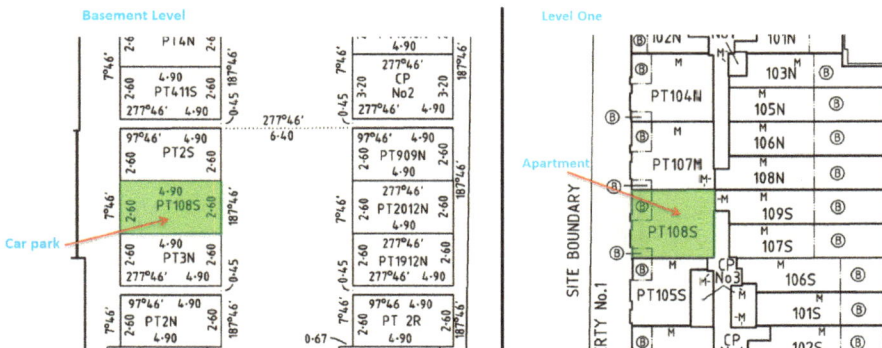

Figure 4. Parts of a volumetric lot.

Common property is defined based on arrangements of indoor or outdoor spaces as well as physical building elements, being more spatially complex than the volumetric lots. In the existing subdivision practices in Victoria, only the spatial extent of communal spaces is represented in 2D floor or cross-sectional plans. Communal building elements are not clearly shown in 2D plans since delineating building elements would result in complexities inside the plans and graphical representation of physical building elements in 2D plans is very difficult. Nevertheless, the notation section of plans describes those building elements which are considered as part of the common property interest. For instance, for common property No. 1 (Figure 5), it is indicated in the notation section of the subdivision plan that "*All walls defining boundaries, floor and ceiling slabs, columns, internal service ducts, conduits, pipe shafts, and electricity consumer mains cables within the building and courtyards are deemed to be part of common property No.1*" (excerpt from a notation section of a subdivision plan).

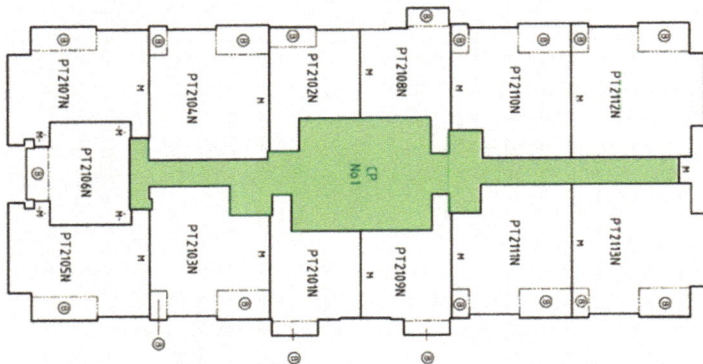

Figure 5. Common property delineated in floor plan diagram.

An easement represents utility networks which provide benefits or pose restrictions on the original land parcel that the building has been developed on (see Figure 6). This legal interest can itself be either physical (visible) elements, such as utility elements outside the buildings and roads, or it can be invisible external spaces surrounding utility elements.

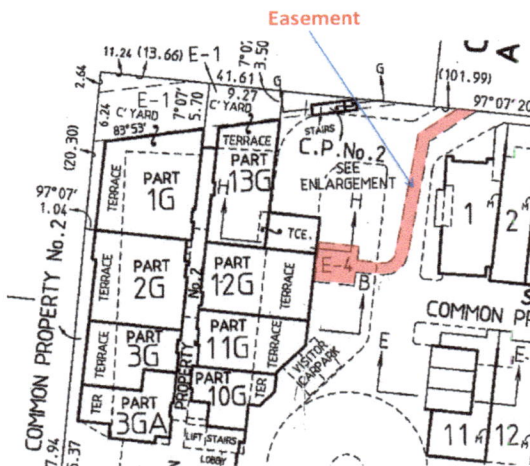

Figure 6. An easement depicted in a plan of subdivision, adapted from [29] (p. 509).

In addition to arrangements of legal interests, legal boundaries defining the spatial extent of each constituting part of the legal interest must be specified. A comprehensive taxonomy of legal boundaries for Victorian cadaster was developed in [2]. This taxonomy reflects an expanded version of the widely known dichotomy between general and fixed boundaries. There are three types of general boundaries, namely structural, projected and ambulatory ones. Structural boundaries are mainly defined through referencing the interior, exterior, or median of the physical structure. As an example of structural boundaries, it can be seen in Figures 4 and 5 that median boundaries are notated using the letter "M". Projected boundaries are mainly defined in balcony and terraced areas of building through extruding walls and ceilings. Ambulatory boundaries are defined by referencing dynamic natural features, such as rivers [37] (p. 360). Fixed boundaries are defined based on precise surveying measurements.

4. Legal Partitioning in BIM-Based Views

In this section, logical definition of legal boundaries and arrangements of ownership spaces in each BIM-based view of buildings will be explained.

4.1. Purely Legal View

Figure 7a represents how legal spaces and their boundaries are modelled within a purely legal BIM-based view. First, the building is partitioned into building levels. After that, each building level is partitioned into its constituent legal spaces. For defining each legal space, at least three legal boundaries are required to ensure that it is a closed legal space. Each legal boundary is associated with one or two legal spaces. The "BoundaryPosition" attribute of "Legal_Boundary" entity is used to semantically distinguish various types of legal boundaries such as interior, exterior and median ones.

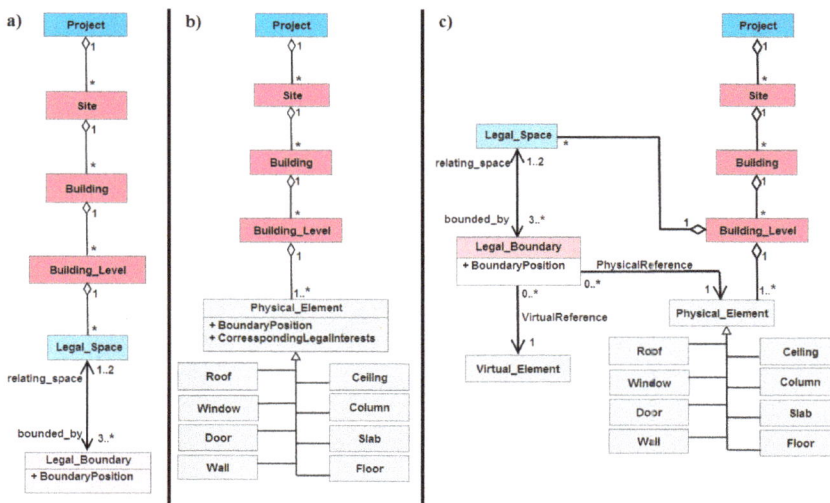

Figure 7. Logical modelling of legal spaces and boundaries in BIM-based views, (a) Purely legal view; (b) Purely physical view; (c) Integrated legal and physical view.

4.2. Purely Physical View

Legal partitioning in purely physical view of BIM environment is underpinned by the diagram shown in Figure 7b. In a purely physical project, the building is partitioned into its levels. Then, each level is constructed based on its constituent physical elements. These physical elements include walls, windows, doors, ceilings, floors, slabs, columns, slabs, and roof. To assign legal interests in the physical model, two new attributes were assigned to each building element: "CorrespondingLegalInterests" and "BoundaryPosition". The "CorrespondingLegalInterests" attribute specifies legal interests which are associated with the building element. The "BoundaryPosition" attribute of "Physical_Element" entity defines whether the exterior or interior face of the building element is used to delineate the legal boundary.

4.3. Intergrated View

Integrated view is composed of both physical elements and legal spaces (see Figure 7c). In these models, after partitioning the building into building levels, both cognitive legal spaces and physical elements are used in legal subdivision of each level. Similar to purely legal views, legal boundaries define the spatial extent of legal spaces in an integrated view of BIM environment. However, it is possible to reference both physical and virtual elements defining legal boundaries. This can be achieved by defining "PhysicalReference" and "VirtualReference" relationships between "Legal_Boundary" entity and "Physical_Element" and "Virtual_Element" entities, respectively. The "BoundaryPosition" attribute of "Legal_Boundary" entity can be used to provide a semantic definition for various boundary types (internal, external, and so on).

5. Assessment of BIM Models

In this section, an integrated BIM model is compared with purely legal and purely physical BIM models. In order to create a purely legal BIM model of the development, subdivision plans for this development were used. Only legal spaces were used to define volumetric lots, common properties, and easements within the legal model (see Figure 8a). Architectural plans were used to construct the purely physical model of the development. Various building elements such as walls, ceilings, floors, doors, and windows were constructed (see Figure 8b). The integrated model is constructed by integrating architectural elements, which are created by 2D architectural plans, and legal spaces as well as legal boundaries, which are defined based on subdivision plans (see Figure 8c).

All BIM models were prepared in Autodesk Revit. Since Revit has its own proprietary and closed data format for storing BIM models, all BIM models were converted into open IFC files which can be visualized and queried in any BIM-based platform. The Solibri Model Viewer was used for visualization of BIM models in IFC format. The reason for selecting Solibri Model Viewer was that a recent investigation of BIM visualization tools found that this viewer performs well when compared to other tools [38]. To perform various queries on BIM models, the xBIM toolkit (More details about xBIM is provided on https://github.com/xBimTeam) was used since it provides the ability to query and manipulate IFC files. The specifications of the workstation used for querying and visualizing BIM models included an Intel Core i7 340 GHz CPU, 4 GB of RAM, AMD Radeon HD 6350 GPU running Windows 7 x64. In order to minimize the effect of other tasks of the operating system on the measurement related to querying and visualization of each BIM model, all user installed processes and applications were shut down. This indicates that the side effects of other tasks were thus minimized and measurements for each BIM model were calculated in similar conditions.

The metrics used for comparing the models included a number of objects and geometry batches, visualization speed (frame per second), query time, modelling legal boundaries, and visual communication of legal boundaries. The number of objects and geometry batches metrics are suggested as appropriate measures for the size and complexity of BIM models [39]. The frame per second (FPS) metric is typically used to measure visualization speed and real-time interaction with BIM models [38]. The query time metric is also an important factor when a user wants to search for and retrieve specific information from BIM models [40]. Modelling and visual communication of legal boundaries metrics are defined to explore the ability of each BIM model in terms of delineating a legal boundary and intuitive understanding of the legal boundary [41,42]. The following subsections provide the measurement results of the assessment metrics for each BIM model.

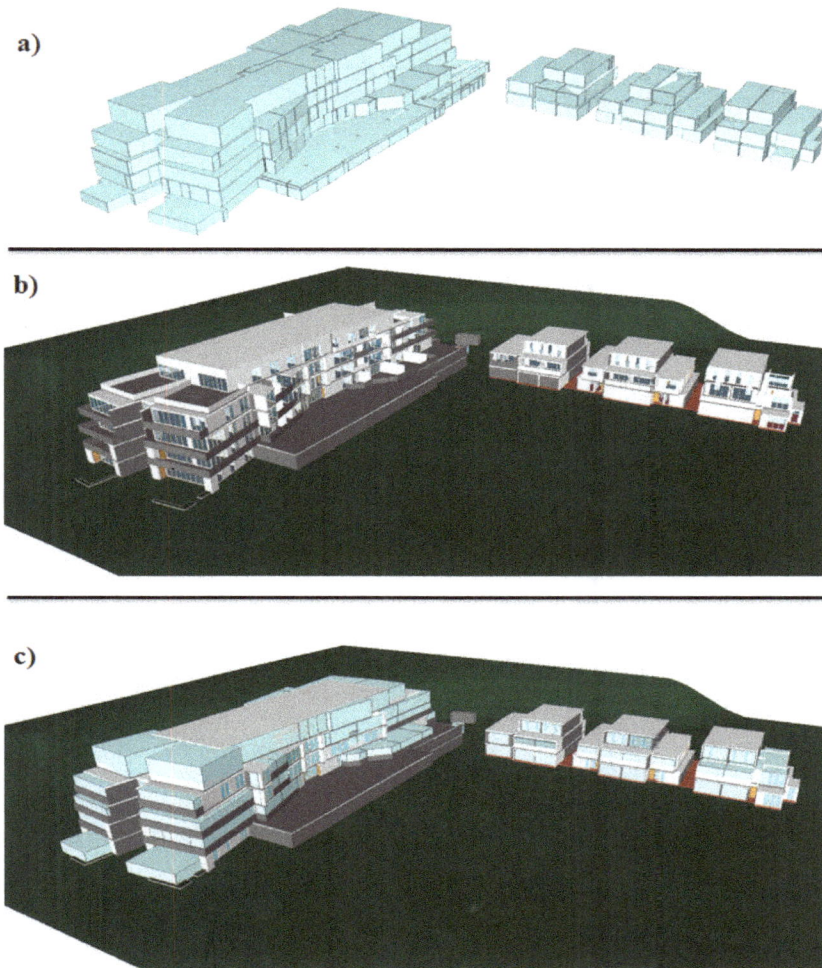

Figure 8. Implemented BIM models: (**a**) Purely legal model; (**b**) Purely physical model; (**c**) Integrated model (In legal and integrated models, the invisible common property space between building and townhouses was not shown for clarity purposes).

5.1. Number of Objects and Geometry Batches

In order to measure the number of objects and geometry batches, the xBIM toolkit was used since it provides functions for counting objects and geometry batches within IFC files. Table 1 compares the BIM models in terms of the number of objects and the number of geometry batches. These metrics are used for measuring the size of BIM models. The "Number of objects" metric refers to all spatial objects contained within each BIM model. A geometry batch refers to number of parts constituting each spatial object. For instance, windows are composed of two batches, one for the frame and the other for the glass, while walls typically include one batch. It can be seen in Table 1 that legal BIM model has the minimum size, whereas the largest size belongs to the integrated model. Another result is that there is a considerable difference between the legal model and the physical model in terms of both number of objects and geometry batches. This stems from the fact that one part of a legal interest, for example an apartment unit of a volumetric lot, can be spatially defined by only one legal space object (IfcSpace), whereas at least six physical objects, including four walls (IfcWall) and two slabs (IfcSlab), are required to define spatial extent of the same part of a legal interest. This indicates that storing legal models would be easier than physical models.

Table 1. Number of objects and geometry batches in each BIM model.

BIM Model	Number of Objects	Number of Geometry Batches
Legal model	173	173
Physical model	1562	1954
Integrated model	1735	2127

5.2. Visualization Speed

The real time rendering performance of BIM tools is fundamental to interacting with them. Frame rate or frame per second (FPS) represents the visualization and rendering speed of graphical applications. In 3D environments, a threshold FPS value of 15 Hz is essential for smooth interaction with 3D models. However, an FPS value of around 60 Hz in highly interactive 3D environments, such as 3D games, is required to provide smooth interaction with the user. Visualizing and rendering BIM models requires less interactivity compared to gaming environments. Literature suggests that an FPS value of 30 Hz is acceptable when interacting with BIM models [38]. Most BIM tools, such as the Solibri Model Viewer, provide the ability to view BIM models from various viewpoints. The common viewpoints supported by BIM tools include top, bottom, front, back, right, left, and isometric. Therefore, the minimum, maximum and average FPS values were measured over a 60 second interaction time with each BIM model from each viewpoint. The results are represented as a series of charts in Figure 9. These charts show trends of visualization speed in BIM models as discussed below:

The rendering performance of all BIM models was the smoothest when the models were represented from top viewpoint. From this perspective, the FPS values fluctuated between 35 Hz and 45 Hz, which is more than the recommended FPS value of 30 Hz.

The isometric viewpoint generated the worst results for all BIM models, resulting in FPS values that were around the essential threshold (15 Hz). On some occasions, FPS values were lower than 15 Hz and interaction with all models was not smooth.

From most viewpoints, the FPS values of the physical model were found to be more than those of the legal model, although the number of objects for the physical model is more than the legal model. The largest difference between the FPS values of the purely physical model and those of purely legal and integrated models was observed from the front viewpoint. The smoothest interaction with purely physical models can be explained by the fact that the volume of the 3D space occupied by building elements is less than volume of the legal spaces. Therefore, interacting with the physical model was smoother than the legal model.

The integrated model was the least interactive model compared with the purely legal and purely physical models since the volume of space occupied by this model is the summation of the volumes of both legal spaces and physical building elements.

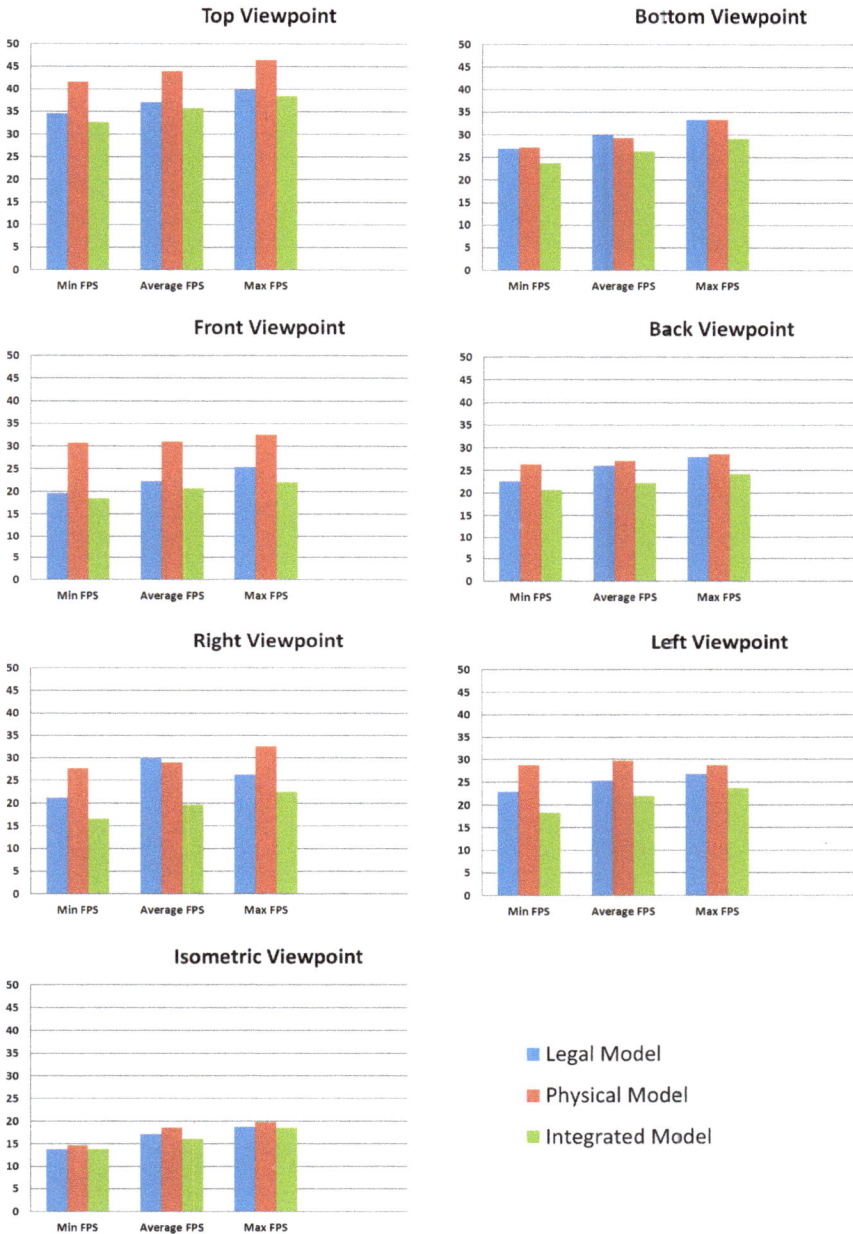

Figure 9. Minimum, maximum and average FPS for each model from various viewpoints in Solibri Model Viewer.

5.3. Semantic Query Time

Since the selected development includes volumetric lots, common properties and easements, semantic queries for finding these legal interests were executed. Initially. there were significant discrepancies in the time measured for these queries. After running queries approximately 30 times, the observed differences between two consecutive runs were insignificant (less than one millisecond). Therefore, there execution of each query was done for another five times. Figure 10 shows the average value of the query execution time for the selected legal interests in each BIM model. Overall, queries in the legal model were executed faster than the physical model. One reason is that because the total number of objects in the legal model was less than in the physical model. Another reason is that while only one legal space is required within the legal model to query a specific part of a legal interest, at least six objects (four walls and two ceilings) are required to query a specific part of a legal interest. Unsurprisingly, querying legal interests took the longest time for the integrated model since it has the largest number of objects. Another result is that the querying time for the common property of the building involves more than strata lots and easements in all BIM models. The reason is that common property is composed of a large number of parts in comparison with strata lots and easements. For example, in an integrated model, a common property includes walls between lots and ceilings between levels, corridors, lobbies, elevators, and stairs, whereas a volumetric lot include a maximum four spaces (apartment unit, storage space, and two car park spaces) and easements typically comprise two parts (such as a utility element and a legal space around that element).

Query time(Milliseconds)

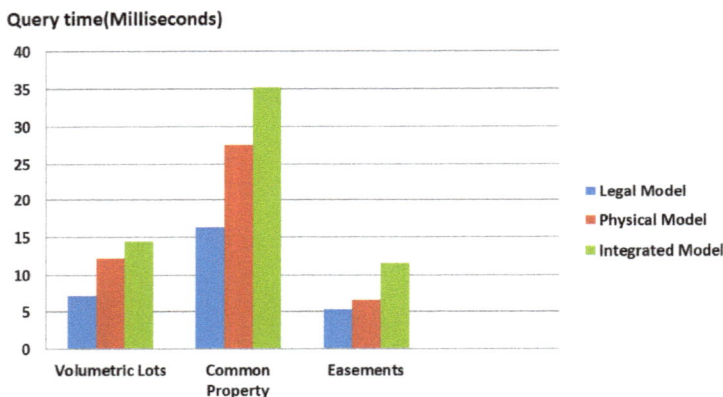

Figure 10. Measured average time for executing semantic queries in BIM models.

5.4. Managing Legal Boundaries

For managing legal boundaries, two distinct concepts should be differentiated from each other: modelling and visual communication. Modelling means that a specific boundary type can be stored and recorded within a 3D model irrespective of the approach used for its visualization. Visual communication can be defined as communicating a specific boundary type merely through graphical or image-based representations. The following subsections explain how modelling and visual communication of each legal boundary can be supported in each BIM model.

5.4.1. Modelling Legal Boundaries

Table 2 shows the capability of each model for storing and recording different types of boundaries. Both legal and integrated models have the ability to model all sorts of legal boundaries. In contrast, the physical model is only able to model interior and exterior structural boundaries as well as ambulatory ones. Median structural boundaries as well as fixed and projected boundaries do not have a physical manifestation. Therefore, it is impossible to model these boundaries within a purely physical model.

Table 2. Feasibility of modelling legal boundaries in each BIM model (–: not supported, +: supported).

Legal Boundary	Physical Model	Legal Model	Integrated Model
Interior structural boundary	+	+	+
Median structural boundary	–	+	+
Exterior structural boundary	+	+	+
Ambulatory boundary	+	+	+
Projected boundary	–	+	+
Fixed boundary	–	+	+

5.4.2. Visual Communication of Legal Boundaries

In order to assess each BIM model in terms of visually communicating each boundary type, a human computer interaction experiment was conducted. There were three groups of human subjects, namely 6 students, 6 land administration experts, and 6 BIM experts. In the experiment, the participants were asked to interact with each BIM model and explore different types of legal boundaries defined inside each BIM model in approximately 10 min. Subsequently, the participants responded either "yes" or "no" if they would be able to visually identify a specific legal boundary.

Table 3 shows the results of participants' responses to evaluating the purely legal model. For students, it was difficult to distinguish various types of legal boundaries in the purely legal view of BIM environment since most students did not have adequate expertise in identifying the location of boundaries in this view. In contrast, land administration experts were able to identify interior, exterior, ambulatory, projected and fixed boundaries due to their knowledge of legal boundary definition. However, median boundaries for the majority of land administration experts were difficult to be discerned visually. For BIM experts, only interior and exterior boundaries were visually clear. This is because BIM experts are familiar with constructing internal and external functional spaces and their boundaries. Some examples of legal boundaries defined in the purely legal model were shown in Figure 11.

Table 3. Responses to visual understanding of legal boundaries in the purely legal model.

Legal Boundary	Students		Land Administration Experts		BIM Experts	
	Yes	No	Yes	No	Yes	No
Interior structural boundary	1	5	4	2	5	1
Median structural boundary	0	6	1	5	1	5
Exterior structural boundary	0	6	5	1	4	2
Ambulatory boundary	0	6	4	2	2	4
Projected boundary	2	4	6	0	1	5
Fixed boundary	1	5	5	1	0	6

Participants' assessment of the purely physical BIM-based view is presented in Table 4. All students and BIM experts were able to identify interior, exterior and ambulatory boundaries. Their identification of these legal boundary types was predicated on using physical elements as references for those boundaries. Land administration experts also identified interior and exterior boundaries easily (see Figure 12). However, most of land administration experts expressed that only physical representation of the dynamic natural feature, which is a river in this study, is not adequate enough to communicate ambulatory boundaries since the location of these boundaries change over the period of time. The vast majority of participants confirmed that median, projected, and fixed boundaries are difficult to be understood in the purely physical view of BIM environment.

Table 4. Responses to visual understanding of legal boundaries in the purely physical model.

Legal Boundary	Students		Land Administration Experts		BIM Experts	
	Yes	No	Yes	No	Yes	No
Interior structural boundary	6	0	5	1	6	0
Median structural boundary	2	4	1	5	0	6
Exterior structural boundary	6	0	6	0	6	0
Ambulatory boundary	6	0	2	4	5	1
Projected boundary	0	6	0	6	0	6
Fixed boundary	0	6	0	6	0	6

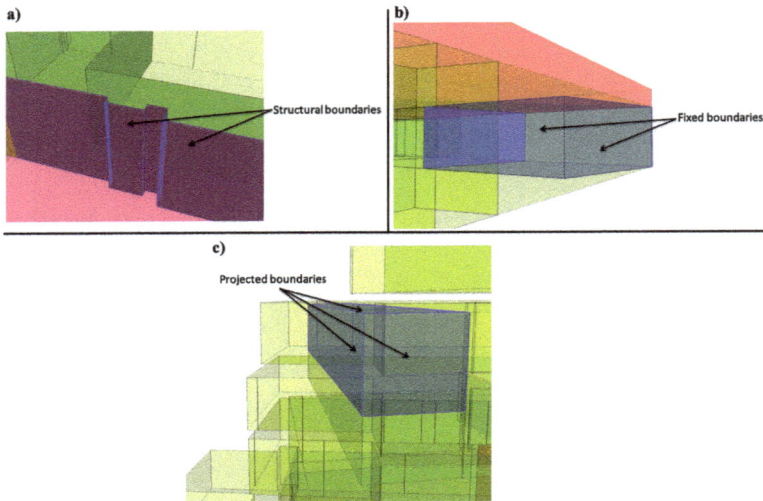

Figure 11. Examples of legal boundaries represented in the purely legal model: (a) Structural boundaries; (b) Fixed boundaries of a parking space; (c) Projected boundaries.

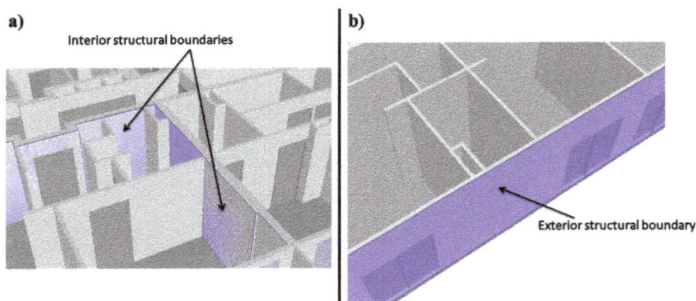

Figure 12. Examples of the legal boundaries represented in the purely physical model: (a) Internal wall boundaries; (b) External wall boundary.

Table 5 provides the results of the subjective assessment of the integrated BIM model. Most participants confirmed that the integrated model provides a clear communication of all legal boundary types since the relationships between physical elements and legal spaces are discernable inside the model. For the vast majority of the subjects, the integration of legal and physical objects was helpful in disambiguating the location of structural boundaries, which is mainly assigned as interior, median or exterior face of the corresponding structural element. Although land administration experts do not require physical elements in understanding project and fixed boundaries, they confirmed that those boundaries can be better communicated when physical elements are leveraged in the BIM environment. Figure 13 provides some examples of legal boundaries shown inside the integrated BIM model.

Table 5. Responses to visual understanding of legal boundaries in the integrated model.

Legal Boundary	Students		Land Administration Experts		BIM Experts	
	Yes	No	Yes	No	Yes	No
Interior structural boundary	6	0	6	0	6	0
Median structural boundary	5	1	4	2	0	6
Exterior structural boundary	6	0	6	0	6	0
Ambulatory boundary	6	0	6	0	5	1
Projected boundary	6	0	5	1	6	0
Fixed boundary	6	0	6	0	6	0

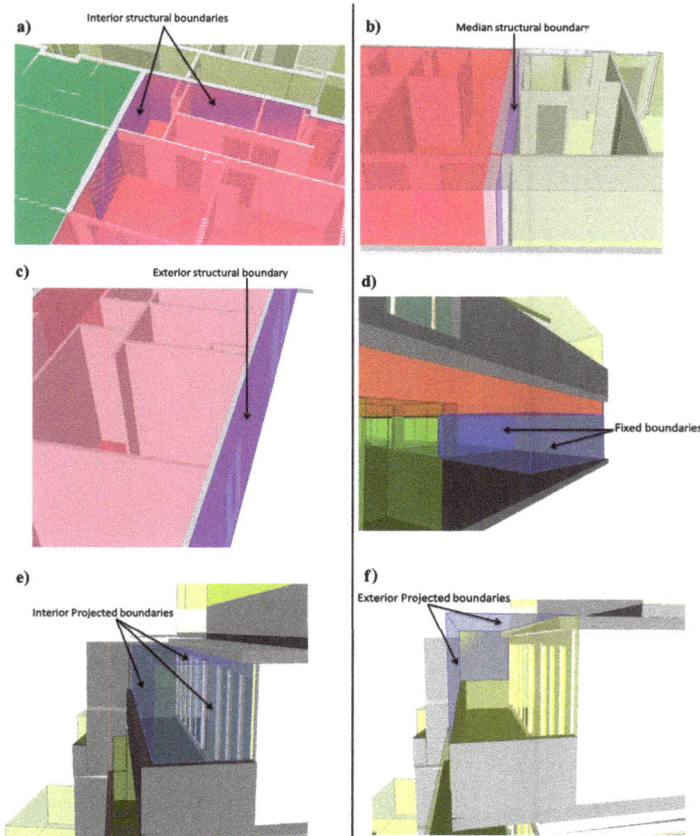

Figure 13. Examples of legal boundaries represented in the integrated model: (**a**) Internal wall boundaries; (**b**) Median wall boundary; (**c**) External wall boundary; (**d**) Fixed Boundaries of a parking space; (**e**) Internal projected boundaries; (**f**) External projected boundaries.

6. Conclusions

This article investigated the performance of three common BIM-based methods, namely purely legal, purely physical and integrated approaches, for storing, managing and communicating legal interests in multistorey buildings. The investigation of current practices in Victoria, Australia showed that a broad range of boundaries and various spatial structures are used to define vertically stratified legal arrangements. The assessment metrics were used to explore the capability of each implemented BIM model. These metrics mainly provided a reflection on the performance of each model in terms of management and communication of legal interests. One finding was that interaction tasks, such as visual inspection and retrieving specific legal interests, in purely legal or physical models was easier to be performed compared to integrated models. It was also found that integrated models provide a more visual communication of the location of legal boundaries, particularly when these boundaries are defined by referencing building elements. This finding indicates that the lay users may find it easier to understand their legal rights in integrated BIM models.

The main focus of this research was on using BIM for multistorey buildings. However, lifecycle information about other built assets, such as roads and bridges, will be considered in future versions of BIM and IFC standard. Therefore, future work could involve investigating the potential capability of the new BIM environment, which comprises both buildings and other urban infrastructure, for managing legal arrangements beyond the spatial extent of multistorey developments. Another important aspect could be investigating the role of legal interests, which are defined in BIM models, for purposes broader than land administration. Since BIM is currently being used in the facility management industry, a BIM model enriched with legal information could pave the way for determining maintenance responsibilities for assets, especially communal facilities, in complex buildings.

Acknowledgments: This work was supported by the Australian Research Council (ARC) [grant number LP110200178].

Author Contributions: This research is a result of collaboration and contribution of all authors. This work is part of the ARC project titled "Land and Property Information in 3D". Abbas Rajabifard is a chief lead investigator of this project. Mohsen Kalantari has a project manager role in the project. Behnam Atazadeh implemented BIM models and conducted the assessments under the supervision and guidance of Mohsen Kalantari and Abbas Rajabifard. Each author had a substantial contribution in the preparation of the manuscript.

Conflicts of Interest: The authors declare no conflict of interest.

References

1. Stoter, J.; Ploeger, H.; Roes, R.; van der Riet, E.; Biljecki, F.; Ledoux, H. First 3D Cadastral Registration of Multi-level Ownerships Rights in the Netherlands. In Proceedings of the 5th International FIG 3D Cadastre Workshop, Athens, Greece, 18–20 October 2016; pp. 491–504.
2. Atazadeh, B.; Kalantari, M.; Rajabifard, A.; Ho, S. Modelling building ownership boundaries within BIM environment: A case study in Victoria, Australia. *Comput. Environ. Urban Syst.* **2017**, *61*, 24–38. [CrossRef]
3. Kolbe, T.; Gröger, G.; Plümer, L. CityGML: Interoperable Access to 3D City Models. In *Geo-Information for Disaster Management SE-63*; van Oosterom, P., Zlatanova, S., Fendel, E., Eds.; Springer: Berlin/Heidelberg, Germany, 2005; pp. 883–899.
4. Lee, J.; Li, K.J.; Zlatanova, S.; Kolbe, T.H.; Nagel, C.; Becker, T. *OGC® IndoorGML*; Open Geospatial Consortium: Wayland, MA, USA, 2014.
5. Liebich, T. *IFC 2x Edition 3. Model Implementation Guide*; Version 2.0.; AEC3 Ltd.: Wycombe, UK, 2009.
6. Çağdaş, V. An Application Domain Extension to CityGML for immovable property taxation: A Turkish case study. *Int. J. Appl. Earth Observ. Geoinform.* **2013**, *21*, 545–555. [CrossRef]
7. Soon, K.H.; Thompson, R.; Khoo, V. Semantics-based Fusion for CityGML and 3D LandXML. In Proceedings of the 4th International Workshop on 3D Cadastres, Dubai, UAE, 9–11 November 2014; pp. 323–338.
8. Heidari, M.; Allameh, E.; de Vries, B.; Timmermans, H.; Jessurun, J.; Mozaffar, F. Smart-BIM virtual prototype implementation. *Autom. Constr.* **2014**, *39*, 134–144. [CrossRef]

9. Azhar, S.; Khalfan, M.; Maqsood, T. Building information modelling (BIM): now and beyond. *Constr. Econ. Build.* **2015**, *12*, 15–28. [CrossRef]
10. Isikdag, U.; Aouad, G.; Underwood, J.; Wu, S. Building information models: a review on storage and exchange mechanisms. In Proceedings of the 24th CIB W78 Conference, Maribor, Slovenia, 26–29 June 2007.
11. Bazjanac, V.; Crawley, D.B. *The Implementation of Industry Foundation Classes in Simulation Tools for the Building Industry*; Building Simulation: Prague, Czech Republic, 1997.
12. *Industry Foundation Classes (IFC) for Data Sharing in the Construction and Facility Management Industries*; ISO16739; International Organization for Standardization (ISO): Geneva, Switzerland, 2013.
13. Atazadeh, B.; Kalantari, M.; Rajabifard, A. Comparing Three Types of BIM-based Models for Managing 3D Ownership Interests in Multi-level Buildings. In Proceedings of the 5th International FIG 3D Cadastre Workshop, Athenes, Greece, 18–20 October 2016; pp. 183–198.
14. Stoter, J.E.; Van Oosterom, P. Technological aspects of a full 3D cadastral registration. *Int. J. Geogr. Inform. Sci.* **2005**, *19*, 669–696. [CrossRef]
15. Guo, R.; Li, L.; Ying, S.; Luo, P.; He, B.; Jiang, R. Developing a 3D cadastre for the administration of urban land use: A case study of Shenzhen, China. *Comput. Environ. Urban Syst.* **2013**, *40*, 46–55. [CrossRef]
16. Stoter, J.; Ploeger, H.; van Oosterom, P. 3D cadastre in the Netherlands: Developments and international applicability. *Comput. Environ. Urban Syst.* **2013**, *40*, 56–67. [CrossRef]
17. Drobež, P.; Fras, M.K.; Ferlan, M.; Lisec, A. Transition from 2D to 3D real property cadastre: The case of the Slovenian cadastre. *Comput. Environ. Urban Syst.* **2017**, *62*, 125–135. [CrossRef]
18. Karabin, M. Registration of untypical 3D objects in Polish cadastre—Do we need 3D cadastre? *Geod. Cartogr.* **2012**, *61*, 75–89.
19. Atazadeh, B.; Kalantari, M.; Rajabifard, A.; Ho, S.; Ngo, T. Building Information Modelling for High-rise Land Administration. *Trans. GIS* **2017**, *21*, 91–113. [CrossRef]
20. Karki, S. 3D Cadastre Implementation Issues in Australia. Master's Thesis, The University of Southern Queensland, Toowoomba, Australia, 2013.
21. Dimopoulou, E.; Elia, E. Legal aspects of 3D property rights, restrictions and responsibilities in Greece and Cyprus. In Proceedings of the 3rd International Workhsop on 3D Cadatres: Development and Practices, Shenzhen, China, 25–26 October 2012; pp. 41–60.
22. Gröger, G.; Plümer, L. CityGML—Interoperable semantic 3D city models. *ISPRS J. Photogramm. Remote Sens.* **2012**, *71*, 12–33. [CrossRef]
23. Kolbe, T.H. BIM, CityGML, and Related Standardization. In Proceedings of the Digital Landscape Architecture Conference 2012, Bernburg/Dessau, Germany, 31 May–2 June 2012.
24. *Geographic Information—Land Administration Domain Model (LADM)*; ISO19152; International Organization for Standardization (ISO): Geneva, Switzerland, 2012.
25. Liebich, T. *IFC4—The New BuildingSMART Standard*; buildingSMART Standard: Geneva, Switzerland, 2013.
26. Schenck, D.A.; Wilson, P.R. *Information Modeling: The EXPRESS Way*; Oxford University Press, Inc.: Oxford, UK, 1994.
27. Benner, J.; Geiger, A.; Leinemann, K. Flexible generation of semantic 3D building models. In Proceedings of the 1st International Workshop on Next Generation 3D City Models, Citeseer, Bonn, Germany, 21–22 June 2005.
28. Ding, L.; Drogemuller, R.; Rosenman, M.; Marchant, D.; Gero, J. *Automating Code Checking for Building Designs—DesignCheck*; University of Wollongong: Wollongong, Australia, 2006.
29. Atazadeh, B.; Kalantari, M.; Rajabifard, A.; Ho, S.; Champion, T. Extending a BIM-based data model to support 3D digital management of complex ownership spaces. *Int. J. Geogr. Inform. Sci.* **2017**, *31*, 499–522. [CrossRef]
30. Groger, G.; Kolbe, T.H.; Nagel, C.; Hafele, K.H. *OGC City Geography Markup Language (CityGML) En-Coding Standard*; Open Geospatial Consortium: Wayland, MA, USA, 2012.
31. Nagel, C. Spatio-Semantic Modelling of Indoor Environments for Indoor Navigation. Ph.D. Thesis, Technische Universität Berlin, Berlin, Germany, 2014.
32. Lemmen, C.; van Oosterom, P.; Bennett, R. The Land Administration Domain Model. *Land Use Policy* **2015**, *49*, 535–545. [CrossRef]
33. Lemmen, C.H.J. *A Domain Model for Land Administration*; TU Delft, Delft University of Technology: Delft, The Netherlands, 2012.

127

34. Dsilva, M.G. A Feasibility Study on CityGML for Cadastral Purposes. Master's Thesis, Eindhoven University of Technology, Eindhoven, The Netherlands, 2009.

35. Li, L.; Wu, J.; Zhu, H.; Duan, X.; Luo, F. 3D modeling of the ownership structure of condominium units. *Comput. Environ. Urban Syst.* **2016**, *59*, 50–63. [CrossRef]

36. Atazadeh, B.; Kalantari, M.; Rajabifard, A.; Champion, T.; Ho, S. Harnessing BIM for 3D digital management of stratified ownership rights in buildings. In Proceedings of the FIG Working Week 2016 Recovery from Disaster, Christchurch, New Zealand, 2–6 May 2016.

37. Williamson, I.P.; Enemark, S.; Wallace, J.; Rajabifard, A. *Land Administration for Sustainable Development*; ESRI Press Academic: Redlands, CA, USA, 2010.

38. Johansson, M.; Roupé, M.; Bosch-Sijtsema, P. Real-time visualization of building information models (BIM). *Autom. Constr.* **2015**, *54*, 69–82. [CrossRef]

39. Amor, R.; Jiang, Y.; Chen, X. BIM in 2007—Are we there yet. Proceedings of CIB W78 Conference on Bringing ITC Knowledge to Work, Maribor, Slovenia, 27–29 June 2007; pp. 26–29.

40. Schweizer, R. Spatial BIM Queries: A Comparison between CPU and GPU based Approaches. Bachelor's Thesis, Technische Universit at Munchen Faculty of Civil, Geo and Environmental Engineering, Munich, Germany, 2015.

41. Shojaei, D. 3D Cadastral Visualisation: Understanding Users' Requirements. Ph.D. Thesis, The University of Melbourne, Melbourne, Australia, 2014.

42. Aien, A.; Kalantari, M.; Rajabifard, A.; Williamson, I.; Wallace, J. Towards integration of 3D legal and physical objects in cadastral data models. *Land Use Policy* **2013**, *35*, 140–154. [CrossRef]

International Journal of
Geo-Information

MDPI

Article

Overview of the Croatian Land Administration System and the Possibilities for Its Upgrade to 3D by Existing Data

Nikola Vučić [1,*], Miodrag Roić [2], Mario Mađer [2], Saša Vranić [2] and Peter van Oosterom [3]

1   State Geodetic Administration, Gruška 20, 10000 Zagreb, Croatia
2   Faculty of Geodesy, University of Zagreb, Kačićeva 26, 10000 Zagreb, Croatia; mroic@geof.hr (M.R.); mmadjer@gmail.com (M.M.); svranic@geof.hr (S.V.)
3   Faculty of Architecture and the Built Environment, Department OTB, GIS Technology Section, Delft University of Technology, P.O. Box 5030, 2600 GA Delft, The Netherlands; p.j.m.vanOosterom@tudelft.nl
*   Correspondence: nikola.vucic@dgu.hr; Tel.: +385-1-6165-439

Received: 31 March 2017; Accepted: 17 July 2017; Published: 20 July 2017

Abstract: This paper explores the laws and other legal acts related to the Croatian 3D cadastre with an emphasis on those which relate to interests in strata, spatial planning, and other regulations that are valid or were valid on Croatian territory. The effects of the application of these regulations on the present situation of registration in cadastre and land register were considered. This paper also explores current legal, institutional, and technical solutions implemented in the Croatian Land Administration System and the possibilities for its upgrade to 3D cadastre. Implementation of any technological option to establish a 3D cadastre is tightly related to legislation. Hence, legislation and technological options are considered to find solutions that will be possible to implement. One suggestion presented in this paper was to use other sources of 3D data such as topographic signs or symbols used to represent topographic objects on 2D maps. In combination with other geodetic and cartographic products, useful information can be obtained, often quite relevant to provide a reference context for a 3D cadastre. Topographic signs on topographic maps and on other geodetic products provide a representation of complex real-world situations (tunnels, bridges, overpasses etc.) that are not usually presented on cadastral maps. This paper presents the possibility of utilizing those topographic signs to achieve the first steps towards establishing a 3D cadastre. Furthermore, this study proposes the establishment of a 3D Multipurpose Land Administration System as the most efficient system of land administration in a time when spatial information is easier to obtain than ever before and traditional real estate registers are subject to frequent and demanding changes.

Keywords: 3D cadastre; 3D real property; land administration; legal; institutional; technical; topographic signs

1. Introduction

A cadastre is generally a parcel-based, up-to-date land information system containing records of real properties and interests (i.e., rights, restrictions and responsibilities). A more extensive use of land administration information began with the development of multipurpose cadastres. Their establishment has proven to be quite demanding, and it is difficult to find efficient multipurpose cadastres in any country. It was only the development of information technologies that really opened up the possibilities for the development of Multipurpose Land Administration Systems (MLAS), as currently, the differences between the systems for registering land and the systems for registering land tenure do not allow a unified approach between countries [1].

The amount and the complexity of the information maintained as per the public authority regulations are constantly increasing, and are strongly related to the development of technology. The possibilities of new technologies are also constantly increasing, thus opening the possibility of collecting new, additional information not previously acquired due to technical difficulties. Cadastral data are basic data for land administration systems. Their availability in a digital form makes them interesting for the increasing number of new areas of human activity and is essential for their further development, which has led to a constant increase in demand for cadastral information. Therefore, countries have to work on improving the information flow by collecting and maintaining 3D cadastral data to keep up with the growing requirements, especially those related to efficient land administration.

The definitions and descriptions of 3D properties often focus on the technical and registration aspects rather than the legal aspects. In the legal category, most studies have addressed national legislation and its practical use. For more than 10 years, Paulsson and Paasch [2] conducted a thorough analysis of the overall field of 3D cadastres and concluded that further fundamental legal research on 3D property was required. The same assumption was also valid for the Croatian case and was confirmed by insight into the general state of registered land administration data.

During the Second International 3D Cadastres Workshop in Delft, The Netherlands (2011), five practical issues were highlighted [3]. By applying and answering these questions, a close impression of the general state of 3D data in the Republic of Croatia was obtained:

- Which types of 3D cadastral objects (3D properties) can be registered? Are these always related to constructions (buildings, pipelines, tunnels, etc.) as in Norway and Sweden, or could they be any part of the 3D space (both airspace or in the subsurface)?
- In cases where infrastructure objects cross 2D parcel boundaries (such as long tunnels, pipelines, and cables networks), should these be divided based on the surface parcels (as in Queensland, Australia), or treated as one cadastral object (as in Sweden and the Netherlands)?
- How do we deal with the fact that the legal status of such objects does not have to be the same for all ground parcels? For example, one construction located on three ground parcels, each on the basis of another type of right (e.g., easement, restrictive covenant, lease).
- For the representation (and initial registration) of a 3D cadastral object, is the legal space specified by its own coordinates in a shared reference system (as is the practice for 2D in most countries), or is it specified with reference to existing topographic objects/boundaries?
- Should 3D registration and visualization reflect the actual dimensions, or is it sufficient to have a visualization of property units in buildings based on standard floor-to-floor heights, as in Spain? What is the legal value of these boundaries? Is an investigation of the source documents (title deed, survey plan) needed to obtain legally binding information?

In this paper, amongst other things, the answers to these questions are given and elaborated from a Croatian perspective. These findings serve as a starting point and general overview of the Croatian Land Administration System. We believe that most data for establishing the 3D cadastre in the Republic of Croatia has already been collected and maintained, but often in a form not suitable for direct implementation. Therefore, the aim of this paper is to show the possibilities of implementing a 3D cadastre by using the data already registered in the existing official registers and databases that are part of the Croatian Land Administration System. The objectives undertaken to achieve this aim can be summarized in several steps, which include an overview of the relevant Croatian legislation, an analysis of the existing data in official registers, and a proposal of the methods for establishing the 3D cadastre by using this data.

The rest of the paper is organized as follows. Section 2 deals with the legal and technical aspects of a Croatian 3D cadastre, while Section 3 describes its institutional aspects. Section 4 lists available sets of data which could be used in establishing a 3D MLAS. A subsection on technical aspects investigates how 3D models can be created from the listed data, but without going into a deeper analysis of these procedures. Section 5 deals with 3D topographical signs, while Section 6 describes the link between topographical signs, cadastral maps, and cadastral registry entries. The paper ends with our conclusions.

2. Legal and Technical Aspects of the Croatian 3D Cadastre

Business and technological systems of public administration are largely based on registers. The terms registry or register are used to denote the organization where information on registered land rights are held. Information on registered land is typically textual and spatial, with the former typically maintained in a registry (or a land register in Croatia) and the latter in a cadastre office. Introducing interoperability into the registers is one of the key drivers for optimizing public administration, starting from the simple automation of existing processes, to the overall transformation of the system, and the construction of modern user-oriented services. An additional incentive for such a transformation was the foreseeable need for interoperability between the national system of registers and the European Union (EU) registers, and potentially in the future with other registers worldwide. The current system of registers in the Republic of Croatia is significantly vertically structured [4].

Spatial data and data on real property in Croatia are managed in multiple registers with many end users. Basic registers include the Cadastre and Land register. The Cadastre register maintains data on the position, shape, and area of real properties, while the Land register registers data on rights, restrictions, and responsibilities. Responsible institutions of public authority are the State Geodetic Administration (for the Cadastre register) and municipal courts (for the Land register). In cadastral offices (20 regional cadastral offices with their 92 branches as well as the Municipal Office for Cadastre and Geodetic Works of the City of Zagreb), real properties are registered based on their technical characteristics. The cadastral data on real properties (cadastral parcels) are the basis for the establishment, renewal, storage, and maintenance of land registers that are kept across 109 land register offices. In land registers, the data on cadastral parcel title holders are associated with the data on cadastral parcels defined by the cadastre. Real property in Croatian real property law is based on the *superficies solo cedit* principle, where a land surface parcel includes everything relatively permanently associated with the parcel on or below the land surface (primarily buildings, houses, etc.). A real property (LA_BAUnit in ISO 19152), in Croatian legislation, may consist of one or more land parcels registered in the land register in the same property sheet, hence they are legally combined in a single body (registered land unit). Grass, trees, fruits, and all valuable commodities provided on the surface of the land are parts of this real property until the land is divided.

The Croatian Land Administration System (LAS) is in the process of registering 3D cadastral objects related to construction (buildings, pipelines, tunnels). In cases where infrastructure objects cross 2D parcel boundaries such as long tunnels, pipelines, and cable networks, the Croatian LAS treats them as one cadastral object. The building and physical planning laws are very strict and do not allow for differences in legal status between the ground parcels where such an object has been built, e.g., one construction situated on three ground parcels, each on the basis of another type of right (e.g., easement, restrictive covenant, lease). For building a new construction in a similar situation, the right of construction can be utilized. The right of construction is a limited real right on someone's parcel, and it entitles the construction right holder to build a construction on the surface of other owner's land or underneath it; and the owner of the land has to agree to this construction. The right of construction in legal terms is equal to the definition of real property. For the representation (and initial registration) of a 3D cadastral object, its legal space is specified by reference to existing topographic objects/boundaries. For the first implementation of the Croatian 3D cadastre and visualization of property units in buildings, it was sufficient to use standard floor-to-floor heights.

2.1. Buildings and Particular Parts of a Real Property

Buildings are registered in the cadastre at the obligatory request of a party. A geodetic report prepared by the authorized surveying company must be supplied with this request. The responsible cadastral office must first review and certify the report. Since 2007, cadastral offices have partially participated in controlling the legalities of the construction of all buildings. As a precondition for registering buildings into the cadastre, and then taking that data into land registers, the geodetic reports must also include the building permit or other relevant documents. Any particular parts of real properties can also be registered in the non-technical (written) part of the cadastral documentation. However, given that all the necessary data cannot be recorded in as much detail as in the Land register, there can be inconsistencies between the cadastral and land register data when describing particular parts of the same real property.

Buildings and other structures are registered in the cadastre with the following attributes: location (2D coordinates), area, intended building use, building name, and house number. A land register takes over two-dimensional data on real property from the cadastre. Real property may be further divided into common and particular parts, and registered in the land register based on the report on the particular part of the real property (Figure 1). In this way, co-owners of a real property have a co-owner relationship on the common parts, and each of them becomes an individual owner of a particular part (for example an apartment or office space). This way of registering particular parts began in 1996, and has yet to be implemented for the majority of real properties, as land register registration is purely voluntary.

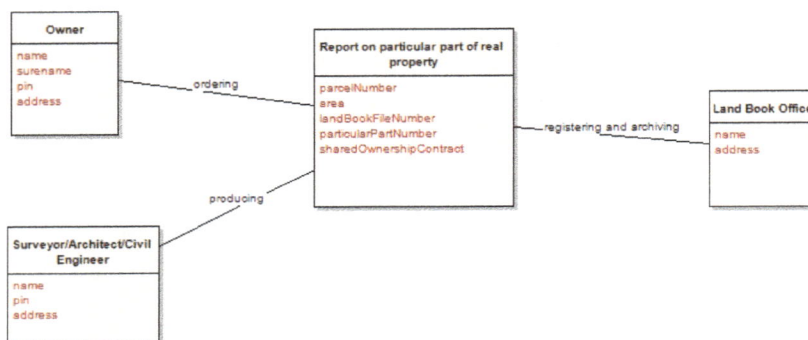

Figure 1. Schematic presentation of creating a report on a particular part of a real property.

Data on buildings are transcribed into land registers based on the data delivered to the land register by the cadastral office. Ownership of a particular part of property (e.g., an apartment or office space) is realized through registration in a land register. These particular parts may be registered if they make independent units of use. Particular parts may include balconies, terraces, basements, and attics, under the strict condition that they exclusively serve a single particular part and that they are clearly separated from other real property parts. The registration of particular parts of a real property in the land register is not possible without a report on the particular part, which in legal terms means the retention of a real property as a single body. This same procedure is commonly used in land registers to formally unify the land (which was often publicly owned) with the building constructed on that land. Partition of the real property establishes ownership of a particular part of a real property (apartment, office space, garage, etc.) that becomes associated with a proportionally shared part on the property. A fair relationship in financing the maintenance of a building is furthermore made possible by establishing the ratio of each party's ownership in the common real property and hence each party's proportional share in the shared ownership of common parts.

The report on a particular part of a real property establishes the size and shape of the common and particular parts of a single real property (apartment, office space, etc.), and draws connections for reference purposes against the real property as a unit. Additionally, data on particular parts must be technically processed with drawings of particular and common parts with the required labels and areas of particular parts. Figure 2 shows two particular parts of a real property differentiated by hatch style. Drawings like these are provided in analogue format. Furthermore, a shared ownership contract must also be provided.

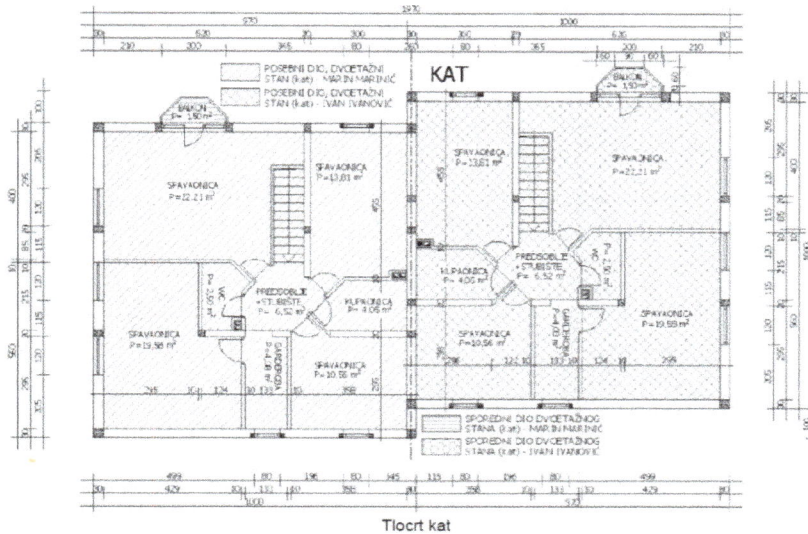

Tlocrt kat

Figure 2. Particular parts of a real property [5].

According to the Regulation of the Connecting Land Register and the Book of Deposited Contracts [6], the legal obligation of the building manager is to start the procedure of connecting the land register with the Book of Deposited Contracts. This regulation was brought about as it was noticed that many buildings built several decades ago were not registered in the cadastre or the land register, and have not been partitioned. According to the mentioned regulation, the description of the particular part of a real property contains the data on intended use (office space, apartment, or other), number of rooms, position of particular part in a building, and the surface of the particular part (for example, a two-bedroom apartment on the first floor, right side, with the usable area of 52 m^2). Furthermore, the description of a particular part of a real property also mentions the additions (for example, a woodshed, a garage, a parking space, a balcony, a garden, a terrace, and similar), along with their area, if there is such data.

It is necessary to standardize the appearance and methods of creating building subdivision plans. A building information modelling (BIM)-based approach was proposed by Atazadeh et al. [7] as a potential solution to overcome 3D cadastral challenges. There exists a rich amount of physical information inside the BIM models; however, ownership data elements are not recorded in BIM models, but it is possible to connect ownership data elements with BIM data by developing a specific application to link those two sets of data.

2.2. Utility Cadastre

Utility cadastre in the Republic of Croatia is defined by the Law on State Survey and Real Property Cadastre [8], and utility cadastre registers remain under the jurisdiction of local government (towns and municipalities). The State Geodetic Administration (SGA) of the Republic of Croatia suggested that the physical registration of utilities should be organized at a national level in the Republic of Croatia, and the law was changed in December 2016 to follow this direction. The newly adopted Croatian

Utility Cadastre (when incorporated in the Croatian Land Administration System) should be able to streamline the provision of essential services such as water, sewerage, electricity, and communication networks. This normative decision saw the Law on State Survey and Real Property Cadastre align with Directive 2014/61/EU of the European Parliament and of the Council on 15 May 2014 on the impacts of reducing the costs of installing electronic high-speed communications networks. The Digital Agenda is a comprehensive plan of the European Commission to stimulate economic growth through the creation of a more competitive and modern digital Europe [9].

The mentioned and accepted changes of the current Law on State Survey and Real Property Cadastre passed all preparatory procedures. Based on the current status of the Utility Cadastre in the Republic of Croatia, the newly implemented Utility Cadastre should achieve the following goals [10]:

- To obtain information about the "occupancy of space" with regard to underground utilities and other infrastructure;
- To prevent infrastructure-related negative publicity, as well as prevent and reduce the cost of direct and indirect damage;
- To manage the infrastructure and implementation of conditions for keeping records of the utility infrastructure;
- To develop this in such a way that infrastructure data will merge with land cadastre data and be made available in the same projected coordinate reference system (Croatian Terrestrial Reference System 96 (HTRS96) for all interested parties).

Until now, most of the data was in an analogue form and therefore not fulfilling its potential. With the adopted changes to the Law, the 3D data of the utility cadastre will be maintained in digital form and as such, be made more suitable for 3D cadastre.

2.3. Technical Aspects: Geodetic Projects, Digital Terrain, and Surface Models

In Croatian official registers, there is insufficient data on the elevation of objects. However, data can be acquired through geodetic projects, and the available data on digital terrain and surface models can be utilized to establish a 3D cadastre.

The production of geodetic projects for construction and physical planning, as well as the production of utility cadastre reports (for public utility infrastructure), require the use of technical specifications to determine coordinates in the coordinate system of the Republic of Croatia. These specifications prescribe the correct ways to measure and store 3D coordinates in both analogue and digital forms for cadastral purposes (land, real property, and utility cadastre) as well as detailed topographic surveying, preparation of surveying maps, and all other georeferenced products.

Digital elevation models (DEM) provide basic quantitative information about the Earth's surface. Most data providers and professional users prefer the term DEM for both the digital terrain model (DTM) and the digital surface model (DSM). A DTM usually refers to the physical surface of the Earth (elevations of the bare ground surface) without objects such as vegetation or buildings, while a DSM describes the upper surface of the landscape, including the height of the vegetation, man-made structures and other surface features, and only gives elevations of the terrain in areas where there is little or no ground cover [11].

Data acquisition for the development of DTM and DSM are from 3D photogrammetric data acquisition from aerial images (stereopairs), as per the principles of the CROTIS (Croatian Topographic Information System).

Based on the principle of reusing public sector information, data on buildings and other man-made structures could be obtained from the subtraction of DTM and DSM as both models are available in the SGA. However, it can only currently be used for general or statistical purposes. For more specific purposes regarding 3D cadastre requiring higher demands of accuracy and precision, this method for obtaining quality 3D data is still not adequate.

3. Institutional Aspects of the 3D Cadastre

Currently in Croatia there are no official records that can provide complete information on all buildings as spatial objects. Cadastre registers (which fall under the jurisdiction of the State Geodetic Administration) and land registers (under the jurisdiction of the Judicial Authority (Ministry of Justice)) are the only official and systematically maintained registers that contain data on real property [12] that also include buildings. The condition, integrity, and structure of the data collected on buildings maintained in these registers do not allow insight into the state and basic characteristics of certain buildings and the overall condition of buildings across the entire country. Therefore, one of the strategic objectives of the State Geodetic Administration is the establishment of a multipurpose cadastre of buildings to provide such information. The SGA has ordered an implementation study of the cadastre of buildings. This study has provided answers on how to establish institutional, legislative, and financial frameworks, and proposed a structure for the data model and technical standards for the information system of such a cadastre. Findings of the study pointed out that the State Geodetic Administration was the responsible institution for a cadastre of buildings. For detailed elaboration of a cadastre of buildings, the Law on State Survey and Real Property Cadastre should also be updated. Regarding the financial resources for the establishment of this new register, it should be provided from the State budget. Furthermore, this study also provided short-term and long-term strategic guidelines regarding system architecture, data model, the specific needs of stakeholders, the required legislation, the benefits delivered by such a system, and the financial resources needed for its establishment and maintenance. Our study also defined the implementation phases of a building cadastre based on the experiences of EU countries who had already introduced similar systems into their daily operations. The study, among other activities, further questioned the needs of the following future key users: the Ministry of Construction and Physical Planning, Tax Administration, Ministry of Justice, Ministry of the Interior Affairs, Croatian Chamber of Economy, National Protection and Rescue Directorate, Croatian Bureau of Statistics, Croatian Office for the State Property Management, and a representative sample of Croatian cities and municipalities. All of this was undertaken to involve the general public in the project and to consider the needs of the users which will bring—after the establishment of the unified multipurpose register of buildings—the added values of more regular spatial planning, property tax collection, the overall development of cities and municipalities, and the overall benefit to the state institutions and society as a whole [13]. Thus, the aim of the SGA was to develop a building cadastre based on the results of this study.

In Croatia, there are number of activities aimed at improving data, business processes, and the organization of land administration, and all of these fall under the National Real Property Registration and Cadastre Program known as Organized Land [14]. One of the project's key objectives is to realize and implement a Joint Information System (JIS) to combine both the land register and a cadastre. The JIS is a unique system which will replace the current different databases, cadastral data models, and associated applications in the cadastral offices of the State Geodetic Administration, as well as the land register databases and applications in the offices of the municipal courts. The SGA implemented the JIS in all cadastral offices in Croatia by November 2016. Today, the JIS provides support for the implementation of all regulated business processes and tasks, as well as transparent monitoring and data reporting from the cadastre and land registers. This system has special values in its administration and functionalities, and is hosted in a highly secure environment. The establishment of the JIS accelerates registration as it: (1) integrates the spatial and legal data of real property in both cadastral and land register systems; (2) raises the level of security in real property transactions; (3) provides better management of both systems; (4) streamlines business processes; (5) improves customer relations; and (6) increases the speed and quality of service. To provide the best speed and quality of services to key users and the general public, the State Geodetic Administration and Ministry of Justice developed (and continue to develop) a public One-Stop-Shop (OSS) web application that represents the link to the cadastral and land register data (or JIS). OSS allows all users to search and access an overview of the basic cadastral and land register data, but also contains a section for registered users who can then view the data, apply for public documents, view their case status, and receive issued documents into an electronic mailbox. Other additional cadastral data functionalities (to be implemented by the end of 2017) such as the electronic exchange of data between cadastral offices and licensed geodetic engineers, and the delivery of digital reports in cadastral offices, will standardize and speed up the review and confirmation of geodetic reports as well as accelerate the process of real property registration. Furthermore, this will also provide citizens with easy and quick access to public documents and data.

Effects of Institutional Issues on the General State of Land Related Data

Recently conducted research in Reference [4] included the analysis of land-related registers in the Republic of Croatia with the purpose of assessing the overall condition of land-related data. These registers primarily included data on land features such as cadastral parcels, buildings, and utilities (which can be found as the source of 3D data), as well as the interests (rights, restrictions, responsibilities) established over those features and spaces.

One of the aims of the abovementioned research in Reference [4] was to determine the level of redundancy between the registers closely related to the domain of land administration. For that purpose, the Land Administration Domain Model ISO 19152 (LADM) was used as it represents the basis of all land administration systems worldwide, including both the legal and the spatial components of land administration. In addition, a detailed analysis of the current legislation regulating the domain of registered data was conducted.

The analysis confirmed a significant amount of data redundancy, even at the level of individual public authorities responsible for governing multiple registers (Figure 3), which is a direct consequence of a non-existing linkage between the official registers. However, situations like this should not occur as the public authority responsible for more than one register has unrestricted access to all of the data, is fully familiar with the underlying data models, and should therefore be able to find simpler solutions for data sharing than when data is kept in different jurisdictions.

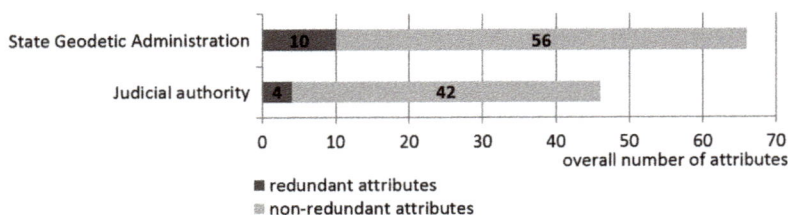

Figure 3. Redundancy at individual public authorities.

Additionally, situations like these should not exist in a mature information infrastructure (II) as part of the e-Government. A non-existing linkage is the primary reason for copying data from one register to another, thus producing redundancies. These data redundancies in the public authorities often lead to situations where data on land features in one register does not match the corresponding data in another register. Therefore, registers have often been viewed as unreliable, which undoubtedly lessens their actual value and creates unwanted consequences for the relevant public authority, as well as for citizens and companies as the end-users of that data. However, there are no excuses for keeping such redundant data. With the introduction of e-services as a form of modern Information and Communications Technology (ICT) [15], an environment has been created where there are no longer any technological barriers for data sharing, only those of administrative character. As a result, there is still no full systematic solution for linking registers at the data level in the Republic of Croatia.

4. Establishment of a 3D Multipurpose Land Administration System in Croatia

Urbanization, which includes the development of high-rise apartments and the advent of complex building structures, has created unique challenges that cannot be met by 2D land and property information. These include interrelated titles and complex plans relating to [16]:

- cadastral parcels; and
- buildings, both internal (indoor plans) and external attributes (roof and façade).

Croatia does not have the complex buildings and structures found in Japan, China, or the Netherlands, but certain 3D situations exist which cannot be modelled in a 2D plane. One of the main problems in Croatia is that building data are scattered in several places including the topographical database, the registry of house numbers, the documentation acquired during building legalization, and so forth. Unfortunately, one direct consequence of such a state is the high level of data redundancy, which further causes additional errors in the data.

4.1. Sources of 3D Data in Croatia

Although data models for 3D data have existed for some time, the collection and generation of 3D data has not reached the level required to use 3D models across the wide variety of use cases as defined in Reference [17].

In Croatia, only the city of Zagreb has published 3D data online [18] for Zagreb and the neighbouring city of Zaprešić (Figure 4). Data can be viewed, and users can obtain basic information on buildings; furthermore, the Zagreb City Office for Strategic Planning and Development uses 3D data in its daily operations for strategic planning and urban development. However, these data cannot fulfil the needs of a 3D MLAS since the objects are modelled on the building level and building part. To attach various rights, restrictions, or responsibilities to building units, a further division of building parts in building units is necessary.

Figure 4. The Croatian National Theatre, Zagreb as displayed in the Zagreb 3D geoportal [18].

The implementation study of the multipurpose building cadastre in the Republic of Croatia is a State Geodetic Administration project, which is one step towards the establishment of a 3D MLAS [13]. One of the main outputs of the study was a conceptual model of a buildings cadastre that considered the available data and the needs of the stakeholders. The basis for the conceptual model was the INSPIRE directive, i.e., the data theme Buildings (Figure 5).

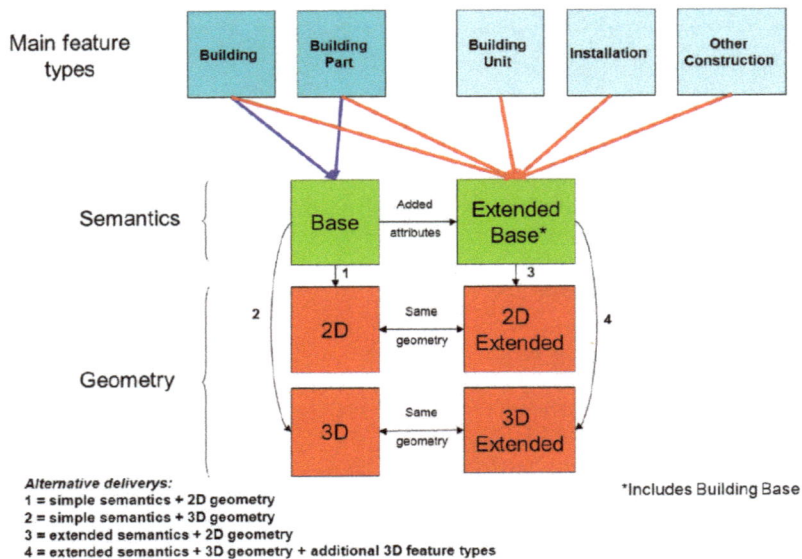

Figure 5. Content and structure of application schemas for the theme Buildings [19].

The INSPIRE Buildings theme uses CityGML for 3D model representation. CityGML defines five levels of detail (LoD): LoD 0, LoD 1, LoD 2, LoD 3, and LoD 4. In addition, data specifications in the Buildings theme can define several profiles regarding the complexity and availability of data in buildings such as:

- 2D/3D Base;
- 2D/3D Extended.

A 2D geometry profile includes 2D or 2.5D, i.e., a 2D footprint and height. A base profile only contains a basic set of building attributes while an extended profile has additional attributes including construction materials, installations, and relationships to other classes such as building units or cadastral parcels, addresses, and so on.

Since a large number of buildings in Croatia does not have any kind of technical documentation, information on these buildings need to be collected from existing data sources and represented in a 2D extended profile using LoD 0. For such buildings, a 3D model (LoD 1) could be created as explained in Reference [20]. Buildings and building parts that have technical documentation (for example, new buildings which have valid construction and usage permits) will be represented in 3D profile (LoD 2), while building units will be represented in 2.5D (footprint and height).

A great source of 3D data could be gained from the building legalization process, which has been in progress since 2012 [21]. More than 800,000 households have initiated the process of building legalization; as most households have more than one building, the number of buildings is much larger. At the time of writing (March 2017), more than 500,000 cases have been resolved. With future legal amendments on the treatment of illegally constructed buildings and other structures, the government of the Republic of Croatia has extended the deadline for submitting applications for legalization to the end of June 2018, and the number of cases for building legalization are likely to increase. For each building (based on size) in the process of legalization, a detailed project (for larger buildings) and footprint with height at the top and at the edges of the roof was provided (for smaller buildings). From that data, LoD 3 (for larger buildings) and LoD 1 or LoD 2 (for simpler buildings) could be created depending on building size.

One of the main problems with such data is its heterogeneity. Since there are no standards for building drawings, each company creates designs based on their in-house rules and standards. Some companies use Computer-Aided Design (CAD) applications, others use more advanced applications such as ArchiCAD. Given the lack of standardization, it is very difficult to implement an automated workflow to create 3D models.

4.2. Technical Aspect of Establishing 3D MLAS

The problems mentioned in the previous sections, such as data redundancy, unrelated data, data heterogenity, and errors in data, make the automatic generation of 3D models a challenging task. Initially, 3D models could be generated from the existing sources previously mentioned, or following the approach described by Biljecki et al. [20], or through the semi-automatic approach explained by Pouliot et al. [22]. Since existing 3D data are heterogenuous, a machine learning paradigm could be used to overcome slight differences in the data models across various sources.

Section 5 provides an overview of how certain 3D situations are represented on 2D cadastral maps with topographical signs. These situations could be combined with 3D data generated without elevation data as explained by Biljecki et al. [20], where it is simple to generate a 3D model from a cadastral map and the number of storeys. The height of the building was estimated by multiplying the number of storeys with the average height of a single storey. Lines within buildings were used to enhance the 3D model generation algorithm.

Figure 6 shows the building footprint (blue line) represented on a cadastral map (a). Within the building polygon is a number which represents the building type. The red dashed line represents passage under the building. The red solid line represents the line which separates construction parts of the building, such as eaves or different building parts. Based on the type of line, it is possible to determine whether the building should be clipped from the top or from the bottom (b). The presumption is that the height of passage under the building is equal to the height of one storey. This presumption is valid for the majority of buildings in Croatia. It must be noted that this example is simplified and does not consider issues such as buildings that are not registered on the cadastral map, and the recognition of information on the number of building storeys from building permits and other documents. Although this idea has the potential to generate 3D models easily and quickly, it requires additional research to test the feasibility and accuracy of the data produced in this manner.

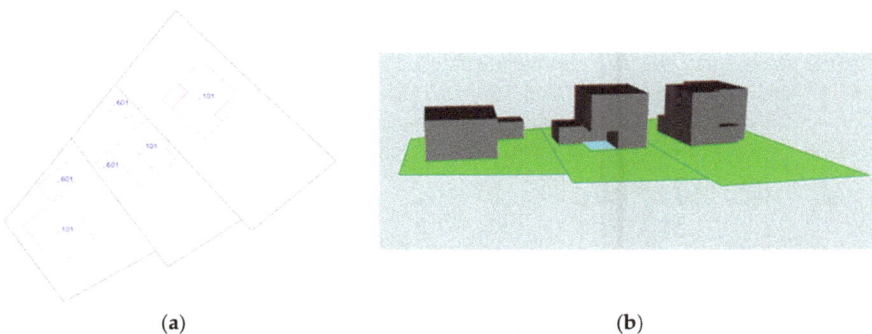

(a)   (b)

Figure 6. 3D model generated from (a) cadastral map, (b) 3D model.

Bennet et al. [23] recognized the criteria that future cadastres should satisfy to serve as a background for quality land administration, which include:

- Improving the integration between the field data collection and execution of transactions on the data;
- Defining and formalizing the types of transactions on the data;
- Describing and formalizing the transaction processes for various types of transactions to outsource a large proportion of the updating process.

The previously listed criteria should be applied to a 3D Land Administration System since the private sector is generally more adaptable to user needs. To improve the modelling manner, storage, and exchange of field survey data, the international standard ISO 19156 can be used. Furthermore, the scientific literature contains publications on modelling land survey data and their integration with 2D cadastral data. Van Oosterom and Lemmen [24] explored the possibilities of modelling land survey data based on ISO standard 19156 (Observations and Measurements—O&M). Vranić et al. [25] went a step beyond and developed a more refined model based on ISO 19156 for Global Navigation Satellite system (GNSS) observations that could be used for the efficient storage and retrieval of land survey data. In his Master's thesis, Soffers [26] proposed a solution for modelling relationships between survey data and resulting cadastral maps, i.e., spatial units based on the Land Administration Domain Model (LADM) sub package Surveying and Representation. Storing survey data in a digital form (that is, in readable form) to a computer that is related to spatial units enables easier searching and usage of the data. Mentioned studies have dealt with the integration of cadastral data with classical land survey data such as tachymetry or GNSS, but none of them address the modelling of 3D data from Light Detection and Ranging (LIDAR) or other sensors that produce 3D data. If 3D models are generated without elevation data (as explained in Biljecki et al. [20]) or from existing data gained from legalization (such as topographical databases, etc.), these sources and their relationship to spatial units should be modelled via the LADM LA_SpatialSource class.

A large proportion of 3D data can be acquired from existing data as explained previously, but the data are heterogenous and of different accuracy; however, this can be improved through the incremental updating of 3D data. If the process of updating 3D data is outsourced to the private sector, it is necessary to summarize the transaction types and their correctness conditions. Matijević et al. [27] performed a basic analysis of transactions in cadastre and planar partition as traditional European parcel-based cadastres use planar partition to manage their geometry. In a later work described in Reference [28], initial research on the correctness of cadastral parcels represented by simple polygons (as defined in international standards ISO 19107 and ISO 19125) was conducted where the authors defined the general criteria requiring satisfaction in order to correct the planar partition of the working area. Later, Vranić et al. [29] defined a complete list of transaction types on cadastral parcels and defined preconditions and postconditions for each transaction type. Although these papers concentrated on the correctness of 2D data, these can also be extended towards defining correctness criteria for 3D data. Initial research on the correctness of 3D data has already been conducted by Karki et al. [30] where they used ISO standard 19152 (Land Administration Domain Model: LADM) as a basis to define several encoding strategies for 3D data to ensure the achievable level. Furthermore, depending on the level of maturity of system validation, criteria can be defined.

5. 3D Topographical Data in Topographic Databases

One consequence of the growing densification in urban land-use is an increase in situations regarding the vertical stratification of rights. As traditional 2D cadastral models are not able to fully handle spatial information on ownership rights in the third dimension, 3D cadastre has attracted researchers wanting to better register and spatially represent overlapping real-world situations [31].

Models of physical objects created from field observations may differ from legal object boundaries, and these differences between facts in the field and facts in the documentation can often become an obstacle in realizing the various rights related to land administration. Therefore, it is important that buildings are registered in the official registers (both in cadastre and land registers) in a way that accurately depicts the actual state of real property. Thus, we have physical 3D objects, legal objects with 3D space, and other categories such as economic objects that can be used in valuations.

5.1. Buildings

In the Croatian Land Administration System, there are special topographic signs for 2D maps showing 3D situations such as buildings overlapping other structures (such as tunnels, roads, or other parcels with building parts crossing the parcel boundaries either above or below ground level).

Additionally, there are also special signs that represent underground buildings on cadastral and topographic maps (Figure 7). Cadastre can register underground buildings only on cadastral maps and in the written part of cadastral documentation, but without area information, which can be documented in the land register. Many new underground buildings were built in Croatia during the last twenty ears, which is the reason behind the consideration of new regulations by the State Geodetic Administration of the Republic of Croatia that will enable the registration of underground buildings with area and other attributes into the cadastre.

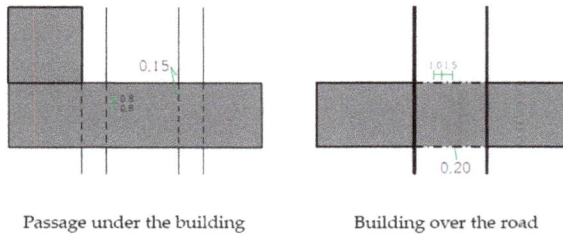

Passage under the building   Building over the road

Figure 7. Example of topographic signs for buildings [32].

5.2. Other 3D Situations from Real Life

Other 3D real-life situations (such as pedestrian passageways) are also regulated by the geodetic and cadastral legislation. Good examples of potential 3D objects where representation can only be found in topographic signs on topographic maps are bridges, tunnels, pipelines, overpasses, etc. Furthermore, there are different topographic signs that serve a particular purpose and apply to different situations. For example, there are a few different types of bridges in the Croatian Land Administration System which are represented by different topographic signs (Figure 8).

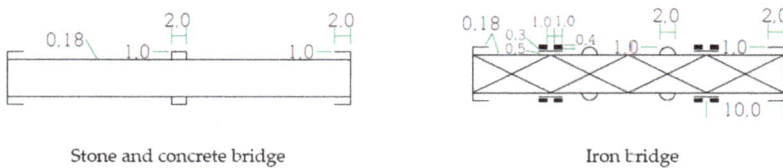

Stone and concrete bridge   Iron bridge

Figure 8. Example of topographic signs for bridges [32].

In these cases, cadastral data are more complete and accurate, but lack important information on 3D objects. In contrast, topographic maps contain this type of information, but with less spatial accuracy. However, by combining cadastral data and topographic maps, it is possible to gain new values in terms of complete information on 3D space.

6. Link between Topographical Data and Cadastral Data

The base register for large-scale topography forms an excellent context for 3D cadastral solutions as 3D cadastral parcels (3D legal spaces) are often related to (planned) physical objects such as buildings, tunnels, pipelines, and other constructions. For reference purposes, 3D legal objects and their 3D physical counterparts should be associated, which has two implications:

- That the 3D physical object descriptions (topographic objects) should exist, but are not obvious as most countries still use 2D representation for large-scale topographic base maps and;
- That the topographic objects should be usable and referenceable even when the data is maintained by other organizations.

LADM supports Spatial Data Infrastructure (SDI) implementations as the information infrastructure requires unique ID numbers for all objects, a full database history (versioning), and the blueprints of external classes such as topographic objects (buildings, tunnels, pipelines, etc.). LADM ensures the semantics of land administration data to other users within the SDI [33], and also provides conceptual descriptions for land administration, including 3D topology. Furthermore, it allows for land-related data to be organized in a standardized and interoperable way to support different types of spatial data. According to the requirements of LADM, topological information alone is not sufficient to describe a 3D spatial unit, and geometrical information must also be associated with each topological primitive either as direct or indirect geometries (via related topological primitives with geometries). For the 3D topology model in LADM as described in the Spatial profiles of Annex E7, there are no overlapping volumes (3D_SpatialUnit) [34].

For each real-life object in the Digital Topographic Database, a unified feature identifier can be assigned which can also be used to link with other cadastral data, for example, it can be implemented in an SGA Geoportal, which is the central place to access spatial data and is one of the basic elements of the National Spatial Data Infrastructure.

The SGA Geoportal consists of the following sets of data:

- 1:5000 digital orthophoto map, recorded and produced during 2011 for the whole country, and recorded and produced in 2014/2015 for some parts of country;
- Croatian Base Map 1:5000 (topographic);
- Topographic Maps 1:25,000, 1:100,000, 1:200,000;
- Digital cadastral maps (with the following contents: cadastral offices, cadastral municipality, cadastral parcels, buildings, house numbers);
- Central Registry of Spatial Units (counties, cities and municipalities, towns, local government);
- Generalized content 1:100,000 topographic map (roads, railways, ferry piers, lakes, waterways);
- Digital Terrain Model (DTM).

In the remainder of this section, two case studies are presented: (1) a road tunnel through a mountain; and (2) a pedestrian tunnel below a road.

6.1. Croatian Topographic Information System (CROTIS)

The Croatian Topographic Information System (CROTIS) version 1.0 data model was created in September 2000 [35] and served to define and standardize the data model and collection of topographic data, processing, accuracy, topological relations, and data exchange. CROTIS consists of the following five documents:

- Book I: Conceptual Data Model—Alphanumeric and Graphic Code System
- Book II: Topological Relations
- Book III: Exchange Data Structures and Exchange Graphic Models
- Book IV: Object Catalogue

- Book V: Fundamental Principles

CROTIS was created as a model for the establishment of the Basic Topographic Database. In 2014, a new data model was developed under the name CROTIS 2.0. In the new data model, feature classes were grouped based on the analysis of identical attributes to ensure simple implementation into the object-relation database. During the development of the feature catalogue for the new CROTIS model, special attention was paid to defining the feature classes, attributes, and values. When defining new feature classes, consideration was made as to whether a feature class referred to, for instance, land cover (meadow, forest, rocks, etc.) or land usage (port, marina, beach, etc.) [36]. The CROTIS 2.0 data model consists of the packages shown in Table 1).

Table 1. List of packages and classes in the Croatian Topographic Information System (CROTIS 2.0).

Package	Class
Structures	Building
	Smaller structures
	Larger structures
	Built barriers
Utility lines	Utility line
	Belonging elements of utility networks
Traffic	Traffic areas
	Track
	Surface elements of traffic
	Line elements of traffic
Cover and land use	Agriculture land
	Areas under the trees
	Other natural areas
	Public areas
	Commercial areas
	Land use
Hydrography	Wide water flow
	Narrow watercourse
	Coast line
	Sea, standing water
	Elements of the watercourse
	Water barriers
Relief	Altitude
	Depth
	Isobath
	Contours
	Relief forms
Geographical names	Geographical name

6.2. Case Study 1: Tunnel

Sveti Rok tunnel is a double-tube tunnel located on the Sveti Rok–Maslenica section of the Zagreb–Split–Dubrovnik Motorway. The length of the left tunnel tube is 5679 m and the length of the right is 5670 m. The traffic runs through both tubes separately for each traffic direction. The northern tunnel portal is placed at 561 m above sea level and the southern portal at 510 m. The tunnel passes through the Velebit Mountain in the corridor of the Mali Alan mountain saddle (Figure 9).

Figure 9. Sveti Rok road tunnel.

An example of tunnel representation in the Croatian Land Administration System is shown in Figure 10. The tunnel, which was built under many cadastral parcels, would, in a vertical sense, belong to those cadastral parcels. However, by functionality, it is permanently connected to the land where the entrance to the tunnel is located, and not to the cadastral parcels that extend above it. Therefore, according to legislation, the tunnel is permanently connected only to the land where the entrance is which makes it one property.

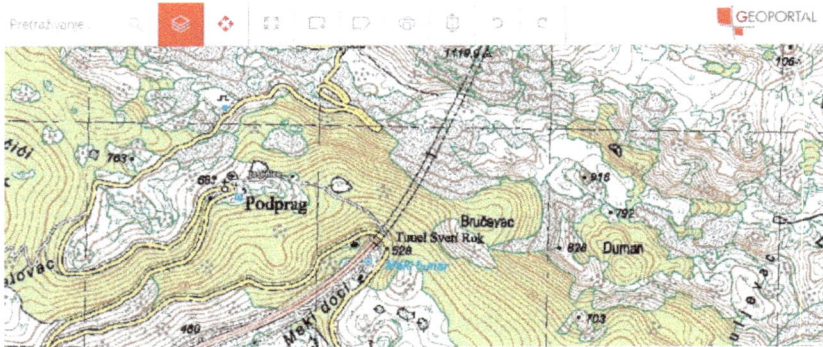

Figure 10. Tunnel as seen on a 1:25,000 topographic map [37].

6.3. Case Study 2: Pedestrian Passage under the Street

In Croatia, pedestrian passages under roads are mainly situated in large cities, and smaller towns do not usually have these 3D situations (Figure 11).

Further development of software and hardware technologies for spatial information has made it easy to combine geodetic and cartographic products in the modern digital environment. Today, all Croatian cadastral offices maintain digital cadastral maps overlapped with orthophoto images, and all other geodetic and cartographic products are available on the Croatian SGA Geoportal (Figure 12). At times, specific details are not shown on the cadastral map (like the underground passage from Figure 11), although there is a legal basis for its representation on the cadastral map. It is thought that this may be due to differences in time period during the map creation, as well as the selective and inconsistent application of the rules that may have caused these problems.

Figure 11. Pedestrian passage under the street in Zagreb, Savica [38].

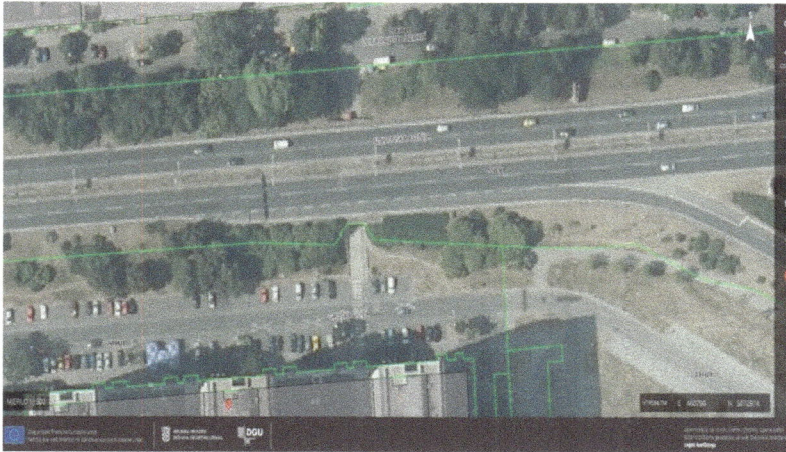

Figure 12. Pedestrian passage under the street in Zagreb, Savica, on the State Geodetic Administration Geoportal (digital ortophoto image overlapped with digital cadastral map) [37].

7. Conclusions

All branches of Croatian government—the legislative, judicial, and executive branches—are responsible for the land and associated land-related policies, each one of them within their own powers and competence. There are a great number of institutions responsible for land and different land-related policies, but each of them within their own scope of competence, and there is very little coordination between them. The scope of activities and competence of the public administration bodies are clearly regulated by law and formally, they should not be mutually overlapping.

However, in the Republic of Croatia (and other countries where cadastre was established a long time ago) many registers and official databases on land and interests were created where certain overlaps between some segments are evident. These have most often been established independently and therefore contain much redundant data. However, their interaction can be used to gain new values and establish Multipurpose Land Administration Systems. As shown in this paper, there are multiple sources of 3D data in Croatia, including reports on a particular part of real property (deposited in the land register), digital terrain models and digital surface models, the Basic Topographic Database, the utility cadastre data, and the topographic content of cadastral and topographic maps (topographic signs). At a technical level, this problem can be solved by harmonizing the data model, for which the LADM can be used as a basis. However, legislative and institutional issues appear to be much more demanding tasks in practice.

It is not imperative that the 3D cadastre is established as soon as possible, instead, it is much more appropriate to develop it gradually and with good prioritization. It is indisputable that larger cities with more complex 3D situations in real life (such as a large number of public utility infrastructure, more underground structures, and many buildings with two or more floors) would have a greater use of the 3D cadastre compared to small towns, especially villages. A cost-benefit analysis of establishing a 3D cadastre in certain areas could be used to determine these priorities. The cost-benefit analysis undertaken in the implementation study of the building cadastre in the Republic of Croatia has shown that it would be feasible and cost effective to implement a building cadastre by the end of 2020, so that it could be used for the taxation of real properties based on their value, and for the 2021 census.

This study, among others, has provided an overview of the available 3D data in Croatia. As previously mentioned, the current data on buildings and other 3D objects are scattered, redundant, and loosely related. However, regulations defining the maintenance of cadastral maps and data collection workflow in a straightforward manner will allow for easier mass usage of the data. The simple test case presented in Section 4 shows how a 3D model could be generated by using a building footprint from a cadastral map and information on the number of floors. Further research will focus on testing the applicability and the usability of the currently available 3D data. Furthermore, various workflows for 3D model generation will also be created, which will include 3D data from multiple sources, and the generated 3D model will be tested in terms of completeness, accuracy, and correctness.

Acknowledgments: This work has been fully supported by the Croatian Science Foundation under project HRZZ-IP-11-2013-7714.

Author Contributions: Nikola Vučić, Miodrag Roić, Mario Mađer, and Saša Vranić contributed to organizing the research in part of the legal and institutional aspects of the Croatian cadastral system, as well as writing the paper. The materials for the paper were prepared by Nikola Vučić, Miodrag Roić, Mario Mađer, Saša Vranić and Peter van Oosterom. All authors have read and approved the final manuscript.

Conflicts of Interest: The authors declare no conflict of interest.

References

1. Roić, M.; Mastelić Ivić, S.; Matijević, H.; Cetl, V.; Tomić, H. Towards Standardized Concept of Multipurpose Land Administration. In Proceedings of the FIG Working Week, Christchurch, New Zealand, 2–6 May 2016.
2. Paulsson, J.; Paasch, J.M. 3D Property Research from a Legal Perspective. *Comput. Environ. Urban Syst.* **2013**, *40*, 7–13. [CrossRef]
3. Ploeger, H. Legal Framework 3D Cadastres. Position paper 1. In Proceedings of the 2nd International Workshop on 3D Cadastres, Delft, The Netherlands, 16–18 November 2011; pp. 16–18.
4. Mađer, M.; Matijević, H.; Roić, M. Analysis of possibilities for linking land registers and other official registers in the Republic of Croatia based on LADM. *Land Use Policy* **2015**, *49*, 606–616. [CrossRef]
5. Tušinec, I. Etažiranjem do vlasničkih ovlasti. *Ekscentar* **2005**, *27*, 23–29. (In Croatian)
6. Official Gazzette of the Republic of Croatia. Pravilnik o povezivanju zemljišne knjige i knjige položenih ugovora i upisu vlasništva posebnog dijela nekretnine (etažnog vlasništva). Available online: http://narodne-novine.nn.hr/clanci/sluzbeni/2013_09_121_2596.html (accessed on 20 July 2017).

7. Atazadeh, B.; Kalantari, M.; Rajabifard, A.; Champion, T.; Ho, S. Harnessing BIM for 3D digital management of stratified ownership rights in buildings. In Proceedings of the FIG Working Week, Christchurch, New Zealand, 2–6 May 2016.

8. Official Gazzette of the Republic of Croatia. Zakon o državnoj izmjeri i katastru nekretnina. Available online: https://www.zakon.hr/z/156/Zakon-o-dr%C5%BEavnoj-izmjeri-i-katastru-nekretnina (accessed on 20 July 2017).

9. E-Savjetovanja. Available online: https://esavjetovanja.gov.hr/ECon/MainScreen?entityId=3020 (accessed on 27 August 2016).

10. Vučić, N.; Roić, M.; Markovinović, D. Towards 3D and 4D Cadastre in Croatia. In Proceedings of the 4th International Workshop on 3D Cadastres, Dubai, UAE, 9–11 November 2014.

11. Varga, M.; Bašić, T. Quality Assessment and Comparison of Global Digital Elevation Models for Croatia. Kartografija i Geoinformacije: Zagreb, Croatia. *J. Croat. Cartogr. Soc.* **2013**, *12*, 20.

12. Roić, M. *Upravljanje Zemljišnim Informacijama—Katastar*; Faculty of Geodesy: Geodesy, Zagreb, 2012; Volume 199, p. 199.

13. State Geodetic Administration. Development of the Study on the Implementation of the Cadastre of Buildings in the Republic of Croatia. Available online: http://www.dgu.hr/assets/uploads/OGLAS/ToR_katastar%20zgrada_HRV.pdf (accessed on 23 June 2017).

14. Organized Land. Available online: http://www.uredjenazemlja.hr/default.aspx?id=17 (accessed on 2 February 2017).

15. Van Oosterom, P.; Groothedde, A.; Lemmen, C.; Van der Molen, P.; Uitermark, H. Land Administration as a Cornerstone in the Global Spatial Information Infrastructure. *Int. J. Spat. Data Infrastruct. Res.* **2009**, *4*, 298–331.

16. Jazayeri, I. A geometric and semantic evaluation of 3D data sourcing methods for land and property information. *Land Use Policy* **2014**, *36*, 219–230. [CrossRef]

17. Biljecki, F.; Stoter, J.; Ledoux, H.; Zlatanova, S.; Çöltekin, A. Applications of 3D City Models: State of the Art Review. *ISPRS Int. J. Geo-Inf.* **2015**, *4*, 2842–2889. [CrossRef]

18. Zagreb Geoportal. Available online: http://zagreb.gdi.net/zg3d/ (accessed on 21 March 2017).

19. Data Specification on Buildings—Draft Technical Guidelines. Available online: http://inspire.ec.europa.eu/documents/Data_Specifications/INSPIRE_DataSpecification_BU_v3.0rc3.pdf (accessed on 21 March 2017).

20. Biljecki, F.; Ledoux, H.; Stoter, J. Generating 3D city models without elevation data. *Comput. Environ. Urban. Syst.* **2017**, *64*, 1–18. [CrossRef]

21. Status of Building Legalization. Available online: https://legalizacija.mgipu.hr/izvjesce (accessed on 26 March 2017).

22. Pouliot, J.; Roy, T.; Fouquet-Asselin, G.; Desgroseilliers, J. 3D Cadastre in the province of Quebec: A First experiment for the construction of a volumetric representation. In *Advances in 3D Geo-Information Sciences*; Springer: Berlin/Heidelberg, Germany, 2011; pp. 149–162.

23. Bennet, R.; Rajabifard, A.; Kalantari, M.; Wallace, J.; Williamson, I. Cadastral features: Building a new vision for the nature and role cadastres. In Proceedings of the XXIV FIG International congress: Facing the challenges: Building the capacity, International Federation of Surveyors (FIG), Sydney, Australia, 11–16 April 2010.

24. Van Oosterom, P.; Lemmen, C.; Uitermark, H.; Boekelo, G.; Verkuijl, G. Land administration standardization with focus on surveying and spatial representations. In Proceedings of the Survey Summit, the ACSM Annual Conference, San Diego, CA, USA, 7–13 July 2011.

25. Vranić, S.; Mađer, M.; Matijević, H.; Bašić, T. Recording the CROPOS GNSS measurements in XML—A step towards digital geodetic technical report. In Proceedings of the 4th CROPOS Conference, Zagreb, Croatia, 22 May 2015; pp. 140–153.

26. Soffers, P. Designing an Integrated Future Data Model for Survey Data and Cadastral Mapping. Master's Thesis, Delft University of Technology, Delft, The Netherlands, 2017; p. 113.

27. Matijević, H.; Biljecki, Z.; Pavičić, S.; Roić, M. Transaction Processing on Planar Partition for Cadastral Application. In Proceedings of the FIG Working Week 2008 (FIG), Stockholm, Sweden, 14–19 June 2008.

28. Matijević, H.; Biljecki, Z.; Vranić, S. Correctness Preconditions For Changes on Geometry of Cadastral Parcels with Analysis of Implementation Options. In Proceedings of the Papers—1st Serbian Geodetic Congress (Republic Geodetic Authority), Beograd, Serbia, 1–3 December 2011; pp. 201–208.

29. Vranić, S.; Matijević, H.; Roić, M. Modelling outsourceable transactions on polygon-based cadastral parcels. *Int. J. Geogr. Inf. Sci.* **2015**, *29*, 454–474. [CrossRef]

30. Karki, S.; Thompson, R.; McDougall, K. Data validation in 3D cadastre. In *Developments in 3D Geo-Information Sciences*; Springer: Berlin/Heidelberg, Germany, 2010; pp. 92–122.

31. Duarte-de-Almeida, J.P.; Liu, X.; Ellul, C.; Rodrigues-de-Carvalho, M.M. Towards a Property Registry 3D Model in Portugal: Preliminary Case Study Implementation Tests. In *Innovations in 3D Geo-Information Sciences*; Isikdag, U., Ed.; Springer: Cham, Switzerland, 2014; pp. 291–320.

32. Official Gazzette of the Republic of Croatia. Pravilnik o kartografskim znakovima. Available online: http://narodne-novine.nn.hr/clanci/sluzbeni/2011_09_104_2118.html (accessed on 20 July 2017).

33. Jeong, D.; Jang, B.; Lee, J.; Hong, S.; Van Oosterom, P.; De Zeeuw, K.; Stoter, J.; Lemmen, C.; Zevenbergen, J. Initial Design of an LADM-based 3D Cadastre—Case Study from Korea. In Proceedings of the 3rd International Workshop on 3D Cadastres: Developments and Practices, Shenzhen, China, 25–26 October 2012.

34. Zulkifli, N.A.; Abdul Rahman, A.; van Oosterom, P. An overview of 3D topology for LADM-based objects. In Proceedings of the The International Archives of the Photogrammetry, Remote Sensing and Spatial Information Sciences, Kuala Lumpur, Malaysia, 28–30 October 2015; Volume XL-2/W4.

35. Biljecki, Z. *Topografsko Informacijski Sustav RH—CROTIS*; State Geodetic Administration: Zagreb, Croatia, 2000.

36. Landek, I.; Marjanović, M.; Šimat, I. Croatian Topographic Information System CROTIS 2.0. Data Model. Kartografija i Geoinformacije: Zagreb, Croatia. *J. Croat. Cartogr. Soc.* **2014**, *14*, 24.

37. State Geodetic Administration. Available online: http://geoportal.dgu.hr (accessed on 23 February 2017).

38. Society of Architects of Zagreb City. Available online: http://www.d-a-z.hr (accessed on 27 August 2016).

International Journal of
Geo-Information

MDPI

Article

Towards 3D Cadastre in Serbia: Development of Serbian Cadastral Domain Model

Aleksandra Radulović *, Dubravka Sladić and Miro Govedarica

Faculty of Technical Sciences, University of Novi Sad, 21000 Novi Sad, Serbia; dudab@uns.ac.rs (D.S.);
miro@uns.ac.rs (M.G.)
* Correspondence: sanjica@uns.ac.rs; Tel.: +381-63-102-8536

Received: 29 August 2017; Accepted: 16 October 2017; Published: 19 October 2017

Abstract: This paper proposes a Serbian cadastral domain model as the country profile for the real estate cadastre, based on the Land Administration Domain Model (LADM), defined within ISO 19152. National laws and other legal acts were analyzed and the incorrect applications of the law are outlined. The national "Strategy of measures and activities for increasing the quality of services in the field of geospatial data and registration of real property rights in the official state records", which was adopted in 2017, cites the shortcomings of the existing cadastral information system. The proposed profile can solve several problems with the system, such as the lack of interoperability, mismatch of graphic and alphanumeric data, and lack of an integrated cadastral information system. Based on the existing data, the basic concepts of the Serbian cadastre were extracted and the applicability of LADM was tested on an obtained conceptual model. Upon obtaining positive results, a complete country profile was developed according to valid national laws and rulebooks. A table of mappings of LADM classes and country profile classes is presented in this paper together with an analysis of the conformance level. The proposed Serbian country profile is completely conformant at the medium level and on several high-level classes. LADM also provides support for three-dimensional (3D) representations and 3D registration of rights, so the creation of a country profile for Serbia is a starting point toward a 3D cadastre. Given the existence of buildings with overlapping rights and restrictions in 3D, considering expanding the spatial profile with 3D geometries is necessary. Possible solutions to these situations were analyzed. Since the two-dimensional (2D) cadastre in Serbia is not fully formed, the proposed solution is to use the 2D model for simple right situations, and the 3D model for more complex situations.

Keywords: LADM; Serbian country profile; 3D cadastre

---

1. Introduction

The term "land administration" refers to "the processes of determining, recording, and disseminating information about the ownership, value, and use of land when implementing land management policies" [1]. Land administration functions may be divided into four components: juridical, regulatory, fiscal, and information management. The juridical component focuses on registration of rights on immovable properties, the regulatory component is mostly concerned with the development and use of land, the fiscal component focuses on the economic utility of properties and supports their valuation and taxation, and the information management component focuses on the development of land information systems [2]. The core of land administration is the cadastral system, which identifies legal boundaries of cadastral parcels through geodetic survey and produces cadastral maps and, at least in our case, records legal rights on properties.

Well-structured and organized cadastral records and cadastral maps are a prerequisite for providing better services in land administration. Many problems have occurred related to the collection and maintenance of cadastral data in Serbia and attempts have been made to overcome these issues. However, one of the main problems still remains and that problem is related to the way in which data are organized and maintained, which affects the data integrity causing an inability to address user needs in a timely manner [3]. The Serbian cadastre consists of two parts: Cadastral records that represent the legal relationship between people and real properties, and the cadastral maps that, beside geometry data and annotations, contain land use components visualized by topographic symbols.

Cadastral records in Serbia have encountered numerous problems in previous years, such as the existence of different institutions responsible for the management of land data and real properties, storing data in multiple places, data storage in analog form, discrepancy in records in relation to the actual situation, separation of alphanumeric and geometric data, complex structure of the records as a result of inheriting data from different sources, poor performance of search and update of data, and lack of a standard format for data exchange. Some of the problems have been resolved but many have not. Problems relating to the lack of uniformity of the data model, as well as to the use of non-relational data model, exist in all municipalities in Serbia. Other problems are related to the use of different software solutions, which are mainly based on outdated technologies, and the use of the concept of immovable property in software solutions as defined by outdated law. The aforementioned problems can lead to data redundancy, re-implementing functionality, and disturbed correctness of the data.

The data model is the core of the system. To avoid problems and to achieve efficient access, sharing, and exchange of cadastral data on the principles of interoperability, creating a domain model according to national legislation and current standards in the field of geospatial data and cadastre is necessary, particularly for the Land Administration Domain Model (LADM) within ISO 19152 [4].

This research was motivated by the problems that have arisen in the Serbian land administration due to the inadequate legacy cadastral information system. Specifically, there is a need in the Serbian cadastre to develop a new system based on a new data model that will support better cadastral data management, provide better integrity of data by establishing relations among data, and enable more advanced search of the data. For example, a user is not able to perform a search to retrieve all the rights of a certain party, so this functionality must be provided. In addition, serious cadastral data integrity problems that are present should be addressed in the modeling phase of the cadastral information system development. Legal data, which are textual, and surveying data, which are mainly spatial, are usually hard to maintain in a consistent state. The source of the problem is because those two parts of the same land administration system were modeled independently.

The aim of this research was to develop a domain model for the Serbian cadastre, with the purpose of developing a model that will support the resolution of the mentioned problems. The model should be based on the LADM [5,6] to provide interoperability and better data quality. A roadmap to adopt the LADM in the cadastral information system has been provided in Kalantari et al. [7], through several stages that should be considered by land administration organizations, in this case, the Serbian Republic Geodetic Authority. The developed domain model will then be proposed as a country profile.

The aim was also to analyze the needs and possibilities for the development of a three-dimensional (3D) cadastre [8] in Serbia, based on current research and development trends [9]. The main idea behind this is that, since a big gap exists between current cadastral information organization in Serbia and current research and technology trends, and the need for change is significant, the cadastre should not be limited to what is currently recorded in the information system, but also to anticipate future requirements. This idea is supported by the fact that there is an increasing use of modern technologies for 3D data acquisition in Serbia, such as photogrammetry, LIDAR, terrestrial laser scanning, and a number of 3D models in CityGML, and other formats for parts of the cities, developed for various purposes and for various organizations, that may provide input for 3D rights registration.

Several country profiles based on a LADM have been proposed including Poland [10], Czech Republic [11], and Croatia [12,13]. The model-driven architecture (MDA) approach can be used to automatically implement the database schema, which can be then put into use by a country [14]. In addition, several initiatives concerning the development of a 3D cadastre are ongoing, including:

the development of a 3D cadastre in The Netherlands [15], a transition from two-dimensional (2D) to 3D cadastre in Slovenia [16], and the extension of the Korean LADM country profile to build a 3D cadastre model [17].

This paper presents a domain model of the Serbian cadastre to propose a LADM-based country profile with the goals of ensuring that the LADM-based country profile reflects the current cadastral registration and the corresponding legal requirements, and to determine the conformance between the proposed Serbian country profile and the LADM, and to determine if an extension into a 3D cadastre is possible.

This paper is divided into four sections. Section 1 is the introduction, in which the motivation and aim of this research is described. Section 2 describes the cadastral information system in Serbia that is currently in use, and emphasizes the need for its improvement. Section 3 presents how to transform data model that is currently in use into a LADM-based domain model for the Serbian cadastre that the authors propose for the country profile. Section 4 presents the research toward the development of a 3D cadastre in Serbia. Conclusions and future work are given afterward.

2. Cadastral Information System in Serbia

The cadastral information system in Serbia is regulated by the Law on State Survey and Cadastre from 2009 [18] ("the Law"), Amendments to the Law in 2015 [19], and the Rulebook on Cadastre Survey and Real Estate Cadastre from 2016 [20]. The Law defines the real estate cadastre as a basic and public register of real properties and the rights to those properties. Real properties that are recorded in the cadastre are divided into: Land (cadastral parcel), ground and underground buildings and separate parts of the buildings that make one structural units, such as apartments, business offices, garages, etc. "Buildings" refer to residential buildings, commercial buildings, cultural objects, sports and recreation, roads, bridges, underground passages, dams, embankments, and others. The data on the underground buildings are connected to the parcel where the main entrance or one of the entrances to the underground building is located. The utility network cadastre is the main register of the lines and rights to them, together with the property owner's rights, and contains information on the following: water supply network, sewage and drainage network, hot water network, electricity network, telecommunications network, pipeline network, gas pipeline network, and common facilities. Article 157 of the Law states the necessity for implementing a unified information system of both the real estate cadastre and the utility network cadastre, but this has not yet been achieved.

In the information system of the real estate cadastre, alphanumeric data are separated from graphic (spatial) data. Geographical information systems (GIS) and Computer Aided Design (CAD) tools are used for the transformation of analog cadastral maps into digital vector format. Ninety percent of cadastral municipalities are digitized [21]. After digitization, the topological and geometric consistency of the content of the digital cadastral map are tested by checking the fulfillment of the rules for creating geometric and topological relationships between objects. The alphanumeric parts of the information system for real properties in Serbia were implemented in a DOS application, based on FoxDbf tables for a significant part of Serbian territory, while a small part of data is stored in a Microsoft Access database. Data models in these two applications differ from each other, with an evident absence of relationships in the FoxDbf database. The data model is based on the real estate folio concept, a legacy paper document that consists of four sheets. Figure 1 displays the database tables that represent the four sheets with the most important attributes. Other attributes are omitted from the figure for better visibility. The "A" sheet contains data about one or more parcels. The "B" sheet contains data about parties which have rights over parcels from the A sheet. The "V" sheet contains data about buildings and parts of buildings, such as apartments and business offices, which are located in parcels listed in the A sheet, together with data about their rights. The "G" sheet contains data about restrictions on real properties defined in the A and V sheets. This is how the Law on State Survey and Cadastre and the Registration of Rights on Real Estates Cadastre (1992) [22] are defined in the real estate folio. However, the current Law defines a real estate folio as data about a single real property together with its rights and restrictions.

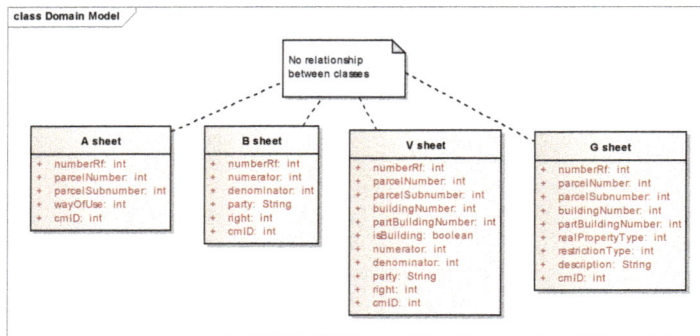

Figure 1. Existing data model.

In January 2017, "The Strategy of measures and activities for increasing the quality of services in the field of geospatial data and registration of real property rights in the official state records" [22] ("the Strategy") was adopted. It contains measures to achieve the implementation of the plan by 2020. The Strategy analyzes the existing state of the cadastral system and lists the following weaknesses in terms of access and use of spatial data in the Republic of Serbia: Unclear data, mismatch of data, the heterogeneity of the information technology systems, multiplication of effort and costs, insufficient use of standards, lack of appropriate digital data, lack of interoperability, poor reliability and confidentiality of data, unclear quality of data, mismatch of graphic and alphanumeric data, and lack of an integrated cadastral information system. The Strategy recognizes the INSPIRE Directive [23] and ISO 19152 standard as world trends in geodetic science that need to be followed and applied. Several attempts to develop a new cadastral information system in Serbia have been made in past years, but those projects failed due to bureaucratic reasons. A new project is underway and it should implement all measures mentioned in the Strategy.

Based on the above, and given the development of information technology, a need exists to improve the current cadastral system and to integrate existing subsystems (alphanumeric and graphical, as well as the utility network subsystem) into one unifying data model that will be completely based on current legislation and standards in the field of spatial data and land administration. According to "the Strategy of establishing a spatial data infrastructure in the Republic of Serbia" [24], the national data infrastructure should be established in accordance with the principles set out in the INSPIRE Directive. The INSPIRE Directive states that all the rules for its implementation should consider the standards that have been developed by the international standardization bodies. Since the cadastral data are part of the spatial data infrastructure, the domain model for the cadastre should be based on the domain model for land administration (LADM) defined within the ISO 19152 standard and the national legislation.

3. Serbian Country Profile for the Real Estate Cadastre

Research was performed using the following steps: analysis of the requirements defined in the Law and the Strategy; analysis of the international standards in the field of research and literature review; analysis of the current cadastral information system, its data dictionary, and data sets; reverse engineering for to transform the physical model to a logical model; and conceptual modeling to capture concepts in the cadastral domain with the aim of developing a standardized domain model.

3.1. Conceptual Model of the Serbian Real Estate Cadastre

The first step in creating a new domain model was the formation of a conceptual model of the real estate cadastre for Serbia. In the current database schema, real properties, rights, and restrictions are grouped by sheet as listed in the real estate folio document, defined by the Law in 1992. Figure 2 shows a conceptual model that represents the contents of A sheet. The A sheet contains data about parcels and parts of parcels, which represent different types of land use within one parcel. Basic attributes for a parcel are the number of the parcel, cadastral municipality, and the number of the real estate folio. According to this, a conceptual model was derived and including the classes of CadastralMunicipality, Parcel, PartOfParcel, and RealEstateFolio. A cadastral municipality is formed as a collection of a certain number of parcels within which the numbers of the parcels range from 1 to n. A cadastral municipality is a territorial unit in which the cadastral survey was completed, covering the area of one city or a village. It is also possible that several villages belong to one cadastral municipality or that parts of one city are in several cadastral municipalities. Parcels are recorded within the real estate folio so that one folio can contain several parcels. A cadastral parcel is a basic cadastral territorial unit and represents a part of the land in the cadastral municipality defined by the border and marked by a unique number on which exists an ownership right.

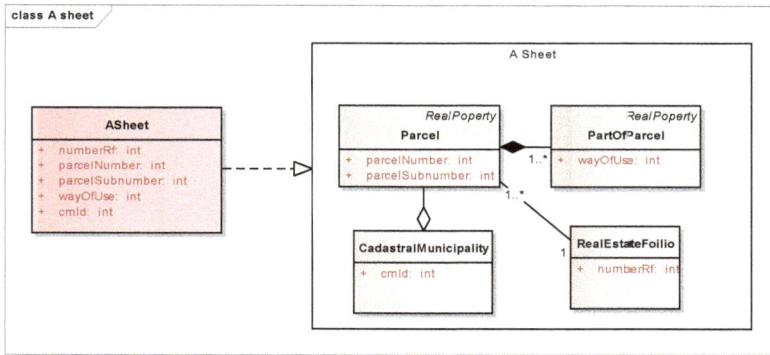

Figure 2. Transformation of the "A" sheet class.

In a similar manner, the other classes in Figure 1 were transformed into the conceptual model. The B sheet was transformed into the classes Party and RightsOverParcel. Party refers to the records of natural and legal parties who have some rights over real properties. RightsOverParcel refers to a specific right bringing together a party, the type of the right, and share of the right, as well as the number of the real estate folio. The V sheet contains data on buildings and parts of buildings together with information about rights to them. A building is identified by a number and the corresponding parcel. This leads to the following classes: Building, PartOfBuilding, and RightsOverBuildingAndPart, which is the rights on a building and its parts. The G sheet contains data about restrictions on real properties. The Restriction class was created for this purpose. Afterward, a generalization of the conceptual model was performed by introducing two abstract classes: RealProperty, representing all three types of real properties, and Rights, which represents the rights on any type of real property.

Finally, the domain model of the real estate cadastre in Serbia was represented as shown in Figure 3. The real estate folio contains one or more real properties from the A and V sheets, one or more rights to the real properties from the B and V sheets, and either no or more restrictions on the real property from the G sheet. One party may have zero or more property rights.

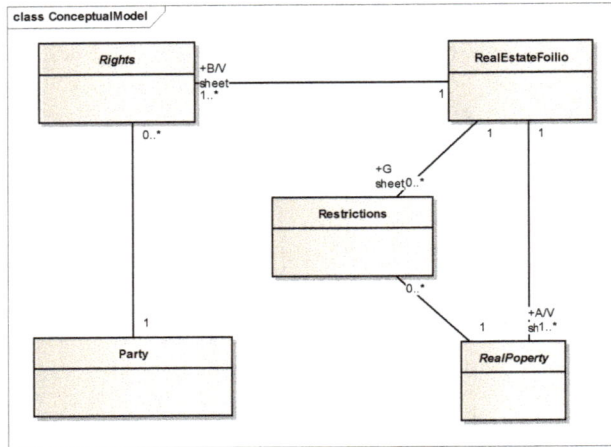

Figure 3. The basic schema of the conceptual model for Serbia.

This model describes the current data structure that is formed under the Law from 1992. According to the Law from 2009, a real estate folio contains a single real property with its rights and restrictions. This affects the previous model by changing the multiplicity on class RealProperty from 1..* to 1.

3.2. Serbian Country Profile for Land Administration

The second step in the establishment of the standardized domain model was to analyze the possibilities of fitting the resulting conceptual model in LADM [25,26]. Figure 4 shows the mapping of the basic LADM classes into the Serbian conceptual model classes [27].

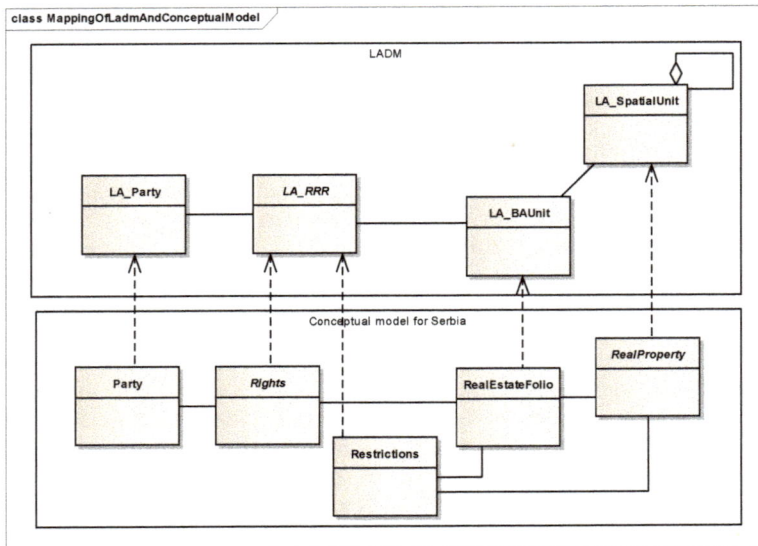

Figure 4. Mapping of basic classes in the Serbian data model to the Land Administration Domain Model (LADM) classes.

Since the LA_Party class describes the right holders, the appropriate class of the conceptual model is the Party class. LA_SpatialUnit describes spatial units, meaning real properties like parcel, building, and part of building, and were mapped to the abstract class RealProperty. LA_RRR describes the rights, restrictions, and responsibilities over a real property. In the conceptual model, two classes were created to describe the rights and restrictions, Rights and Restrictions, and both were derived from the LA_RRR class. The LA_BAUnit class represents a basic administrative unit, which is a set of rights, restrictions, and responsibilities of one or more real properties so that the sum of shares is equal to one. In the conceptual model for Serbia, this is equivalent to the real estate folio, so the corresponding class is RealEstateFolio.

Since the applicability of basic LADM on the Serbian conceptual model has been proven, it was possible to create basic classes corresponding to the Serbian profile (Figure 5). All classes from the Serbian country profile have the prefix RS. The Law defines a real estate folio as data on one real property together with its rights and restrictions. By analyzing the association between RS_SpatialUnit and RS_BAUnit, we concluded that with the current legislation, the cardinality of connection should be one to one. With this change, we accomplished a maximum of one real property being defined in one real estate folio.

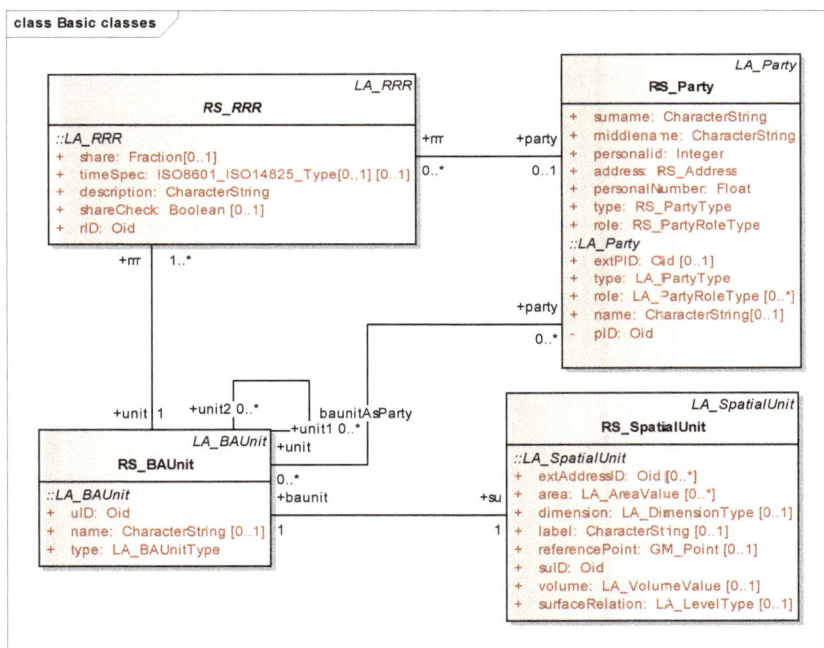

Figure 5. Basic classes in the Serbian profile.

Figures 5–7, 10 and 14 contain a complete country profile for Serbia, followed by instance diagrams (Figures 8, 9, 12, 13 and 19) with examples for characteristic situations. Classes for every package and subpackage are displayed together with code lists. The VersionedObject class was used in the development of country profile to manage and maintain historical data. Inserted and superseded data are given a time-stamp. In this way, the contents of the database could be reconstructed as they were at any historical moment. In the following diagrams, this class will not be shown to save space.

3.2.1. Party Package

The basic class in the Party package is RS_Party, used to describe holders of rights in the cadastral system (Figure 6). In the code list, RS_PartyType refers to the basic administrative unit ("baunit"), group, natural persons, and non-natural persons. In the code list, RS_PartyRoleType includes all the various roles, such as state agency, citizen, notary, etc. Additional attributes in the class RS_Owner are surname, middle name, address, personal identification (personalid), and ID card number (personalNumber). The class RS_GroupParty is a group party which can be "baunit", an association, or a family (RS_GroupPartyType). Group party is used, for example, in a common ownership on the real property in which several parties have the same right with an indefinite share so they act as one party.

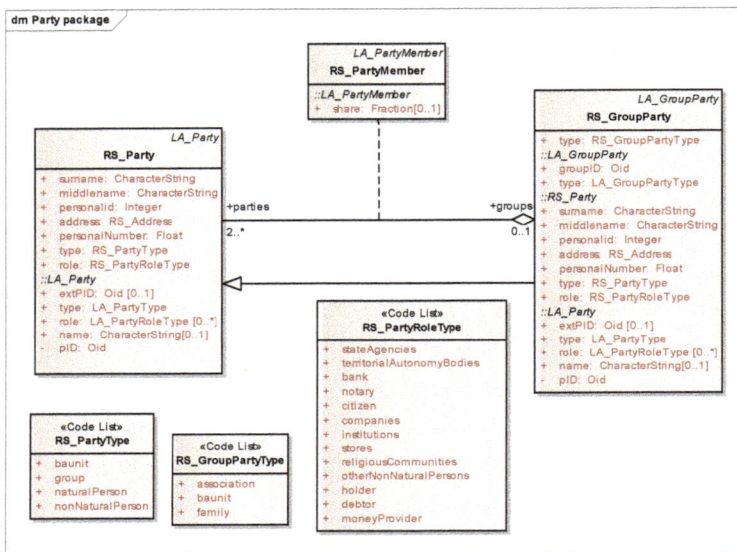

Figure 6. Party package of the Serbian profile.

3.2.2. Administrative Package

Rights, restrictions, "baunits", and administrative sources are defined within the administrative package (Figure 7). RS_RRR defines an abstract class that represents the rights, restrictions and responsibilities, and subclasses RS_Restriction and RS_Right as a such as of this class. The Serbian cadastre does not maintain data about responsibilities like monument maintenance, cleaning a ditch, keeping a snow-free pavement, etc. Such responsibilities are defined within the "Decision on the communal order" for every city itself and are handled by public utility companies, other legal entities, entrepreneurs, and natural persons. To introduce the class LA_Responsibility into the profile, the Law would need to be changed and would require long-term data recording of responsibilities for each real property, so this class will remain outside of the profile. Rights are described by the share and type of the right, which can include ownership, right to use, lease, etc. To describe the type of a share, including a whole right, ideal part, real part, or common, the code list RS_RightShareType was added. The attribute Ownership shows whether a right is an ownership or not. If it is, a type of ownership is selected from the code list RS_OwnershipType (private, public, social, etc.).

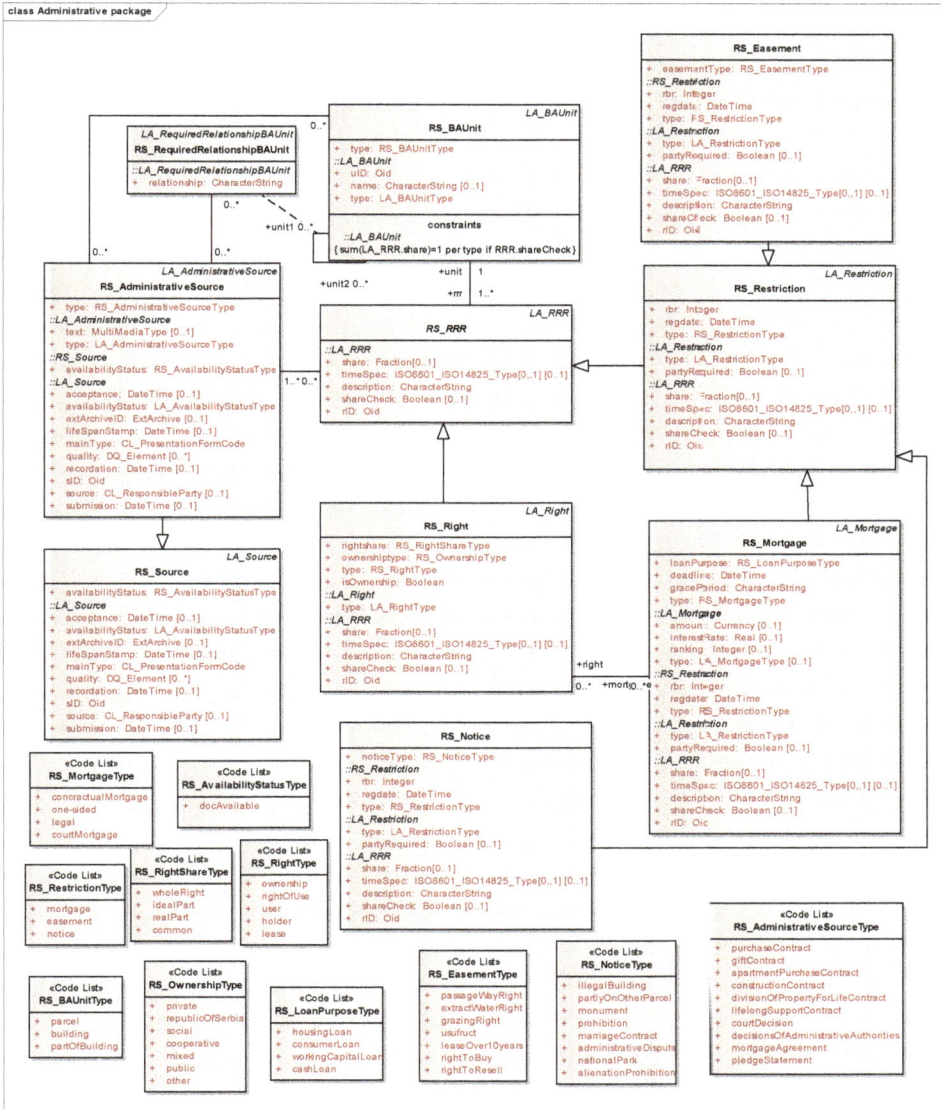

Figure 7. Administrative package of the Serbian profile.

The Law on Property Relations in Serbia [28] recognizes the concept of common ownership, which is an ownership of two or more parties whose shares in the ownership are not determined nor can, as a rule, be alienated while the joint property regime is in place. In addition, for the common parts of the building, such as elevators, roof, attic, stairs, and corridors, owners of apartments, garages, and business offices have the right of common indivisible ownership. Owners of apartments and special parts of the building have the right to use common parts of the building according to their purpose, regardless of the size of their particular part of the building. Such situations are recorded in the Serbian cadastre as common ownership without a defined share of the right. Persons who have common ownership act as a group and can only sell their real property by a common agreement. In the event that an agreement cannot be reached, the division of property is carried out by a court procedure. Figure 8 shows an example of the common ownership of a vineyard by two persons. Class RS_Parcel relates to a spatial unit and is explained in the Section 3.2.3.

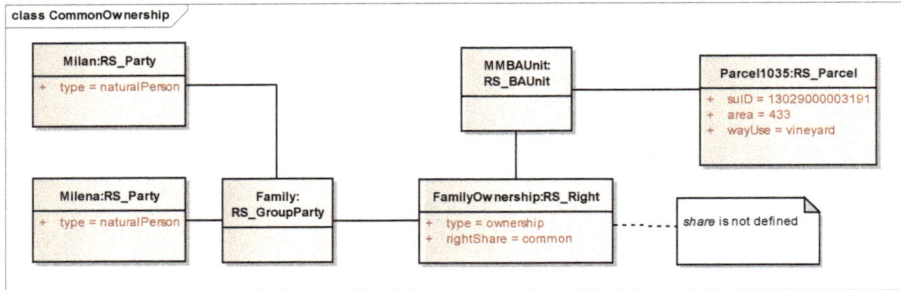

Figure 8. Instance diagram of a common ownership of a vineyard.

Restrictions on real properties in Serbian cadastre are described with ordinal number, description and a type. There are three types of restrictions, mortgages, easements and notices, and therefore three classes, i.e. RS_Mortgage, RS_Easement and RS_Notice.

A mortgage is a special restriction of the ownership right. It involves conveying an interest in a property by a debtor to a creditor, as a security for a financial loan, with the condition that the property is returned, when the loan is paid off. To fully describe a mortgage, additional attributes for deadline registration and a grace period were added. The type of a mortgage is defined in the RS_MortgageType code list. The RS_LoanPurposeType code list contains values for different types of loan purposes. Figure 9 shows a mortgage on an ownership right with a party as the debtor and a bank as the creditor. The mortgage is verified by mortgage agreement created by the notary as a responsible party.

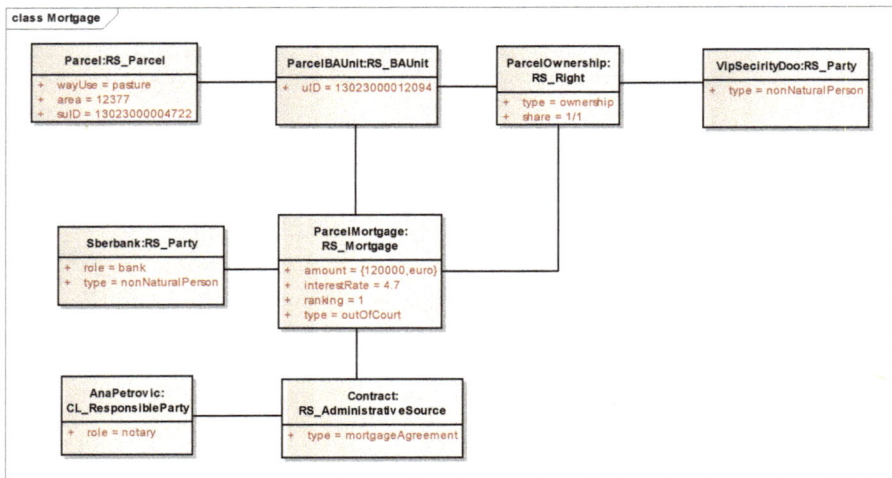

Figure 9. Instance diagram of mortgage on ownership.

Easement is the right of the owner of a real property to use the neighboring real property in a certain way or to require the owner of the neighboring real property to keep from certain behavior. Easement is enrolled as usufruct, a right of passage, as well as other rights prescribed by the Law governing the rights to real properties. Easements are recorded on the real properties in favor of a particular right holder or for the benefit of another real property, regardless of the party with ownership right on that real property. Types of easement are defined in the RS_EasementType code list.

A notice relates to the registration of facts that are of importance for the establishment, modification, termination, or transfer of real rights to real properties, relating to the characteristics of the holder of the right, on the real property or on legal relations regarding the real property. The registration of a notice does not prevent further entries on the real property to which the registration refers. From the moment of registration of the notice, all registrations in the real estate cadastre of the holder, which are contrary to the purpose of the recorded notice, are conditional and depend on the outcome of settling the rights to the real property. Types of notices are defined in the RS_NoticeType code list. A notice is registered when a building is built without all necessary permits, or a part of building lies on another parcel, or when there is a marriage contract between parties, etc.

The RS_BAUnit class represents a set of rights and restrictions on real property so that the total sum of shares is equal to one. In the Serbian cadastre, the concept of "baunit" corresponds to a real estate folio. Since there are three types of real estate folio for parcels, buildings, and parts of buildings, the RS_BAUnitType code list has three corresponding values.

Registration of rights and restrictions on real properties is based on a private or public document, which states if the property is suitable for registration. Registration of real property rights is done on the basis of contract (purchase, gift, purchase of apartment, construction, division of property for life, lifelong support, etc.), court decisions (judgments and solutions), and decisions by the competent administrative bodies. Division of property for life is a special form of gift contract in which the ancestor can renounce and share their property according to their wishes. However, in order for this contract to be valid, all descendants who, according to the law, would be called upon to inherit the ancestry must agree and sign its contents. Registration of a mortgage is done on the basis of a mortgage contract, a pledge statement, or court decisions. These documents are described with the RS_AdministrativeSource class and the RS_AdministrativeSourceType code list.

3.2.3. Spatial Unit Package

Th spatial unit package is based on the RS_SpatialUnitGroup and RS_SpatialUnit classes (Figure 10). The RS_SpatialUnitGroup class is related to a set of spatial objects and can be used to specify the administrative units within the cadastral system, including country, administrative municipality, cadastral district, and cadastral municipality, which are defined in the RS_SpatialUnitGroupType code list.

The RS_SpatialUnit class describes spatial objects in the cadastre, including parcels, buildings, and parts of buildings, as well as utility networks, which are part of the utility network cadastre. A parcel is described by number and sub-number within one cadastral municipality (RS_Parcel). The type of land is determined by the RS_PuposeParcelType code list and refers to whether it is urban construction land, agricultural land, forest land, or public building land. One parcel may consist of several parts with different uses; hence, the RS_PartOfParcel is defined. Since rights and restrictions are defined over the entire parcel, part of the parcel class is a spatial unit within the partOfParcel level, while the parcels, buildings, and parts of the buildings are defined by the ownership level (RS_Level). The types of parcel use include field, garden, orchard, meadow, pasture, the land under the building, the city's green areas, embankments, a pond, an artificial lake, etc. (RS_WayOfUsePartParcel).

The RS_Building class that represents the building requires additional attributes to describe the number of the building, the date of construction, the number of entrances, number of floors underground, on the ground floor, above the ground, and in the attic, legal status (code list RS_LegalStatus), and the building use according to the RS_WayOfUseBuilding code list value (residential buildings, commercial buildings, auxiliary buildings, etc.). For maintaining data about the legal aspects of the parts of buildings, the country profile adds the RS_LegalSpaceBuildingUnit class, with specialization RS_PartOfBuilding, such as an apartment, garage, or business space. A part of a building is described with its number, date of registration, number of rooms, information about the part of the building is located (basement, ground floor, first floor, second floor), and of the type of use of the business space.

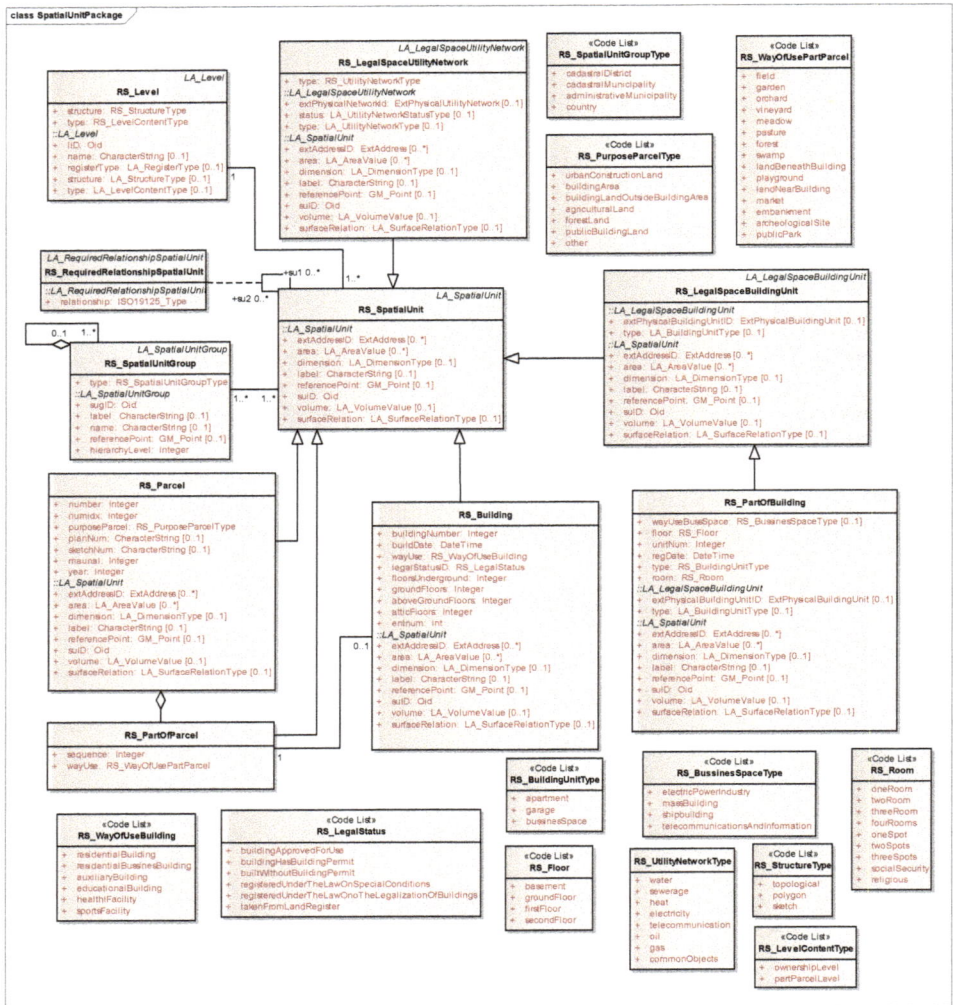

Figure 10. Spatial unit package of the Serbian profile.

The country profile allows the joining of the real estate and utility network cadastre data as required by the Law by introducing the RS_LegalSpaceUtilityNetwork class. The utility network cadastre is a register of the utility network lines together with their rights and restrictions. The types of utility networks are defined within the RS_UtilityNetworkType code list. One type refers to common utility network objects, which are in the current utility network cadastre, used to keep data about underground passages that belong to the real estate cadastre. With the proposed country profile, this can be corrected.

In the real estate cadastre, buildings are normally located within the parcel. Topological relationships between parcel and building can be verified by implementing the ISO 10125-2 standard. Figure 11a shows an example of parcel 1030/2 with one building. An instance of the RS_RequredRelationshipSpatialUnit class was created for the purpose of achieving a topological relationship between the parcel and the building (Figure 12).

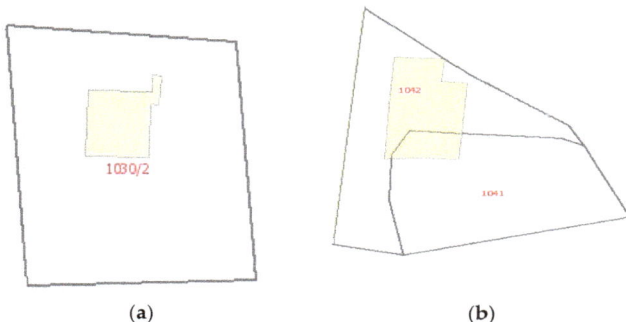

(a)   (b)

Figure 11. (a) Building within a parcel; and (b) building partly on another parcel.

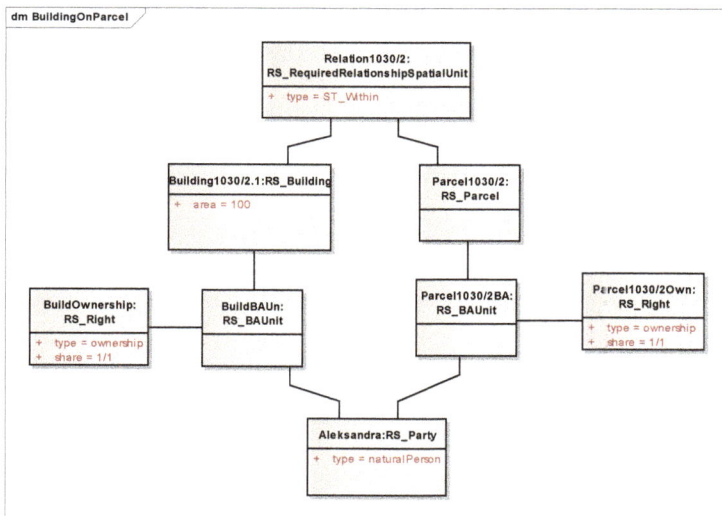

Figure 12. An instance diagram of a building within a parcel.

In practice, it is not uncommon that a building is located on two or more parcels, which would occur when a building was built without all permissions, a building taken from the land register, etc. Figure 11b shows a building which is on parcel 1042, but part of the building crosses parcel 1041. For the building, a notice is recorded that a part of it is located on parcel 1041. In this example, topological relationships cannot be verified by the ST_Within operation, but the connection between the basic administrative units of the building and the other parcel can be defined, using the RS_RequiredRelationshipBAUnit class (Figure 13).

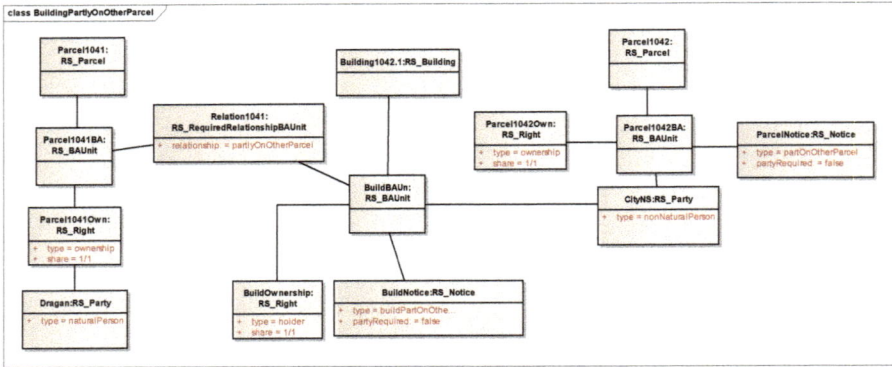

Figure 13. Instance diagram of a building partly on another parcel.

3.2.4. Surveying and Representation Subpackage

Classes in the surveying and representation subpackage are RS_Point, RS_SpatialSource, and RS_BoundaryFaceString (Figure 14). In the cadastral survey procedure, geodetic measurement of real property was carried out according to the actual field situation. The subject of geodetic measurement is the boundary points at the border lines of cadastral municipalities, parcels, parts of the parcel, according to the use type, and buildings, as well as other construction facilities. Data on the geometries of spatial objects can be acquired by using the polar, photogrammetric, global navigation satellite system (GNSS) method, the laser scanning method, and their combination. The individual points or complete spatial units are associated to the RS_SpatialSource class which represents a spatial source. A survey was documented with spatial sources. The types of spatial sources were defined within the RS_SpatialSourceType code list. The spatial profile for Serbia is 2D topological, so the RS_BoundaryFaceString class with GM_MultiCurve type was used for geometries.

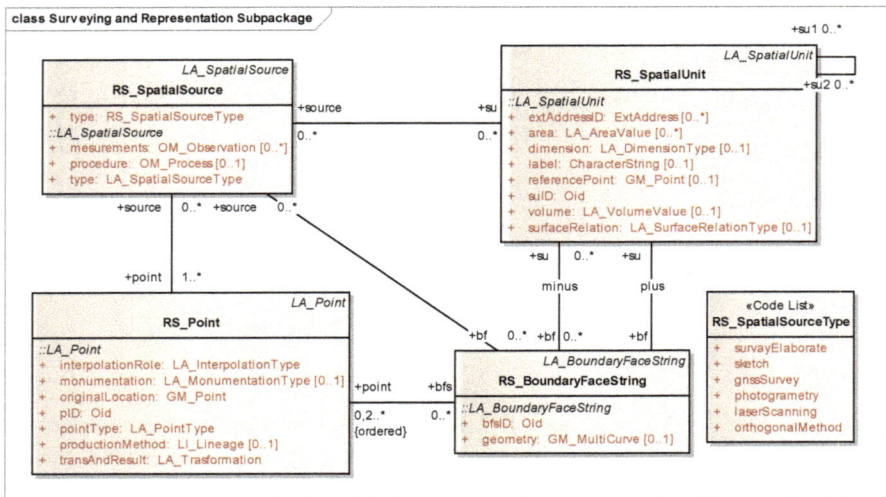

Figure 14. Surveying and representation subpackage of Serbian profile.

3.2.5. Conformity to LADM

The abstract test suite, defined in Annex A of the ISO 19152 standard, specifies the requirements that the development of the country profile has to meet in order to conform to this standard. Three conformance levels are specified per (sub)package: Level 1 (low level), Level 2 (medium level), and Level 3 (high level). A low level of conformance implies that there is at least one class in the country profile that is conformant with the definition of the following classes: LA_AdministrativeSource, LA_Party, LA_RRR, LA_Right, LA_BAUnit, and LA_SpatialUnit. The medium level refers to classes LA_GroupParty, LA_PartyMember, LA_Restriction, LA_SpatialUnitGroup, LA_Level, LA_Point, LA_SpatialSource, LA_BoundaryFaceString, and the corresponding code lists. The high level involves the introduction of classes LA_Responsibility, LA_RequiredRelationshipBAUnit, LA_LegalSpaceBuildingUnit, LA_LegalSpaceUtilityNetwork, LA_RequiredRelationshipSpatialUnit, and LA_BoundaryFace.

Table 1 shows the mapping of the Serbian conceptual model, LADM classes, and the resulting Serbian country profile classes. The level of conformance to LADM is also shown. The Serbian country profile conforms to all classes from the low and medium levels, and to four of the six classes from the high level of conformance.

Table 1. LADM conformance and mapping table for the Serbian country profile.

Serbian Conceptual Data Model	LADM Class	Serbian Profile Class	Conformance Level
Case	LA_Source	RS_Source	1
Party package			
Parties	LA_Party	RS_Party	1
-	LA_GroupParty	RS_GroupParty	2
-	LA_PartyMember	RS_PartyMember	2
Administrative Package			
-	LA_RRR	RS_RRR	1
Rights	LA_Right	RS_Right	1
Restrictions	LA_Restriction	RS_Restriction	2
Restrictions	LA_Restriction	RS_Notice	2
Restrictions	LA_Restriction	RS_Easement	2
-	LA_Responsibility	-	-
RealEstateFolio	LA_BAUnit	RS_BAUnit	1
Restrictions	LA_Mortgage	RS_Mortgage	2
Case	LA_AdministrativeSource	RS_AdministrativeSource	1
-	LA_RequredRelationshipBAUnit	RS_RequredRelationshipBAUnit	3
Spatial Unit Package			
RealEstate	LA_SpatialUnit	RS_SpatialUnit	1
Parcel	LA_SpatialUnit	RS_Parcel	1
PartOfParcel	LA_SpatialUnit	RS_PartOfParcel	1
Building	LA_SpatialUnit	RS_Building	1
CadastralMunicipality	LA_SpatialUnitGroup	RS_SpatialUnitGroup	2
PartOfBuilding	LA_LegalSpaceBuildingUnit	RS_PartOfBuilding	3
Pipeline	LA_LegalSpaceUtilityNetwork	RS_LegalSpaceUtilityNetwork	3
-	LA_Level	RS_Level	2
-	LA_RequredRelationshipSpatialUnit	RS_RequredRelationshipSpatialUnit	3
Surveying and Representations Subpackage			
-	LA_Point	RS_Point	2
Case	LA_SpatialSource	RS_SpatialSource	2
-	LA_BoundaryFaceString	RS_BoundaryFaceString	2
-	LA_BoundaryFace	-	-

4. Discussion and Possibilities for Implementation of 3D Cadastre in Serbia

Current law and the national strategy are looking at measures to overcome the mentioned problems in the information system and do not mention any possible introduction of a 3D component to the cadastre. From the perspective of the Law, our cadastral registration method is sufficient for registering the ownership rights of the buildings in its 2D structure. However, the 2D registration system is facing limitations due to the increasing construction of buildings with more complex structures and the need is growing for the registration of complex ownership rights. In Serbia, many such cases exist, for which different workarounds are used in order for the complex rights to fit into the 2D cadastre. In some cases, neither buildings nor rights have been recorded in the cadastre. In other cases, the data in 2D differ from the appearance of the building in reality, and only a notice has been recorded to indicate the existence of a real property. The overview of some of these situations will be given in the following text.

Figure 15a shows the underground passage below the road in the center of Novi Sad. A number of shops and stores exist in the underground passage. The real estate cadastre only keeps the data about road above the underground passage and the rights to it, while the data about the passage and its rights are not recorded, as shown on the cadastral map (Figure 15b). The data on the underground passage are recorded in the utility network cadastre. According to the Law, the data on all underground buildings should be a part of the real estate cadastre.

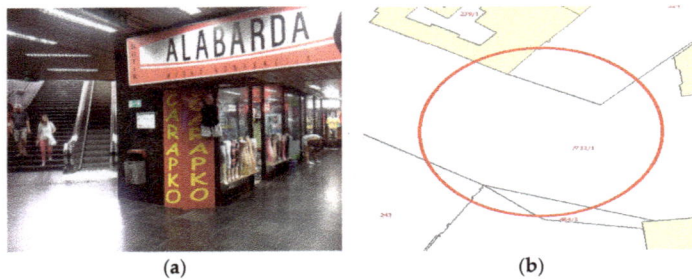

(a)   (b)

Figure 15. (**a**) Underground passage; and (**b**) the cadastral map for the area of the underground passage.

Figure 16a shows skyscrapers, named Geneks Towers in Belgrade, which are connected on the 26th floor with a two-story passage. A rotating restaurant, currently in the reconstruction stage, is located on the top of one of the towers. On the 2D cadastral map, the towers are shown as two separate buildings, and they are recorded as such in the information system (Figure 16b). The notice on one of the towers says that a business space above 26th floor exists, while parts of the building are not listed.

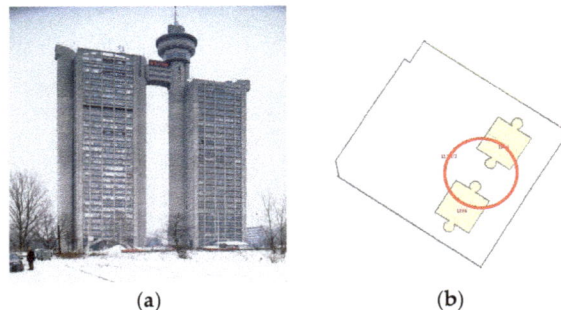

(a)   (b)

Figure 16. (**a**) Geneks Towers (adapted from [29]); and (**b**) the cadastral map for Geneks Towers.

Another example of overlapping rights that the 2D cadastre cannot properly record is the Petrovaradin fortress. It is located on the bank of the Danube River near the city of Novi Sad. The real estate cadastre records the parcels and buildings that are located on the fortress. Figure 17a shows the fortress itself and Figure 1 shows the cadastral map of the entire area of the fortress. It is divided into many small parcels on top of the fortress, while the fortress itself is not located on the parcel, and it is not recorded as an object (building). However, this is an excellent example of the complexity of the concept of a "building".

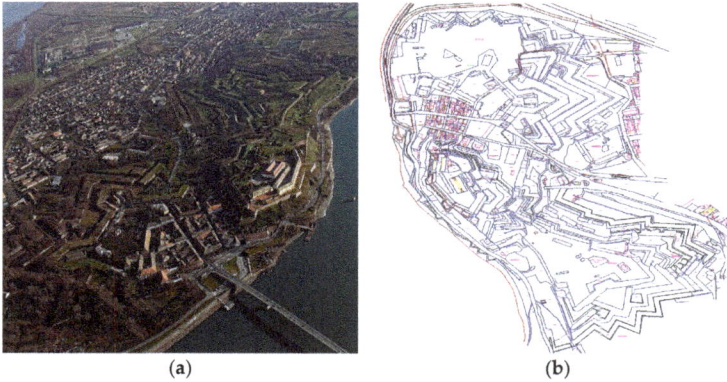

Figure 17. (a) The Petrovaradin fortress (adapted from [30]); and (b) cadastral map of the Petrovaradin fortress.

The Petrovaradin fortress, an exceptional achievement of the fortification architecture of the eighteenth century, is one of the most complex, largest, and best-preserved baroque artillery bastion fortresses in Southeast Europe. Today's fortress extends over an area of over 105 ha, with the total length of the outer defense line of 5500 m, and a complex system of underground military tunnels and galleries. The underground galleries are made in four levels, interconnected by a series of horizontal and vertical communication corridors. The length of these communications and underground chambers is over 16,000 m. The Petrovaradin fortress was completely demilitarized in the mid-twentieth century. It was proclaimed to be a historical monument and a cultural asset of great importance, was placed under state protection, and assigned to civilian use. On the fortress and in the underground tunnels and chambers, the Museum and Archives of Novi Sad, hotels and restaurants, ateliers and galleries of fine artists, an arts academy, astronomical observatory, planetarium, etc. are located [31]. Figure 18 shows a sketch displaying how the tunnels and chambers are distributed within the fortress (Figure 18a), followed by images of the underground military galleries (Figure 18b), and art studios (Figure 18c), that are now located inside the fortress. The fortress is owned by the city of Novi Sad, while the real properties on it and in it are leased. The leases are initiated by a lease contract, and according to the Law, the lease should be registered in the real estate cadastre. Due to the lack of records on the real property itself within the fortress, the legal right of tenants to enter their contract into cadastral records is denied.

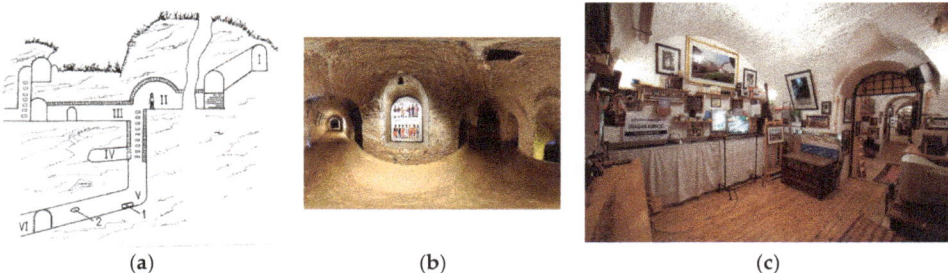

Figure 18. (a) Sketch of an underground tunnel and chambers (adapted from [32]); (b) underground tunnels (adapted from [33]); and (c) underground art gallery of Dragan Kuručić (adapted from [34]).

In the case of the Petrovaradin fortress, the problem of recording the rights to the tunnels and chambers can be solved by using the RequiredRelationshipBAUnit class in the country profile. Due to the inability of the 2D cadastre to record overlapping rights in 3D, and because the area of the fortress is divided into numerous parcels, data on the fortress itself, and its internal parts, are not recorded in the cadastre. If the land on which the fortresses is located is declared a large parcel, then duplication of the cadastral area would occur because the area on top of the fortress is divided into many small parcels within which the hotels, restaurants, galaries, and other buildings are located. Merging all the small parcels into one big parcel is also not a good solution, since there are many buildings located on the fortress. A solution can be reached if the fortress is presented as a building, and the tunnels and chambers as parts of the building. For each of them, the basic administrative unit can be defined. The fortress can be associated with the "baunits" of all the small parcels by using an instance of the RS_RequiredRelationshipBAUnit with the partlyOnOtherParcel type. In this way, the total area of the cadastral municipality remains the same, and the problem of recording rights on the fortress and its special parts is solved. However, the problem of recording data on the fortress in the cadastral map remains. Figure 19 shows the registration of ownership and the lease of the underground chambers.

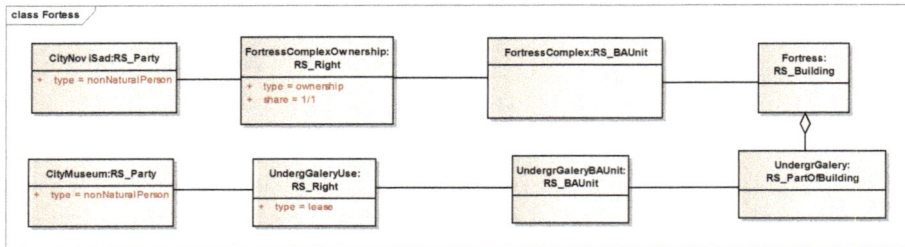

Figure 19. Instance diagram of registering rights on underground gallery.

Based on the previous cases, we concluded that a need for the introduction of 3D into the Serbian cadastre exists. The International Federation of Surveyors (FIG) joint commission 3 and 7 Working Group on 3D Cadastres created a set of questionnaires and sent them to the working group participants to make a world-wide inventory of the status of 3D cadastres at the current moment and the plans for the future. From the questionnaire filled by Serbian experts [35], we concluded that the work on the 3D cadastre would not be planned by 2018. However, if the measures of the Strategy are implemented, it would be possible to consider it in the near future. The FIG Working group on 3D Cadastre proposes several options for creating a 3D model of the cadastre, so that different countries can choose different optimal solutions, depending on the actual situation and conditions: minimalistic, topographic, polyhedral legal, non-polyhedral legal, and topological [36]. The criteria for applying these solutions are based on four key areas of the 3D cadastre: Legal aspects of 3D, initial registration of the 3D parcel, 3D data management, and visualization of 3D objects. These criteria should be further analyzed to chose the right solution for the Serbian cadastre.

ISO 19152 provides support for 3D representations and 3D registration of rights, so the creation of a country profile for Serbia is a starting point toward the development of a 3D cadastre. Given the existence of buildings with overlapping rights and restrictions in 3D, due to their shape or position, considering the expansion of the spatial profile with 3D geometries in such situations is necessary. In Serbia, the 2D cadastre is still not fully formed but it should not be an obstacle to introducing 3D into the cadastre. The availability of production and data collection technologies enables collecting 3D data for these buildings in CityGML format. It is possible to map LADM, especially its representation of legal spaces, to and encode as a CityGML Application Domain Extension (ADE) [37]. Building Information Modeling (BIM) is used for planning, designing, construction, and management of modern residential and business buildings and infrastructure. These data could also be used in the cadastre for the registration of real properties, rights, and restrictions [38,39]. One way of doing this is to introduce external links, with which a 3D view of data could be shown. For example, by creating an association between the RS_Building class from the country profile and AbstractBuilding from the CityGML conceptual model [40]. Another and more complex way would be to expand the Serbian country profile with the RS_BoundaryFace class, and to transform 3D data in this model. In both ways, using the 2D model for simple rights situations, and 3D model for more complex situations and buildings which already have geometries in 3D, would be possible. Whether 3D buildings are stored in the database or not, the consistent maintenance of 3D data should be considered a high priority.

5. Conclusions

Considering the cadastral information system in Serbia is outdated and not satisfactorily functional, there is a requirement, at the country level by the Government, to replace it and develop a new one. We propose that the new system be based on LADM, and have proven our proposal by developing a LADM-based domain model of the cadastre in Serbia. The current Serbian cadastral data organization fits into a LADM-based data model without losing any relevant information, and the conformance with LADM was determined. Using LADM will not only provide interoperability, but also international experience with the goal of providing better service to end users by providing more accurate data in a timely manner, which is not the case with the current system.

The authors analyzed the possibilities of developing the aspects of a 3D cadastre. Future work will include a thorough analysis of all relevant use cases, data acquisition techniques, and possible formats for creation, submission, and registration of 3D spatial units and their rights, storage, and analyses of the 3D data, 3D cadastral visualization, semantics, etc. The aim is to lay the foundation for the development of a legal framework and formal procedures that will enable the use of modern geospatial technologies, which are gaining momentum in this country, but are not yet fully accepted in regular practice.

Author Contributions: Aleksandra Radulović conceived the outline of the article and is the first author of the article. She developed the domain model based on the experience gained through collaboration with Serbian Republic Geodetic Authority on the cadastral information system development. Dubravka Sladić is the co-author of Sections 2–4 and author of Sections 1 and 5. She also helped develop the model through thorough analysis of the data model and system currently in use. Miro Govedarica was the project lead and influenced the work by suggesting the key points.

Conflicts of Interest: The authors declare no conflict of interest.

References

1. Land Administration Guidelines. United Nations: New York, NY, USA; Geneva, Switzerland, 1996. Available online: http://www.unece.org/fileadmin/DAM/hlm/documents/Publications/land.administration.guidelines.e.pdf (accessed on 25 August 2017).
2. Dale, P.; McLaughlin, J. *Land Administration*; Oxford University Press: Oxford, UK, 1999.
3. Sladić, D.; Radulović, A.; Govedarica, M. Cadastral Records in Serbian Land Administration. In Proceedings of the FIG Working Week 2017, Helsinki, Finland, 29 May–2 June 2017.
4. ISO 19152:2012 Geographic information—Land Administration Domain Model (LADM). Available online: https://www.iso.org/standard/51206.html (accessed on 24 August 2017).
5. Lemmen, C.; Van Oosterom, P.; Bennett, R. The Land Administration Domain Model. *Land Use Policy* **2015**, *49*, 535–545. [CrossRef]
6. Van Oosterom, P.; Lemmen, C. The Land Administration Domain Model (LADM): Motivation, standardisation, application and further development. *Land Use Policy* **2015**, *49*, 527–534. [CrossRef]
7. Kalantari, M.; Dinsmore, K.; Urban-Karr, J.; Rajabifard, A. A roadmap to adopt the Land Administration Domain Model in cadastral information system. *Land Use Policy* **2015**, *49*, 552–564. [CrossRef]
8. Stoter, J.; van Oosterom, P. *3D Cadastre in an International Context. Legal, Organizational, and Technological Aspects*; CRC Press: Boca Raton, FL, USA, 2006.

9.   Van Oosterom, P. Research and development in 3D cadastres. *Comp. Environ. Urban Syst.* **2013**, *40*, 1–6. [CrossRef]

10. Bydłosz, J. The application of the Land Administration Domain Model in building a country profile for the Polish cadastre. *Land Use Policy* **2015**, *49*, 598–605. [CrossRef]

11. Janečka, K.; Souček, P. A Country Profile of the Czech Republic Based on an LADM for the Development of a 3D Cadastre. *ISPRS Int. J. Geo-Inf.* **2017**, *6*, 143. [CrossRef]

12. Vučić, N.; Markovinović, D.; Mičević, B. LADM in the Republic of Croatia-making and testing country profile. In Proceedings of the 5th FIG International Land Administration Domain Model Workshop 2013, Kuala Lumpur, Malaysia, 24–25 September 2013.

13. Mađer, M.; Matijević, H.; Roić, M. Analysis of possibilities for linking land registers and other official registers in the Republic of Croatia based on LADM. *Land Use Policy* **2015**, *49*, 606–616. [CrossRef]

14. Psomadak, S.; Dimopoulou, E.; Van Oosterom, P. Model driven architecture engineered land administration in conformance with international standards—illustrated with the Hellenic Cadastre. *Open Geospat. Data Softw. Stand.* **2016**. [CrossRef]

15. Stoter, J.; Ploeger, H.; van Oosterom, P. 3D cadastre in the Netherlands: Developments and international applicability. *Comp. Environ. Urban Syst.* **2013**, *40*, 56–67. [CrossRef]

16. Drobež, P.; Fras, M.K.; Ferlan, M.; Lisec, A. Transition from 2D to 3D real property cadastre: The case of the Slovenian cadastre. *Comp. Environ. Urban Syst.* **2017**, *62*, 125–135. [CrossRef]

17. Lee, B.; Kim, T.; Kwak, B.; Lee, Y.; Choi, J. Improvement of the Korean LADM country profile to build a 3D cadastre model. *Land Use Policy* **2015**, *49*, 660–667. [CrossRef]

18. Official Gazette of the Republic of Serbia. The Law on State Survey and Cadastre. 2009. Available online: http://paragraf.rs/propisi/zakon_o_drzavnom_premeru_i_katastru.html (accessed on 24 August 2017).

19. Official Gazette of the Republic of Serbia. Amendments to the Law on State Survey and Cadastre. 2015. Available online: http://www.rgz.gov.rs/web_preuzimanje_datoteka.asp?LanguageID=3&FileID=1793 (accessed on 24 August 2017).

20. Official Gazette of the Republic of Serbia. The Rulebook on Cadastre Survey and Real Estate Cadastre. 2016. Available online: http://www.rgz.gov.rs/web_preuzimanje_datoteka.asp?LanguageID=3&FileID=1840 (accessed on 24 August 2017).

21. Official Gazette of the Republic of Serbia. The Strategy of Measures and Activities for Increasing the Quality of Services in the Field of Geospatial Data and Registration of Real Property Rights in the Official State Records. 2017. Available online: http://www.rgz.gov.rs/web_preuzimanje_datoteka.asp?LanguageID=3&FileID=2444 (accessed on 25 August 2017).

22. Official Gazette of the Republic of Serbia. The Law on State Survey and Cadastre and Registration of Rights on Real Estates. 1992. Available online: http://notarisrbija.rs/wp-content/uploads/2014/09/zakon_katastar_nepokretnosti.pdf (accessed on 24 August 2017).

23. Official Journal of the European Union. Directive 2007/2/EC of the European Parliament and of the Council of 14 March 2007 Establishing an Infrastructure for Spatial Information in the European Community. Available online: https://inspire.ec.europa.eu/documents/directive-20072ec-european-parliament-and-council-14-march-2007-establishing (accessed on 25 August 2017).

24. Official Gazette of the Republic of Serbia. The Strategy of Establishing a Spatial Data Infrastructure in the Republic of Serbia for the Period 2010–2012. Available online: http://www.geosrbija.rs/DownloadFile.aspx?fileID=51 (accessed on 25 August 2017).

25. Radulović, A. Domain and Service Model for Real Estate Cadastre Geoinformation System. Ph.D. Thesis, Faculty of Technical Sciences, Novi Sad, Serbia, 16 July 2015.

26. Radulović, A.; Sladić, D.; Govedarica, M. Serbian Profile of the Land Administration Domain Model. In Proceedings of the FIG Working Week 2017, Helsinki, Finland, 29 May–2 June 2017.

27. Sladić, D.; Govedarica, M.; Pržulj, Đ.; Radulović, A.; Jovanović, D. Ontology for real estate cadastre. *Surv. Rev.* **2013**, *45*, 357–371. [CrossRef]

28. Official Gazette of the Republic of Serbia. The Law on Property Relations in Serbia. Available online: http://paragraf.rs/propisi/zakon_o_osnovama_svojinskopravnih_odnosa.html (accessed on 25 August 2017).

29. Photo by Błażej Pindor. Available online: https://commons.wikimedia.org/w/index.php?curid=7686393 (accessed on 27 August 2017).

30. Photo from the Official Page of the Institute for the Protection of Cultural Monuments is Novi Sad. Available online: http://www.zzskgns.rs/wp-content/uploads/2014/07/IMG_0048xcxxVECAxxcisto.jpg (accessed on 27 August 2017).

31. Institute for the Protection of Cultural Monuments is Novi Sad. Available online: http://www.zzskgns.rs/ (accessed on 25 August 2017).

32. The Official Site of Researcher and Inventor Veljko Milković. Available online: http://www.veljkomilkovic.com/Images/Podzemlje_clip_image002.jpg (accessed on 27 August 2017).

33. The Official Site of Panacomp Wonderland Travel. Available online: http://www.panacomp.net/wp-content/uploads/2016/03/Vojne-galerije-site-1.jpg?x23386 (accessed on 27 August 2017).

34. Photo by Dragan Kurucić. Available online: http://www.panacomp.net/wp-content/uploads/2016/05/featured-S-17-2.jpg?x23386 (accessed on 27 August 2017).

35. Questionnaire 3D-Cadastres: Status September 2014 Serbia. Available online: http://www.gdmc.nl/3dcadastres/participants/3D_Cadastres_Serbia2014.pdf (accessed on 25 August 2017).

36. FIG Joint Commission 3 and 7 Working Group on 3D Cadastres. Available online: http://www.gdmc.nl/3dcadastres/realization/ (accessed on 25 August 2017).

37. Rönsdorf, C.; Wilson, D.; Stoter, J. Integration of Land Administration Domain Model with CityGML for 3D Cadastre. In Proceedings of the 4th International Workshop on 3D Cadastres, Dubai, UAE, 9–11 November 2014.

38. Oldfield, J.; van Oosterom, P.; Quak, W.; van der Veen, J.; Beetz, J. Can Data from BIMs be Used as Input for a 3D Cadastre? In Proceedings of the 5th International FIG 3D Cadastre Workshop, Athens, Greece, 18–20 October 2016.

39. Ho, S.; Rajabifard, A.; Roić, M. Towards 3D-enabled urban land administration: Strategic lessons from the BIM initiative in Singapore. *Land Use Policy* **2016**, *57*, 1–10. [CrossRef]

40. Góźdź, K.; Pachelski, W.; van Oosterom, P.; Coors, V. The Possibilities of Using CityGML for 3D Representation of Buildings in the Cadastre. In Proceedings of the 4th International Workshop on 3D Cadastres, Dubai, UAE, 9–11 November 2014.

International Journal of
Geo-Information

MDPI

Article

INTERLIS Language for Modelling Legal 3D Spaces and Physical 3D Objects by Including Formalized Implementable Constraints and Meaningful Code Lists

Eftychia Kalogianni [1,2,*], Efi Dimopoulou [1], Wilko Quak [2], Michael Germann [3], Lorenz Jenni [4] and Peter van Oosterom [2]

[1] School of Rural and Surveying Engineering, National Technical University of Athens, 15780 Athens, Greece; efi@survey.ntua.gr

[2] Faculty of Architecture and the Built Environment, Delft University of Technology, 2600 GA Delft, The Netherlands; c.w.quak@tudelft.nl (W.Q.); P.J.M.vanOosterom@tudelft.nl (P.v.O.)

[3] infoGrips Informationssysteme GmbH, 8005 Zürich, Switzerland; michael.germann@infogrips.ch

[4] BSF Swissphoto AG, 8105 Regensdorf-Watt, Switzerland; lorenz.jenni@bsf-swissphoto.com

* Correspondence: efkaloyan@gmail.com; Tel.: +30-6944325903

Received: 31 August 2017; Accepted: 16 October 2017; Published: 21 October 2017

Abstract: The Land Administration Domain Model (LADM) is one of the first ISO spatial domain standards, and has been proven one of the best candidates for unambiguously representing 3D Rights, Restrictions and Responsibilities. Consequently, multiple LADM-based country profile implementations have been developed since the approval of LADM as an ISO standard; however, there is still a gap for technical implementations. This paper summarizes LADM implementation approaches distilled from relevant publications available to date. Models based on land administration standards do focus on the legal aspects of urban structures; however, the juridical boundaries in 3D are sometimes (partly) bound by the corresponding physical objects, leading to ambiguous situations. To that end, more integrated approaches are being developed at a conceptual level, and it is evident that the evaluation and validation of 3D legal and physical models—both separately and together in the form of an integrated model—is vital. This paper briefly presents the different approaches to legal and physical integration that have been developed in the last decade, while the need for more explicit relationships between legal and physical notions is highlighted. In this regard, recent experience gained from implementing INTERLIS, the Swiss standard that enables land information system communications, in LADM-based country profiles, suggests the possibility of an integrated LADM/INTERLIS approach. Considering semantic interoperability within integrated models, the need for more formal semantics is underlined by introducing formalization of code lists and explicit definition of constraints. Last but not least, the first results of case studies based on the generic LADM/INTERLIS approach are presented.

Keywords: 3D cadastre; land administration; legal and physical space; INTERLIS modelling language; constraints; code lists; spatial data modelling; Land Administration Domain Model (LADM); ISO 1952; Model Driven Architecture (MDA)

1. Introduction

1.1. Background

Over the past couple of decades, rapid urbanization has led to an increasing demand and pressure for land development, resulting in the division of property ownership so that different owners can have property interests in limited space on, above or below the ground surface. This means that the built environment is becoming more and more spatially complex. Currently, land administration practices mainly rely on 2D-based systems to define the legal boundaries of legal interests. In response to the challenges regarding registration of multi-level properties, the feasibility of 3D information and 3D digital models is being investigated, and their need is often underlined.

In the context of the significant developments in computer graphics in terms of modelling and rendering 3D models of urban structures in different formats and Levels of Details (LoDs), the research on the integration of legal and physical notions of objects has received significant attention. Legal information refers to legal interests, legal boundaries and legal attributes, while it is also a prerequisite for the management of Rights, Restrictions and Responsibilities (RRRs) maintained by cadastral data models (LADM, ePlan). On the other hand, data associated with physical objects is characterized by geometric and semantic information in various LoDs, maintained by physical data models (CityGML, BIM/IFC, IndoorGML), which do not support legal or cadastral information. Researchers investigate the integration between legal and physical data models to simultaneously manage both legal and physical dimensions of 3D RRR data, as presented in Section 2.2.

Moreover, automatic and semi-automatic methods based on Conceptual Schema Languages (CSL) offering direct implementable model descriptions of the corresponding conceptual models have been developed. The transformation of the logical model into a physical database provides a better understanding of the model at a conceptual level, while revealing its strengths and limitations. Over the years, a large number of projects have been conducted—mostly within the academic discourse, but also in the industry—in which the technical implementation of LADM country profiles has been investigated. In the last couple of years, several projects have suggested describing the LADM country profiles with the conceptual schema language INTERLIS, as they both share the same Model-Driven Architecture principles [1], as also described in Sections 2.3, 4.1 and 4.3.

An important step in this direction is the formulation of constraints, which are often initially described in natural language; however, experience has shown that this results in ambiguities. Formal description of constraints is a complex task that needs to be undertaken in the early stages of conceptual design in order to avoid higher costs, as it has negative effects when added too late. The consistency of (spatial) data can be checked if the underlying constraints are properly modelled and enforced [2]. UML diagrams, together with Object Constraint Language (OCL) notation, have been found to be a suitable tool for expressing the designed constraints. However, constraints defined in OCL cannot be automatically converted into implementations.

Various formalization approaches applicable to (3D) geo-constraints do exist today, and they pave the way for extending the current state-of-the-art of constraint modelling into a higher dimension; a brief review of them is presented in Section 2.4. Furthermore, the importance of semantics in land administration, mostly used to further provide explicit meaning to code list values in a more refined manner, is constantly underlined in various research projects, as briefly presented in Section 2.5.

It should be noted that this paper is part of a wider research study, some of the results of which have already been presented in previous publications and conferences [3–9]. The paper covers a literature review of the whole spectrum of related work, emphasizing the role and use of the INTERLIS language at an international level. Additionally, it explores the possibilities of linking 3D legal RRR spaces, modelled with Land Administration Domain Model (LADM) [10], with their physical counterparts, described as 3D objects.

To this end, standardization techniques are used in order to explicitly define data models including their constraints, and also to test the performance of advanced technological tools in terms of consistency and integrity.

1.2. Scope and Methodology

The need to link the legal and physical notions of objects in order to support 3D applications in the land administration domain, reflecting the interrelations of those two aspects, is presented in this paper. As this topic is broad enough, the paper is structured in a way to cover the wide range of research objectives that need to be addressed; i.e., current best practices and approaches towards this investigation, use of standards in order to realize them, formalisation of constraints, semantically enriched definition of code lists, etc.

Focus is placed on the INTERLIS conceptual language and the corresponding software packages, as well as the implementation of INTERLIS formalism in the LADM standard, specifically in terms of the standard's country profiles for Switzerland, Greece and Colombia. Finally, the paper addresses an approach for enabling the representation of the relationships between physical and legal boundaries, and presents some preliminary results. In order to address the afore-mentioned research objectives, the methodology illustrated in Figure 1 was followed.

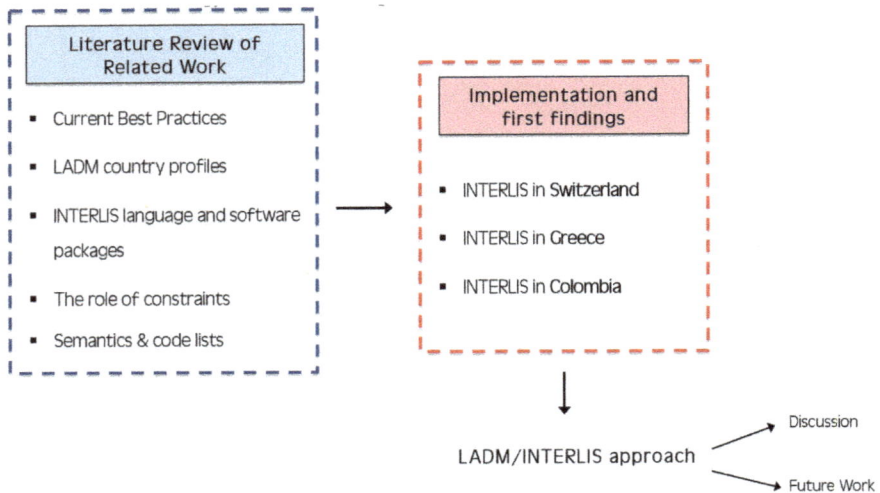

Figure 1. Methodology used to address research objectives.

Following those steps, the rest of the paper is organized as follows: in the next Section, a literature review is presented covering the whole spectrum of necessary background of research information and practices. Section 3 presents INTERLIS, the Swiss standard for land information systems communication and the corresponding software package. The integrated LADM/INTERLIS approach, as well as the knowledge gained from its implementation approaches in three countries—Switzerland, Greece and Colombia—is discussed in Section 4. The proposed structures for semantically enriched code lists, formalization of constraints and 3D data types in INTERLIS are analysed in Section 5, together with some preliminary results from the prototype implementation of the generic LADM/INTERLIS approach. Finally, conclusions are presented in Section 6, focused on addressing limitations and opportunities that arise from this research, while the last Section is dedicated to discussion and recommendations for future work discussing future trends in this domain.

2. Related Work

The necessary background information to this research is briefly presented in the following subsections. The need for the third dimension in the field of land administration is underlined in Section 2.1, while current approaches to and initiatives for the integration of legal and physical concepts that have been carried out in recent years are briefly presented in Section 2.2, followed by examples of MDA-based applications implementing LADM profiles from different countries (Section 2.3). The role of constraints in data modelling is highlighted in Section 2.4 and, in the last subsection, the significance of semantics in land administration is presented with reference to the use and management of code lists.

2.1. The Third Dimension in the Field of Land Administration

The inherent challenges of the third dimension (3D) in the field of geo-technology, as well as the growing importance of 3D modelling in the complete development of land administration activities, are constantly being underlined, along with their need to be managed, resolved and their benefits maximized. As stated in [11], "3D reality needs 3D design, engineering and analyses"; thus, multiple approaches for 3D data acquisition and modelling, 3D data representation, 3D data storage (in spatial databases), 3D file formats for data exchange, and web services that support 3D features have been developed. Indicatively, in 2004, the United States Department of Labour [12] predicted geo-technology to be one of the three "mega-technologies" of the new millennium that promised to drive radical changes in society. Additionally, smart cities are being mapped directly in 3D with buildings being represented at several LoDs; hence, it is worth considering that such data could potentially be reused for land administration purposes.

While registrations and RRRs are necessary elements in land administration systems, the central element is the land parcel, which defines a 3D volume of legal space, including its physical features. For developments above and below the ground, a 2D land parcel is no longer an appropriate basic spatial component for cadastral models, as it cannot adequately represent 3D properties [13]. The increasing complexity of infrastructures and densely built-up areas requires proper registration of the legal status of properties and their spatial components. As stated in [14], "the insight into the third dimension of physical objects helps to understand the location and size of the legal spaces, as well as it is relevant in the context of developing multipurpose cadastral systems".

3D cadastres represent the division of 3D geo-space, and many urban features—such as buildings, open spaces, transportation infrastructure, under/above ground constructions, overcrossing buildings, etc.—are considered 3D cadastral entities [15]. As stated by the authors, a 3D cadastral entity (land, marine, air, underground, etc.) is a synthesis of geometry, attributes and social and legal semantics, which is built by processing constructed 3D objects and topological reconstruction, and incorporating semantic information of both the legal space and the physical component.

Furthermore, 4D and 5D modelling are on the horizon, if not already here, which include the time dimension in both spatial and legal aspects, as well as adding different scales of geo-data, especially in the case of physical features. The research of Oosterom et al. [16] introduced a conceptual full partition of 3D space, including time and scale without overlaps or gaps, and realized in a true 5D generic model, providing a sustainable and solid foundation for the Geo Information Infrastructure (GII).

2.2. Legal and Physical Reality: Towards Integrated Approaches

Standards have been widely used, since they provide efficiency and support in communication between organizations and countries, as well as facilitating system development and data exchange based on common terminology. In the field of land administration, multiple standards have been developed nationally, regionally and internationally in the form of domain-specific standardization, which is necessary for capture the semantics of this field. Integrating such models in order to support registrations and land administration procedures utilizing Geographic Information Systems, along with Data Base Management Systems and applications, for the purpose of implementing land policy measures, is vital.

Legal cadastral data models (defining legal spaces), such as LADM [10] and ePlan [17], are mainly designed to manage and maintain property interests, and do not integrate their physical counterparts [18]. Furthermore, INSPIRE Data specifications on Cadastral Parcels (D2.8.I.6) and INSPIRE Data specifications on Buildings (D2.8.III.2) have been prepared in a way that supports compatibility with LADM. However, Cadastral Parcels focuses on the geometric aspect, without taking into consideration the parties and the associated RRRs applied to it. Moving from land to the sea, the proposed S-121 Maritime Limits and Boundaries Standard [19], built upon ISO 19152, provides a detailed conceptual model description for marine administration based on common national and international requirements.

On the other hand, purely physical models usually manage and store spatial and, sometimes, semantic information at various LoDs; the most frequently used physical data models in the geoinformation field are IFC (ISO 16736:2013) [20,21], OGC CityGML [22], OGC LandInfra/InfraGML [23,24], LandXML [25], and OGC IndoorGML [26]. Among those models, CityGML and IndoorGML provide a comprehensive set of entities, most of which could potentially be used for mapping legal interests within indoor environments [27].

Principally, physical data models provide their own extension mechanisms for incorporating legal objects, whereas legal data models can be connected to physical objects through external linkages [27]. At the research of Stadler et al. [28] it is stated that since legal and physical objects are often maintained separately, confrontation inescapably will lead to geometrical inconsistencies (as will be noticed in the GII setting). Semantic information can help to reduce the ambiguities of geometric integration, provided it is coherently structured with respect to geometry.

This integration has been investigated in recent years, and various approaches are being developed. The need for integrating legal objects with their physical counterparts has also been underlined by [18], who proposed the "3D Cadastral Data Model" (3DCDM), which defines an application schema of GML, and introducing the concepts of Legal Property Object and Physical Object. The model was developed based on the core cadastral data models, and was extended to support urban physical objects, consisting of packages serving different user requirements and applications.

How LADM can be mapped and encoded with the CityGML schema has been examined, mostly suggesting that ADE mechanisms could be an appropriate solution. Dsilva [29] developed a preliminary ADE for managing legal interests within CityGML for cadastral purposes in the Netherlands, called "*KadasterApartment*"; Çağdaş [30] developed a CityGML extension enabling the modelling of cadastral parcels and condominiums to support the requirements of the immovable property taxation system in Turkey; while Rönsdorff et al. [31] proposed two options for creating an ADE using LADM to represent the legal space, and examined how this could be mapped onto the CityGML standard in the context of 3D cadastres. The first option is to develop a jurisdiction-specific profile for LADM, and then implement it as an ADE for CityGML. The second option is to directly implement the fundamental concepts inside LADM, without customizing them for a specific jurisdiction, as a general CityGML ADE. This ADE proposes three core feature classes, namely "*Parcel*", "*LegalSpace*", and "*LegalSpaceGroup*", which are implemented based on the relevant LADM classes.

Gózdz et al. [14] proposed a jurisdiction-specific implementation of LADM as a CityGML ADE within the context of the Polish cadastre. The main objective was to elaborate the possibilities of applying CityGML for cadastral purposes, drawing particular attention to the three-dimensional representation of buildings. In order to link legal objects to their physical counterparts, relationships between the proposed legal classes and "AbstractBuilding" class (from CityGML) were also defined within the ADE mechanism.

Moreover, Li et al. [32] developed a comprehensive LADM-based CityGML ADE as a feasible approach for describing the ownership structure of condominium units in Chinese jurisdictions by proposing a legal and a physical hierarchy. Based on the legislation in China, the proposed ADE facilitates the management of associations between legal and physical notions, and represents the ownership structures of various privately and publicly owned condominium units, defined as multi-level buildings [27].

Gózdz et al. [14] proposed that further research should aim at the investigation of other possible alternatives of combining LADM and CityGML standards, including:

- embedding the selected CityGML classes into a (broader) LADM framework; and
- introducing a link between both domain models (in an SDI or GII setting) using references between object instances.

The authors concluded that introducing semantic representation for land administration within CityGML would be advisable.

Isikdag et al. [33] proposed that integrating 3D RRR spaces with 3D physical models (IFC or CityGML) could provide significant benefits for the valuation and taxation of properties. Moreover, cadastral extension of the Unified Building Model (UBM) was investigated by [34], examining the capability of both the IFC and CityGML standards for dealing with 3D cadastral systems. The authors proposed that UBMs could be extended to include boundaries without physical objects or counterparts, which are necessary for representing above- and below-ground RRR spaces in the context of the Swedish jurisdiction.

Oldfield et al. [35] suggested that space objects (IfcSpace), and the grouping of these spaces as legal zones (IfcZone), could underpin the utilization of BIM models in 3D cadastres. It was also reported that the boundaries of legal spaces could be modelled by "IfcRelSpaceBoundary". The presented workflow described how cadastral data requirements could be efficiently communicated between project initiators and authorities, which would, in turn, facilitate procedures for obtaining legal spaces from BIM models.

Kim et al. [36] proposed a framework for a 3D underground cadastral system with the ability to register various types of properties and manage RRR information using indoor mapping for as-built BIMs associated with 3D properties located underground in the context of the Korean jurisdiction.

Recently, Atazadeh et al. [27] has investigated the feasibility of BIM for urban land administration and, in particular, 3D digital management of legal interests in multi-storey building developments. A BIM data model was extended with legal data, and a prototype BIM model for a multi-storey building development was implemented to demonstrate the viability of the extended IFC data model for 3D digital management and the visualization of data related to complex legal arrangements. Relevant entities, suitable for modelling legal information, were identified and proposed in the IFC standard. These entities have been extended to model legal information with the minimum change possible in the current IFC data structure. The adopted approach for extending relevant IFC entities with legal data elements has mainly been predicated on using the extension mechanism provided within the current schema of the IFC standard.

An approach for linking RRR information to indoor environments was introduced by Zlatanova et al. [26], proposing an LADM-based extension for IndoorGML. For this first approach, it was argued that the subdivision of LADM space on the basis of properties and rights-of-use could be used to define semantically and geometrically available and accessible spaces, and could therefore enrich the IndoorGML concept. The second step of the authors' research, described in [37], was to define an external linkage mechanism for associating IndoorGML entities with LADM entities, or vice versa.

Last but not least, LandInfra is a conceptual model developed to model information about land and infrastructure facilities [23,24]. It is an OGC standardization approach in which some concepts from the cadastral data models (LADM, and LandXML) have been adopted for modelling legal objects, while some physical entities (adopted from IFC and CityGML) have been utilized for modelling physical objects. LandInfra's encoding standard is InfraGML.

Table 1 summarizes the above-mentioned approaches to legal and physical integration.

It is noted that, recently, a comprehensive review of integrated 3D spatial data models was also presented in [27]. Finally, it is evident that, while scientific publications—as also presented in the next subsection—represent one type of research output, the next step is to the transfer these into applied software.

Table 1. Legal and physical integration approaches.

Legal	Physical	Integrated Model	Via	Purpose	Reference
LADM	CityGML	-	ADE	Chinese jurisdiction 3D cadastre Polish cadastre Dutch jurisdiction Turkish jurisdiction	[32] [31] [14] [29] [30] *
ePlan	CityGML & 3D LandXML	-	LADM OWL	Semantic harmonization	[38,39]
LADM	IndoorGML	-	-	assigning RRR information to indoor spaces	[26,37]
LADM & LandXML concepts	IFC & CityGML concepts	LandInfra	-	model information about land & infrastructure facilities	[23,24]
-	-	3DCDM	Legal Property Object & Physical Object	Integrated approach managing legal & physical dimensions of 3D RRR spaces	[18]
-	IFC & BIM	Cadastral extension for UBM	-	3D Cadastral system using IFC and BIM	[34]
-	IFC/CityGML	-	-	Valuation and taxation enhancement	[33]
-	BIM	Integrated BIM model	-	BIM for Urban Land Administration	[27]
-	IFC (IfcSpace, IfcZone)	-	"IfcRelSpaceBoundary"	Obtaining legal spaces from BIM models	[35]
-	BIM (as-built)	-	-	managing RRR information associated with 3D properties located underground in the context of Korean jurisdiction	[36]
LADM	INTERLIS	-	-		[2,3,5]

* The proposed extension has been modified and adopted in OGC's land and infrastructure (LandInfra) conceptual standard [23,24].

2.3. LADM Implementations Based on Model-Driven Architecture Applications

The Land Administration Domain Model (LADM) is one of the first spatial domain standards within ISO TC 211, and aims to support *"an extensible basis for efficient and effective cadastral system development based on a Model Driven Architecture (MDA)"* and to *"enable involved parties, both within one country and between different countries, to communicate based on the shared ontology implied by the model"* [40]. It is a descriptive standard, rather than a prescriptive one that can be expanded, which also meets the need for efficient interoperability, required for fulfilling the principle of legal independence, in which each institution assumes the responsibility for the management of its own legal land objects [41].

As stated by Kaufmann et al. [41], the principle of legal independence stipulates that *"legal land objects, being subject to the same law and underlying a unique adjudication procedure, have to be arranged in one individual data layer; and for every adjudicative process defined by a certain law, a special data layer for the legal land objects underlying this process has to be created."* Thus, the principle of legal independence is clearly supported within its structure, while MDA facilitates the structuring of specifications and thematic specializations in the modelling process. The model is flexible and widely applicable; and therefore, a plethora of LADM-based country profile implementations have been developed since the approval of LADM as an ISO standard in 2012.

Automatic and semi-automatic methods based on Conceptual Schema Languages (CSL) offering direct implementable model descriptions of the corresponding LADM-based conceptual models have grown the research interest in this field. The rest of this Section briefly presents some approaches based on the MDA concept, in which LADM profiles have been implemented, some of these being collaborations between universities in different parts of the world, working together with manufacturers in putting research into practice.

To start with, Hespanha [42] proposed the description of the initial model in UML packages using Enterprise Architect (EA) software, and then exported and parsed the model into Eclipse UML 2.0 class models and diagrams, enabling its implementation in a PostgreSQL/PostGIS database. The final result was a Java abstract layer, accessible to other applications running under the Eclipse Integrated Development Environment (IDE) and the corresponding database schema, which could be further populated with data. In another approach, Hespanha et al. [43] used IBM's Rational Software Architect for implementing the proposed LADM country profile for the Philippines.

Additionally, research was carried out by Delft University of Technology [44], considering the development of a Computer-Aided Software Engineering (CASE) tool that performs a model transformation where the initial UML model, expressed as an XMI, is translated into an SQL file with a set of DDL (Data Definition Language) commands. However, it is a prerequisite that the UML model (i.e., the LADM-based model) be transformed into a Platform Specific Model (PSM), therefore requiring decisions to be taken regarding the hardware and software platform on which the model will be implemented.

Based on the Social Tenure Domain Model (STDM), as the so-called "developing country profile" of LADM, a pro-poor land rights recording system was developed based on free and open-source geographic software products (PostgreSQL/PostGIS, QGIS Python plugin), with the source code freely available from GitHub [45]. LADM was also the starting point for the UN FAO Open-Source Software Project FLOSS Solutions for Open Land Administration (SOLA); an example implementation of LADM (IT System Specification) was released in 2011 using PostgreSQL [46], while LADM was used to develop a concrete feature catalogue for Addis Ababa in 2012 [47].

In recent years, several projects [2,3,5] have suggested describing the developed LADM country profiles with the conceptual schema language INTERLIS. Both INTERLIS and LADM share the same Model Driven Architecture principles [3].

In particular, INTERLIS has been successfully applied in the Swiss Cadastre System for several decades, and became a Swiss standard in 1998. Since 2007, it has been part of the Swiss Federal Act on Geoinformation, and all data models of the Swiss NSDI have to be described with the standard by law. In 2004, the Swiss Land Management Foundation started an initiative to facilitate and speed up LADM implementation by describing the LADM with INTERLIS. The first steps towards the implementation of the Swiss LADM country profile in INTERLIS were introduced by [3]; these are further elaborated in Section 4.1.

Additionally, INTERLIS was selected as the modelling language to obtain a prototype implementation of a proposed Multipurpose Land Administration System (MLAS) for Greece [7]. The implementation of the proposed LADM-based model with INTERLIS followed the methodological steps discussed in [3,4], drawing particular attention to the explicit formulation of constraints, code lists and enumeration values, as well as the initial description of a 3D volumetric primitive.

Finally, institutional stakeholders in Colombia, with technical assistance from a Swiss cooperation (SECO) project, have recently developed a Colombian LADM profile, applying an in-depth planned modelling process. Experts working on the project suggested implementing the developed conceptual profile using INTERLIS and thus, an INTERLIS-based COL-LADM data model was developed, which will be applied in World Bank-financed pilot projects related to a new Multipurpose Cadastre [5]. The Model-Driven Approach (MDA) defined by the Object Management Group (OMG) [1] was suggested, with LADM being the core standard because of the complex institutional setting for land administration in Colombia and, consequently, the need for improved data interoperability and legal independence, as it is defined in [40].

2.4. The Role of Constraints in (Spatial) Data Modelling/Modelling (Geo-) Constraints

There are various ways to model constraints, such as ontology and OCL. Ontology is defined as a *"formal, explicit specification of a shared conceptualization"* [48]. The constraints in ontologies are defined by the components *Rules* and *Axioms*. Rules are statements in the form of an "if–then" sentence describing the logical inferences that can be drawn from an assertion in a particular form. On the other hand, axioms are assertions (including rules) in a logical form that together comprise the overall theory that the ontology describes in its domain of application [49].

Moreover, Object Constraint Language (OCL) is a notational language used to build and analyse software models, usually found as part of the UML. Every expression written in OCL relies on the types (i.e., the classes, interfaces, properties, relationships) that are defined in UML diagrams. It is a constraint and query language designed for object-oriented modelling, which enables the automated parsing, processing and implementation of OCL constraints, referring to classes, attributes, associations, and operations [50].

Furthermore, there are multiple database mechanisms that allow the realisation of constraints, via triggers and procedures, which execute the procedural code after hitting the trigger (that is, failing a certain check when data is inserted, updated or deleted). Triggers are used to avoid insertion of invalid data, and therefore make use of the spatial extensions of database management systems [51].

Constraints establish rules with which data must comply and, hence, should be part of the conceptual model (object class definition), as they provide additional semantics to the model. The implementation of constraints (whether at the front-end, database level or communication level) should be driven automatically by these constraints' specifications within the model [52].

Approaches for modelling and enforcing different types of constraints are many and diverse [51–55]. As stated by [52,53], the main types of traditional constraints include domain constraints, key and relationship structural constraints, and general semantic integrity constraints; and these have been extended with new ones: topological, semantic and user-defined constraints [53]. In particular, the classification of constraints based on relationships between objects can be characterized as follows [55]: thematic constraints, temporal constraints, and spatial constraints. The last can be further categorized into topological constraints, direction constraints, and distance constraints. Mixed constraints occur when the fundamental types of relationship constraints are mixed.

In this paper, the constraints are defined in UML models using OCL, which enables the expression of the constraints at a conceptual level in a formal way, and in a platform/vendor independent manner. Enterprise Architect (EA) software, which supports OCL syntax, as well as the ability to validate the OCL statement against the model itself, was used in the first steps to model the UML diagrams, meaning that verification that the OCL statement is expressed correctly in terms of actual model elements beyond just the syntax was possible, and that the kind of validation syntax that it used corresponded to the actual data types defined for these elements: numbers, strings, collections, etc. The definition of the constraints tested during the LADM/INTERLIS implementation is described in Section 5.2.

It is noted that difficulties are faced in generating an implementation (SQL/DDL) for Enterprise Architect's UML/ OCL models. An indicative example is that there is no normative way to translate UML code lists into SQL expressions, as the corresponding enumerated types of the PostgreSQL are static and cannot be extended.

2.5. Semantics and the Meaning of Code Lists in the Land Administration Domain

Semantic technologies (ontologies, RDF, SKOS, etc.) can be used in land administration and other domains to further provide explicit meaning to code list values in a more refined manner than just a hierarchy [56]. In the context of Spatial Data Infrastructures (SDIs) or Geographic Information Infrastructures (GIIs), reference tools for sharing and maintaining code lists and their definitions are necessary. Hence, it is vital to establish mechanisms to bridge any difficulties in reaching a common understanding of code lists while, at the same time, allowing machine-readability.

In this domain, research has been carried out in recent years concerning the field of land administration at an international and European level. More specifically, the European Land Information Service (EULIS) Glossary is an initiative of the European Economic Interest Group, an online European portal enabling access to land registries across European countries, each with their own land administration legislation. Its goal is to assist better understanding of the local environment, not only literally, but also with regard to terminology [57], enabling the user to display a term and compare its definition with that in another land registry.

Moreover, according to [56] "firstly, it is possible that a European country may be compliant both with INSPIRE and with LADM and secondly, it is made possible through the use of LADM to extend INSPIRE specifications in future, if there are requirements and consensus to do so". It is noted that the European Directive (INSPIRE) does not include Parties and their associated RRRs in the definition of cadastral parcels, and thus, information regarding RRRs is not included in the code lists.

The approach implemented by INSPIRE is presented with regard to the hierarchical structuring of code list values, and means of managing these [58]. The information model used by INSPIRE corresponds to ISO 19135 "*Procedures for item registration*" [59] for managing and disseminating code lists, and in order to reference the identifiers, all unique resource identifiers (URIs) are used.

Additionally, the Web Ontology Language, developed by W3C, is a computational logic-based language, a Semantic Web language designed to represent rich and complex knowledge about things, groups of things, and relations between things. OWL (Web Ontology Language) was developed as a vocabulary extension of RDF, while OWL documents, known as ontologies, can be published on the Web and may refer to or be referred to by other OWL ontologies [60].

Soon [38] presented a formalization of domain ontology for land administration from natural language definitions in the standard, emphasizing user roles. In 2014, Soon et al. [39] proposed an extension of the already-developed LADM Web Ontology Language (OWL), i.e., a semantics-based fusion framework for integrating CityGML with 3D LandXML, adopting ePlan as the conceptual model.

Through this framework, it is expected that a computer system will be able to perform reasoning and inference in the OWL ontology, as well as to retrieve the geometries of a building's legal space or physical space, or both [30,61]. Subsequently, the authors' intention was to utilize the best of all available concepts (i.e., CityGML, LandXML and OWL) without affecting the existing schemas, which have been comprehensively developed for different applications [39].

Furthermore, Simple Knowledge Organization System (SKOS) is a common data model, and provides a framework for developing specifications and standards to support the use of Knowledge Organization Systems (KOS) such as thesauri, classification schemes, subject heading systems, and taxonomies, within the framework of the Semantic Web [62]. It is designed for use as the domain modelling schema when the aim is to represent controlled vocabularies (e.g., taxonomies and thesauri) that organise domain concepts only through hierarchical and associative relationships [62]. It provides a standard way of representing knowledge organization systems using the Resource Description Framework (RDF), allowing interoperability between computer applications. Using RDF also allows knowledge organization systems to be used in distributed, decentralised metadata applications [63].

Moreover, the Cadastre and Land Administration Thesaurus (CaLAThe) provides a controlled, standardized representation of structured vocabularies encoded using Simple Knowledge Organization Systems (SKOS) developed by the World Wide Web Consortium (W3C). CaLaThe is inspired by and derived from ISO 19152 LADM [10], and covers terms regarding real estate, cadastre, land administration, and LADM code lists and classes [63]. A graphical overview of the CaLAThe term collection related to land is presented in Figure 2.

Recently, a group of researchers has initiated the development of a valuation component for LADM as a draft extension module [64] concerning the fiscal parties involved in the valuation practices and fiscal real property units that are the objects of valuation. This valuation information model provides a template for the specification of valuation databases used for recurrently levied property taxes. This initiative is supported by the development of a domain vocabulary and thesaurus [65] to define the semantics of valuation information encoded through SKOS specifications for the standardized representation of structured vocabularies.

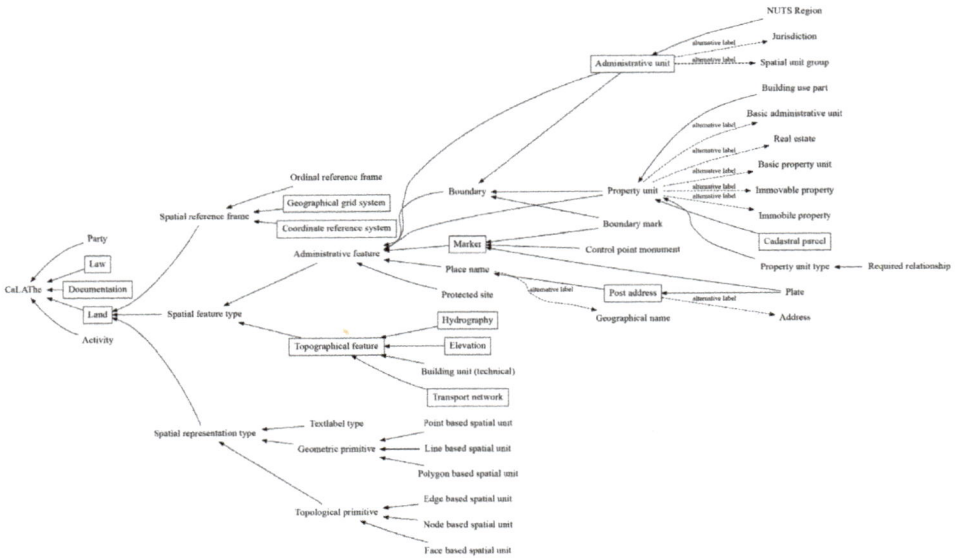

Figure 2. Graphical overview for CaLAThe terms collection relating to land [66].

Last but not least, to further provide explicit meaning to code list values, Lemmen et al. [56] provided input for future LADM development, in that they introduced an extended classification of the LADM class of rights, restrictions and responsibilities. As described by the authors, in the current version of LADM, code lists are in the informative part of the standard, and not in the normative part; hence, their values are indicated only by their name, without any definition. They concluded that two aspects of updating LADM code lists can be identified: organizational (who is responsible for managing a register, who can be involved and have access, what are the roles and responsibilities of the involved parties) and technical (systems for the code list registers/database, web services for accessing and updating code list values, etc.).

Lemmen et al. [56] were the first to propose the use of semantic technologies, ranging from hierarchically structured code lists to the RDF vocabulary, while for structuring and maintaining LADM code lists, ontologies were proposed [65]. The authors mentioned that it is possible to extend LADM and its code lists by using the Legal Cadastral Domain Model [67] and the Social Tenure Domain Model [45] to make it possible to represent RRRs on a more detailed level, including informal rights, restrictions and responsibilities. Figure 3 illustrates an ontology diagram showing land-use relations.

Code lists in LADM, and the rest of the ISO standards on which it is based, are used to express a list of potential values with the aim of enabling the use of local, regional and/or national terminology [68]. In contrast, by means of enumeration, all values admissible for this type are pre-determined, and cannot be extended without creating a new data type, whereas code lists are more open and flexible.

In terms of the life cycle of a model and its components, the model has a long life cycle, code lists have shorter life cycles, and real data has the shortest life cycle, meaning that it should be changed more frequently. Code lists are considered to be neither part of the model, nor real data; and they need to be updated more often than the model, and less often than the data.

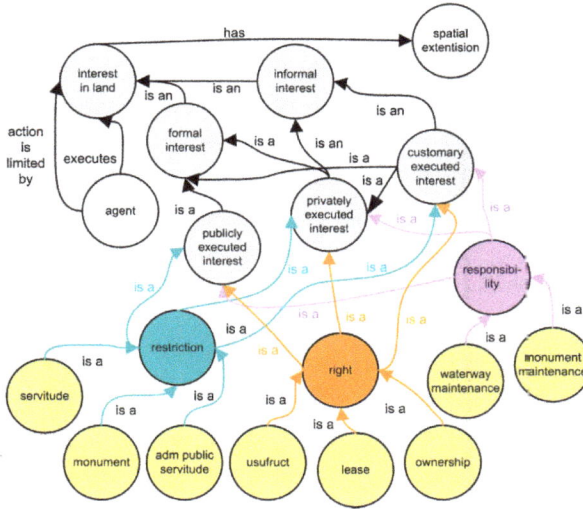

Figure 3. Ontology diagram showing land use relations, exemplified with RRRs listed in LADM Annex J of ISO 19152:2012 [68].

3. INTERLIS—Swiss Standard for Land Administration

This Section briefly introduces the INTERLIS concept and its characteristics as a modelling language, while the INTELIS tools are presented in Sections 3.1–3.7.

INTERLIS is a well-established Swiss national standard (SN 612030 for INTERLIS 1 and SN 312031 for INTERLIS 2) tailored to geospatial applications and, in particular, to geoinformation exchange, and the modelling and integration of geo-data, allowing cooperation between information systems and, especially, geographic information systems [69].

INTERLIS has been part of the Swiss Federal Act on Geoinformation since 2007 [70], and more than 170 data models of the Swiss National Spatial Data Infrastructure (NSDI) have been described in this language. At the same time, INTERLIS is an Object Relational modelling language, which is very precise and highly standardized at the conceptual level assuring a strict separation of model descriptions and data exchange formats [69] and supporting methodological freedom by taking a system-neutral approach. The "duality" of INTERLIS (data model and exchange) is presented in Figure 4.

INTERLIS is a Conceptual Schema Language offering the necessary complement to the UML graphic description language [69]; and therefore, INTERLIS-described models are precise, unequivocal, and can be interpreted without misunderstanding. Data transfer between several databases via a common data model (data schema) described in a common data description language is provided by INTERLIS, as also described in Figure 5.

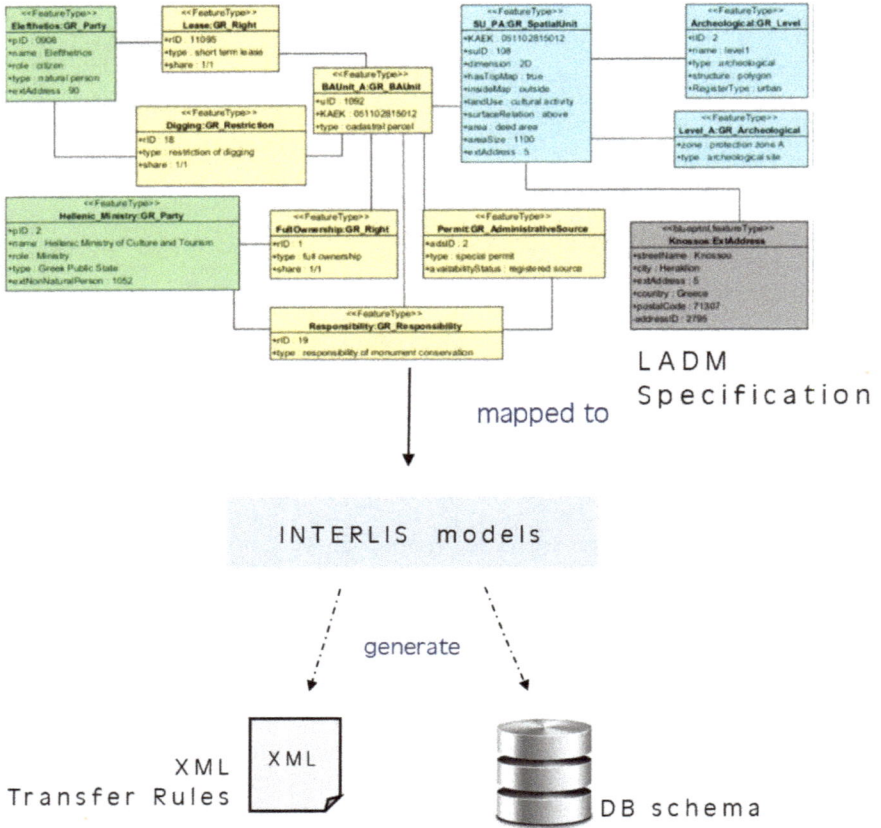

Figure 4. INTERLIS "duality" (adopted from [6], edited).

As has already been mentioned, it follows MDA principles, enabling the utilization of data modelling in close connection with system-neutral (XML-based) interface services [8], and provides strict separation of transfer and modelling tasks. As a result of the experience with INTERLIS and its corresponding software packages, some incompatibility problems between the UML/INTERLIS Editor and other (commercial) modelling software, such as Enterprise Architect have been found. More specifically, this tool does not use OMG XMI, but, rather, Eclipse XMI, which can sometimes hinder communication with other software. It is noted that this is an issue of the UML/INTERLIS Editor, but not of INTERLIS itself, as INTERLIS does not actually rely on UML, although it supports it.

Among the advantages of INTERLIS are the formal description of constraints using an OCL-like language and the ability to quality-check INTERLIS data against INTERLIS data models using tools enabling automated validation of data. This makes INTERLIS development and use unique, as a system-independent means of providing quality-control mechanisms for models described at a conceptual level. Because of this, the quality-checking of data (and more explicit semantics via models and constraints) is becoming more and more important, as there is a growing number of users within the Geo Information Infrastructure further removed from the data producers (i.e., who may not know the data well, and may therefore need explicit documentation/explanation). In INTERLIS 2, four geospatial primitives are defined: coord (coordinate or point types, described by their axes), polyline (describing linear elements), surface, and area (the two types for polygons, defined similarly to polylines, but implying that the linestring is closed).

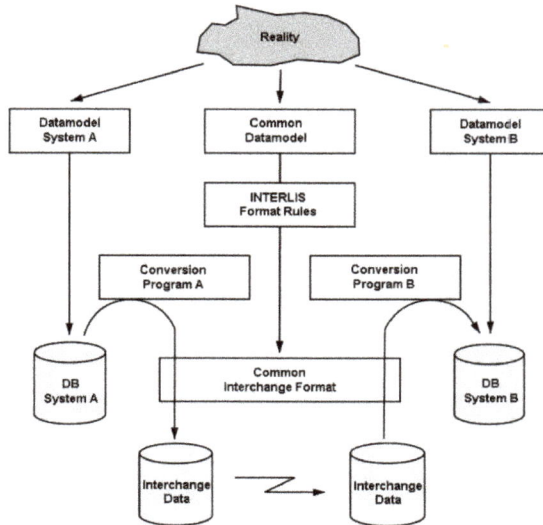

Figure 5. Data transfer between several databases via a common data model (data schema) described in a common data description language [69].

Further information regarding the implementation of conceptual models (LADM and related models) by means of the INTERLIS language is presented in Section 4, INTERLIS is vendor-independent, and a tool chain (Java programs) that can be used to automatically generate implementation components for specific environments has been developed, and is briefly presented in the following paragraphs. The tools can be categorized into three main phases (1. data modelling and exchange format definition, 2. database schema and model conform data generation, and 3. data validation phase), as illustrated in Figure 6. A fourth alternative phase refers to the import of data from external data sources.

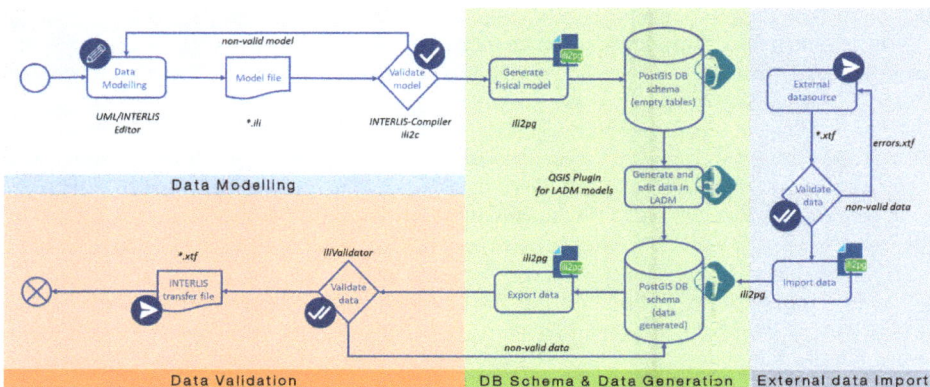

Figure 6. INTERLIS tools and workflow ([6], edited).

3.1. UML/INTERLIS Editor

The UML/INTERLIS Editor tool is used for the following procedures:

- graphical representation of existing INTERLIS data models as UML diagrams, providing a better understanding; and
- description of new models (e.g., LADM country profiles) with UML diagrams.

As is mentioned by [6], the INTERLIS model file comprises an ASCII file, as well as a data file *.xtf; hence, it can be opened in any ASCII editor, and due to the development of plugins, there are editors (jEdit and Notepad++) that support syntax highlighting.

A specific feature of the tool is the import functionality for class diagrams of UML models in XMI format. It is noted that this import is limited to models in XMI Rational Rose format (Rational Rose™UML), as XMI files generated, for instance, in EA software cannot be imported, because they use different XMI versions. However, it is possible to generate an XMI from the UML/INTERLIS Editor and import it into EA.

3.2. INTERLIS Compiler

INTERLIS Compiler (ili2c) validates the syntactical correctness and semantic compliance of INTERLIS data models. More specifically, the tool checks for all kind of syntactical errors: missing declaration of data types, syntax errors, wrongly defined attributes or classes, etc. The compiler reads and writes INTERLIS models, and examines whether or not the models are in accordance with the syntactic and semantic conditions of INTERLIS [69]. Amongst others, the compiler can generate XML schemas, XML-based exchange format (XTF files), GML schemas and HTML tables. It is noted that this toll is a model/schema checker and not data checker.

3.3. INTERLIS Checker

Among other things, a big advantage of INTERLIS is the possibility of validating the model compliancy of the transferred data against its data model [3]. INTERLIS Checker (igchecker2) is the official (licensed) software tool used to quality-check INTERLIS XML data against INTERLIS data models (including specific constraints). The input is a INTERLIS data exchange file in XTF (readable by any GIS that recognizes the GDAL-OGR library for vector formats), and the output log files report whether there are errors or not in the input file. The toll is used by the Federal Cadastre Directorate as well as by almost all of the cantons.

3.4. INTERLIS Validator

The iliValidator tool was developed with funding assistance from two cantons of Switzerland, and the afore-mentioned Swiss cooperation project in Colombia, and which will be further described in Section 4.3. INTERLIS in Colombia. For this implementation, Java was chosen as the programming language, and the tool can be used with a simple Graphic User Interface (GUI) or the command line. Errors found during validation are logged in a simple ASCII log file or in an INTERLIS transfer file based on a simple INTERLIS ("shape file level") data model. With a configuration file, the user is able to switch the constraints on/off, or downgrade them to warnings instead of errors [5]. iliValidator offers the possibility of validating complex user-defined constraints, by calling functions developed in Java from the model and executing them from within the tool.

As underlined by the authors, iliValidator can be used as a programming library, offering the possibility of integrating the data-validation phase into existing software, web services and/or existing processes.

3.5. INTERLIS Loader for Relational Databases

In order to facilitate the translation of the object-oriented INTERLIS data models to relational databases, Object-Relational mappings (O/R mapping) were introduced, and are deployed for the implementation phase, using the tools *ili2pg* (INTERLIS 2 loader for PostgreSQL/PostGIS), *ili2ora* (INTERLIS 2 loader for Oracle) and *ili2gpgk* (INTERLIS loader for OGC Geopackage). Recently, the *ili2sqlserver* (INTERLIS 2 loader for Microsoft SQL Server) was developed. All tools mentioned are part of the ili2db project.

As an example to describe the functions of the tools, the *ili2pg,* can translate INTERLIS data model definitions to a PostgreSQL/PostGIS database, import INTERLIS data to the database created, and export it again to the XML-based INTERLIS exchange format (XTF). The other tools work similarly for their corresponding database management systems.

3.6. QGIS Project Generator Plugin

The Project Generator plugin is a tool developed in Python for QGIS version 3, and built on top of ili2db to centralize the process of generating physical models from INTERLIS models and capturing, importing, editing and exporting data to INTERLIS transfer files (XTF). Users with access to INTERLIS models and QGIS software can produce valid INTERLIS transfer files using QGIS as the data editor, and PostgreSQL\PostGIS as the database (the addition of GeoPackage as an alternative for data storage is currently under construction).

Project Generator downloads ili2db tools if needed, runs ili2db commands to create a physical model, and makes use of such models to configure a QGIS project ready for capturing data. Based on database objects generated by ili2db, as well as on INTERLIS metadata from the original models, Project Generator configures the QGIS layer tree, a form for each layer with appropriate edit widgets for attributes, units, constraints, value lists and relations among layers. With a QGIS project configured in this way, users can capture and edit data according to the rules defined in the INTERLIS models and generate valid data effortlessly. Project Generator can be installed from the official QGIS plugin repository.

3.7. INTERLIS Reader/Writer to FME

INTERLIS Reader/Writer to FME (ilii2fme) is a free tool that provides the Feature Manipulation Engine (FME) with access to INTERLIS 2 and INTERLIS 1 transfer files, supporting the rich geometry model of FME. The tool enables reading INTERLIS 1 and 2 data, and is necessary in order to manually define appropriate ili2fme parameters, read and write INTERLIS models, and write GML data. It is noted that FME is a commercial tool.

4. LADM Implementation in INTERLIS

This Section presents recent experience gained from implementing INTERLIS in LADM-based country profiles, and suggests an integrated LADM/INTERLIS approach. Sections 4.1–4.3 present the experience gained from INTERLIS implementation in three countries: Switzerland, Greece and Colombia.

The generic LADM/INTERLIS approach proposed may be implemented in any LADM-based model, in order to produce a platform-independent exchange format linked to the conceptual model, allowing for automated quality control with promising results. Figure 7 illustrates the ISO standards on which LADM is based, and which have been described in the INTERLIS language, as well as the country profiles that have been translated into INTERLIS up until now.

The second version of the core classes and associations of ISO 19152 LADM modelled with INTERLIS 2 was released in 2016, with full 2D and 3D support, in the context of the Project *"Modernization of Land Administration in Colombia"*. This version is more complete and coherent; it should be particularly noted that some code lists and structures (e.g., Image, ExtArchive) have now been completed, while more constraints have been added to various classes (e.g., the LA_SpatialUnitGroup class, and the LA_BAUnit class, as will also be presented in Section 5.3).

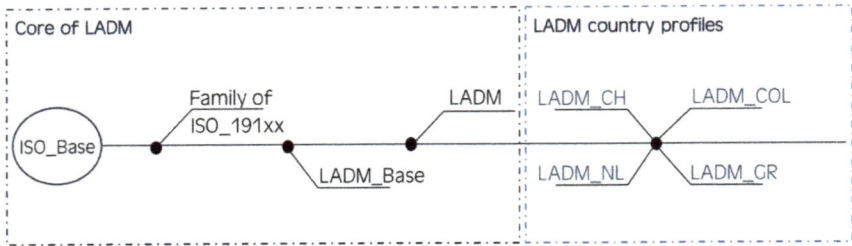

Figure 7. INTERLIS models used to describe LADM country profiles; country profiles that have been translated into INTERLIS language up until now.

An interesting approach in the case of the Colombian project is the modular approach applied: following the principle of legal independence, and in consideration of the MDA, the COL_LADM model is modularized around a Core or Minimum Model, containing the common elements that define the profile. The core model is implemented by the institutions that are responsible for each thematic area of data, tailoring it according to their specialized needs (Figure 8). Thus, the modular approach described implies that the Colombian LADM profile will be formed by a core model and its extended or specialized models for each of the thematic areas or regionalizations (for instance, a specialized model on the Department/District level).

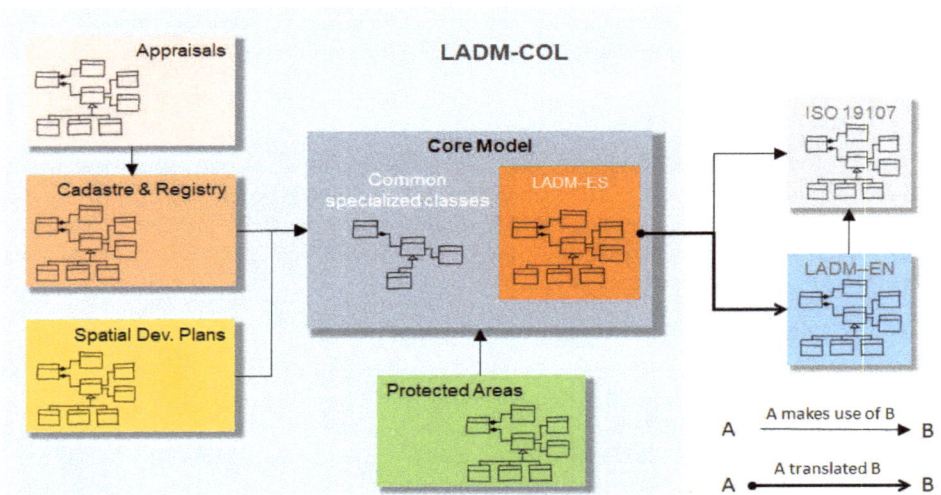

Figure 8. Modularity of COL_LADM model.

This generic "modular" approach—i.e., modelling a core profile, which can be extended by multiple specialized thematic profiles closely linked to each other, reflecting the complex institutional land administration setting—can be implemented in any jurisdiction, as part of an integrated LADM/INTERLIS approach.

4.1. INTERLIS in Switzerland

In Switzerland, the requirement for a clearly defined data model that can be adapted in flexible ways resulted in the development of a conceptual schema and object-oriented language, INTERLIS. INTERLIS is widely used in Switzerland, where the cadastral core data model, as well as many other models (i.e., utility services, land registry, urban planning, etc.), have been defined with INTERLIS [3].

In [5], it is stated that, in the case of Switzerland, the Act on Geoinformation [70] was an important milestone for applying the MDA approach for the entire Federal Spatial Data Infrastructure. The Swiss Land Management Foundation (SLM), a legal entity that is the result of close cooperation between the Swiss private sector and the Swiss government, started an initiative to facilitate and speed up LADM development by describing the standard with INTERLIS. The core work was completed in February 2014, the first results were presented to Dutch Kadaster International. Using this integration, the INTERLIS tool chain can be employed to handle and implement LADM-based country profiles in a computer-assisted manner. Exchange of LADM data between IT systems will be easily possible with the XML-based INTERLIS transfer mechanism, thereby improving implementation efficiency and reducing cost.

It is noted that the LADM/INTERLIS approach developed by the SLM was initially implemented for the Netherlands LADM country profile (NL_LADM) in 2014. One year later, the first version of the LADM country profile for Switzerland (CH_LADM) was introduced, which has since been updated to a new version.

4.2. INTERLIS in Greece

The starting point of the development cycle of INTERLIS implementation for Greece was a model describing a proposed 3D Multipurpose Land Administration System (MLAS) based on LADM. Kalogianni [7] proposed a model for the LADM Greek country profile (GR_LADM hereafter), which considers the current registration of objects in 2D in Greece while, at the same time, being future proof and covering the requirements for future registration (including 3D). It is noted that this approach was developed at a research level.

This model is considered to be an effort to overcome current shortcomings based on international standards, including the representation of a wide range of different types of spatial units—in 2D and 3D—and aiming to establish an appropriate basis for the National Spatial Data Infrastructure (NSDI) of Greece [8]. Given the particular nature of different types of spatial units, and the legislative framework in Greece, as well as the necessity of an integrated structure, an attempt was made to cover all Greek land administration-related information, currently maintained by different organizations, and to harmonize it all in one conceptual model [7].

Consequently, objects (more spatial units) other than those currently comprising the model of the Hellenic Cadastre (HC), were categorized and included in the proposed model, aiming at the creation of a multipurpose land administration system for Greece. Therefore, a wide range of spatial units—including areas of archaeological interest, buildings and unfinished constructions, utilities (legal spaces), 2D and 3D parcels, mines, planning zones, Special Real Property Objects (SRPO) usually found in the Greek islands (anogia, yposkafa), and marine parcels—are supported by the model.

As stated in [71], through the LADM concept, it is possible to describe a common denominator, or a pattern that can be observed in land administration systems. In this context, the concept of "*LA_Level*" in ISO 19152 is a novelty among the other standards related to land administration, as it is defined as a collection of spatial units with a geometric and/or topologic and/or thematic coherence. Additionally, the class "*LA_SpatialUnit*" provides various representations of RRRs and legal spaces; thus, a specialization of the "*LA_Level*" class, the proposed "*GR_Level*" class, is used to organize the spatial units of the GR_LADM, allowing the flexible introduction of spatial data from different sources and of different accuracies. The country profile also includes the content of various code lists, which are an important aspect of standardization. Figure 9 depicts INTERLIS tools and the sequence followed during the implementation of the prototype for Greece.

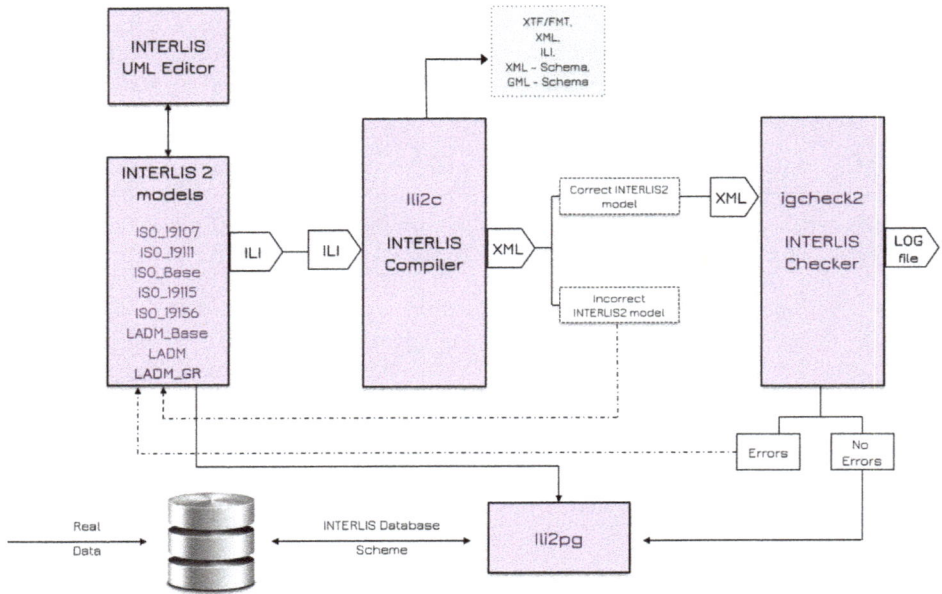

Figure 9. The LADM/INTERLIS approach developed and followed for the INTERLIS implementation in Greece ([8], edited).

Therefore, this model was initially described with UML diagrams using Enterprise Architect (EA) software, and then translated into the INTERLIS modelling language (GR_LADM). Next, INTERLIS tools were used to automatically generate implementation components for specific environments (database schema, exchange format, GUI/editor). Real datasets were gathered from Greek authorities involved in land administration procedures and used to populate the database.

The prototype system data refers to the majority of the levels created in the spatial part of the proposed model, mostly concerning condominiums, 2D and 3D rural parcels, archaeological spaces, special real property objects and planning zones. Although the currently available Greek datasets do conform with the structure of the proposed model, as new elements were added, sample data were generated for terms of completeness; i.e., the class *GR_Archaeological* is a new proposed class for which data was created in accordance with the Greek legislative framework. The database thus populated was used for the assessment. Analysis of the loading and querying of the database provided feedback, and the necessary changes were made to the database schema and the initial conceptual model in order to better represent the reality.

Once a valid INTERLIS model file (.ili format) is available, a database schema can be created in the PostgreSQL/PostGIS database using the ili2pg tool. Additionally, another possible output from ili2c is XML schema, and this can be used as exchange format. This INTERLIS file is then checked, and if there are no errors it can be imported into the database using ili2pg.

4.3. INTERLIS in Colombia

The support from the Swiss government is closely tied to the Peace Accords signed in 2016 between the Colombian government and the Colombian Revolutionary Armed Forces (FARC) [72], which contain an agreement on integral rural reform, including important aspects of establishing a new Multipurpose Cadastre [5,6]. Within this context, the INTERLIS-based Colombian profile of LADM

(COL_LADM hereafter) is part of a set of product specifications that must be applied by the operators of World Bank-funded pilot projects, executed on behalf of the National Planning Department (DNP).

Given the complex institutional framework in Colombia, and the need for appropriate inter-institutional exchange of land information, the process of applying LADM as the basis for a Multipurpose Cadastre was carefully planned and implemented, and involved key stakeholders in land administration, as presented in [6].

Figure 10 illustrates the UML diagram of the Colombian cadastre-registry model—the COL_LADM profile model—which includes classes of ISO 19152, and classes of the core and the specialized cadastre-registry model. It is noted that the model is in Spanish. The figure reflects the cadastre-registry model as it is formally described in INTERLIS, described using the UML/INTERLIS Editor, and implemented using ili2db.

One of the results of the project in Colombia is a web system completely based on a FOSS architecture with several interrelated modules, where the proposed LADM-based model is structured around a Core or Minimum Model with various specialized models formed for each of the common thematic areas (e.g., Cadastre and Registry, Valuation, Spatial Planning, Social Mapping for formalization purposes, Protected Areas, etc.) [6]. The core model is implemented by the institutions involved, who are responsible for each thematic area of data, tailoring it according to their specialized needs through specific classes, relationships, attributes, sets of values and constraints. Thus, in such a modular scheme, what is represented in a specialized model as a BAUnit on the basis of the legislation of a specific institution, may, in another model, be a restriction.

For the project in Colombia, the tool chain that was available for the implementation of the COL_LADM model in INTERLIS is being further enhanced; for example, the different tools mentioned in Section 3 are being integrated into a software called iliSuite, which has a modern GUI where all of the configuration options of tools can be selected, making the tool chain easier to use in general. The software has so far been translated into the English, German and Spanish languages.

The development of a generic web-based system for receiving, validating and storing INTERLIS data, and which is usable by any land administration institution that has to deal with receiving "Model-conform" data produced by third parties, is almost concluded. The whole system architecture is structured around a web interface, including a dashboard of statistical indicators, a GIS for visualizing the data and publishing it to different users, and a document repository with the aim of managing all documents that serve as the administrative or spatial sources of [6]. The core module of the system allows the validation of INTERLIS data through a web interface. It can be used by operators and supervisors of the cadastre surveying projects, as well as—as mentioned above—by other institutions that generate data in compliance with the developed data model (or other official models). This implies a very important conceptual and technological advance, since, with approximately 14 million registered properties in Colombia, only a web-based—and, therefore, automated—validation process would allow this volume of information to be treated in an adequate and timely manner. The validation service takes advantage of the model repository module, in which all official INTERLIS data models are registered, and where the corresponding data model can be selected (or uploaded if not an official model) when proceeding to validate a data set. The use of FOSS, and the modularity of the developed tools, provides great flexibility in terms of integrating them into hybrid GIS environments.

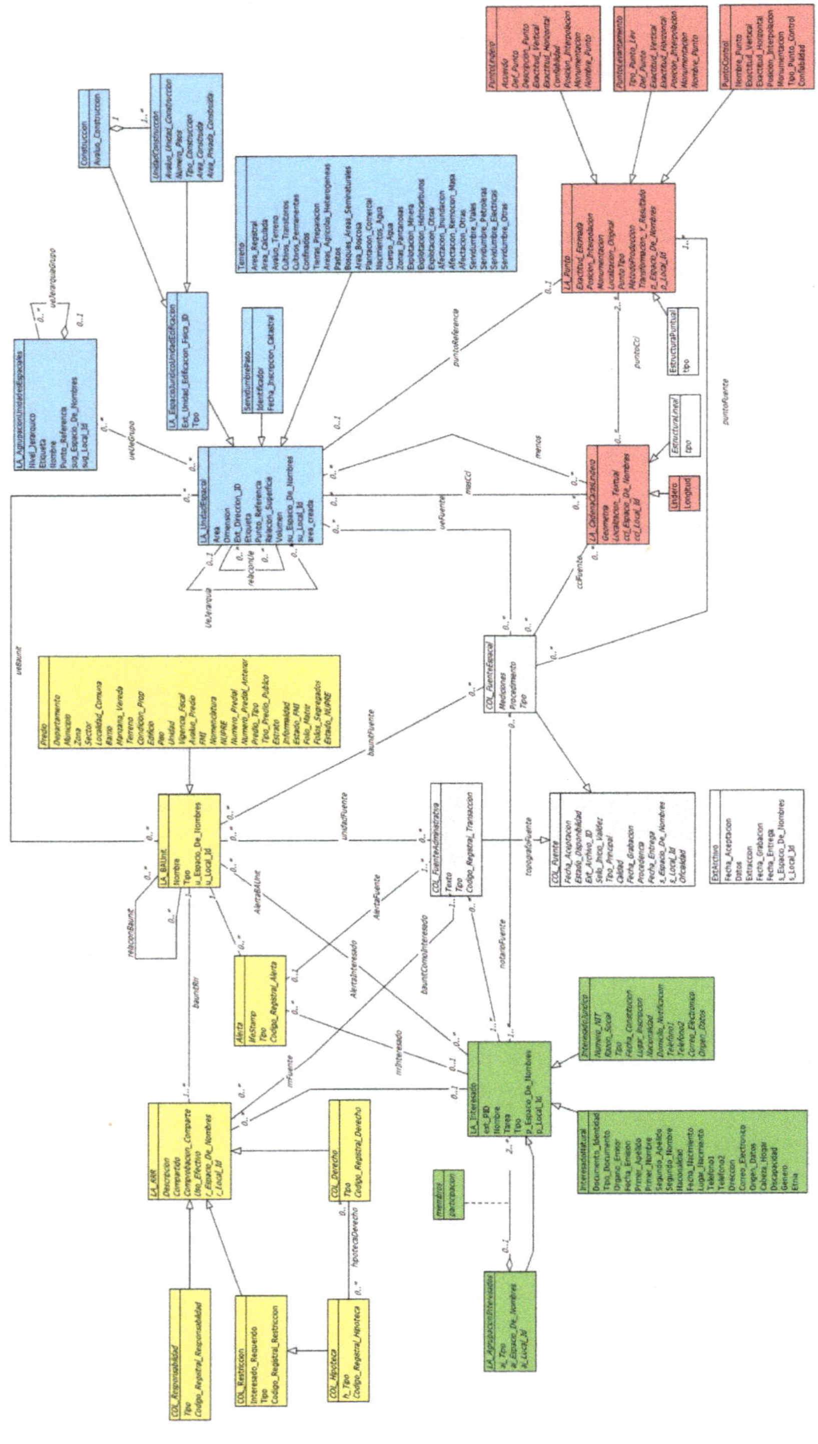

Figure 10. The UML of the COL_LADM profile.

Figure 11 illustrates the web-based system developed for the project in Colombia, which allows for data reception of any INTERLIS-based LADM data, and its integration with a PostgreSQL/PostGIS database and the visualization of any LADM conform data through a GIS Viewer.

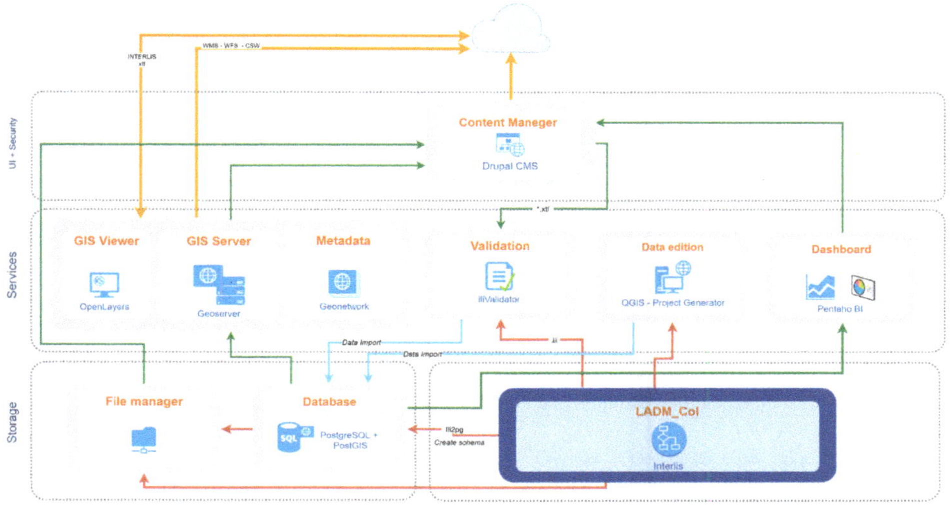

Figure 11. Integration of tools in the Web-Based System developed for the project in Colombia completely based on a FOSS architecture and developed with an MDA approach [6].

5. Implementation

Considering semantic interoperability within integrated legal and physical models, the need for more formal semantics is underlined in Section 5.1, while the proposed formal specification of constraints is presented in Section 5.2. Finally, Section 5.3 presents the 3D data types that are supported by INTERLIS.

5.1. Code Lists and Enumerations

During the implementation of the Greek country profile in INTERLIS, it was decided that, in case of fixed values, the values of the model would be defined as enumeration types, while for values that can be extended, a catalogue table with referential integrity would be used to express code lists.

Furthermore, INTERLIS offers both options; enumerations can be set as lists of values (and even nested) within the model itself, while code lists from external catalogues can be referenced from the model and imported into the database. The main weakness of the latter approach is the potential lack of synchronization between code lists stored in the database, and their corresponding source. For example, two organizations could be using the same model, but with different list values from the same external catalogue (or even different versions of the catalogue).

5.1.1. Semantics in Code Lists

Consistency in the language and terms used is vitally important to semantic interoperability. Glossaries, dictionaries and ontologies, as presented in Section 2.5, support the coherent development of code lists, improve their consistency, and allow for better understanding of the model, the data, and the code lists. Code list values are similar to normal data content, and may often possess some kind of structure that carries meaning; e.g., hierarchical classification codes.

UML code lists are just lists with values without any structure, and therefore, for the INTERLIS GR_LADM model, code lists were designed as structures with attributes, and given a hierarchical structure, which makes them semantically more meaningful, and also extensible [9].

As also proposed by [73], each code list is implemented in the database with one single table. The table name has the extension *"Type"* after the code list name of the conceptual model. It consists of a unique identifier (cID) for each code list and description attributes. The advantage of this type of code list is that its value can be updated, and it can also be versioned when adding the attributes *"beginDateTime"* and *"endDateTime"*.

In the GR_LADM model, the hierarchy is added as a reference to a parent code, the attribute *"parent_cID"*, referring to the "parent" of the code list from which it is inherited, indicating the top-level code list. The attribute should be "NULL" for root codes, as there is no parent (e.g., internationally agreed-upon code lists, as suggested in ISO standards, etc.), while, for the rest, codes are refinements, and have descriptions, comments, etc. Theoretically, the user is not limited to giving only one parent to each code-list value; it could be that the multiplicity can be more than one. However, it was decided, in the context of this research, that for the attribute that indicates the parent of the code lists, the multiplicity would be exactly one (strict hierarchy).

An example from the implementation of a Greek code list is given below in INTERLIS language:

```
STRUCTURE GR_PartyRoleType EXTENDS LADM.Party.LA_PartyRoleType =
    cID: MANDATORY Oid;
    parent_cID: Oid REFERENCE TO LADM.Party.LA_PartyRoleType.cID;
    begin_Date_Time: XMLDate;
    end_Date_Time: XMLDate;
    MANDATORY CONSTRAINT
    end_Date_Time>=begin_Date_Time
    description: CharacterString;
    !! Possible code list values:
    (lawyer,bank,notary,citizen,institution,tax_office,church,surveyor,
    insurance_organization,metropolis,parish,court,courtof_appeal,
    high_court,state_council,legislative_authority,local_authority,
    experoperation_committee,ministry, urban_planning_authority,other);
END GR_PartyRoleType;
```

With the proposed structure, the values take on a hierarchical structure, which provides some semantics, as terms higher in the hierarchy are internationally defined and agreed-upon, facilitating the communication between different countries and organizations.

Every code list has, in theory, the same structure as the one presented above; and therefore, all code lists could be maintained in a single table with an extra attribute to indicate the actual code list to which the code list value belongs.

5.1.2. Enumeration Types

In the GR_LADM INTERLIS model, there are more enumeration types than code lists. This means stronger typing but, at the same time, fixed sets of values that do not allow the extension of the admissible values for a type. This is preferable in some cases—e.g., in the case of dimensions: 0D, 1D, 2D, 3D—but is not appropriate for code lists that may require extension at some point in the future. The structure of the enumerations, which include a fixed set of values, is not (always) simply linear, but features a hierarchical tree-like structure, as described in [69]. The leaves of this tree (but not its branches) form the set of admissible values. This method of creating extensible hierarchical enumerations was used to extend some of the enumerations that have already been introduced in the core LADM model described in INTERLIS, enriching them with the Greek values.

An example of the extensible hierarchical enumerations that were created is given below. In the core LADM model, the enumeration for the LA_StructureType attribute of the LA_Level class was formed as follows:

```
LA_StructureType = (point, line, polygon, other);
```

INTERLIS offers the possibility of extending the parent domain and inheriting the entire enumeration, while making it possible to overwrite its values (this approach has been applied in the COL_LADM and GR_LADM models), as presented below:

```
COL_StructureType EXTENDS LA_StructureType = (
                    other (text, topological, drawing,
                    unstructured));
```

5.2. Formal Specification of Constraints

Constraints in the INTERLIS language can be defined on an object level (MANDATORY CONSTRAINT) or a class level (SET CONSTRAINT, UNIQUE CONSTRAINT) [8]. Therefore, some of the constraints/invariants of the LADM model can be directly expressed by the INTERLIS constraint language.

5.2.1. "Hard" and "Soft" Constraints

Constraints can be subdivided into requirements that must be met by all objects (*"hard"* constraints) and those based on intentions, guidelines, or regulations, which in rare cases can be violated (*"soft"* constraints). Therefore, hard constraints must always be true, otherwise the transaction will be cancelled, and an error given to the user. For instance, the following represents a hard constraint: "the end date should be after the start date".

On the other hand, soft constraints may not always be true, and in the real world there may be some exceptions. Three different ways of dealing with soft constraints are presented in this Section: (1) exception list; (2) percentage of allowed violations; and (3) conditions when the constraint does not apply.

Firstly, if the constraint is true, then the transaction can be continued; if it is violated, then it should be checked whether the case belongs to an exception list. This means that it is a binary decision on the constraint and its presence in the exception list. The exception list can be defined as a way of officially marking entities as an allowed exception. If a soft constraint is violated and case is not included in the exception list, then it is considered to be violating as a hard constraint. The exception list should be created and maintained by the corresponding authorities.

Secondly, in INTERLIS, it is possible to define constraints that only apply as a general rule, and to indicate what percentage of instances of a class must normally comply with the constraint. Such constraints, which can be characterised as "soft" constraints, can be modelled using the rule PlausibilityConstraint.

Finally, adding a condition defining in which cases the constraint has to be checked represents the third type of soft constraint. For instance:

```
CLASS LA_BAUnit EXTENDS VersionedObject =
    name: CharacterString;
    type: MANDATORY (basic_propery_unit, right_unit, other);
    uID: MANDATORY Oid;
    constraints
    {sum(RRR.share)=1 per type if RRR.shareCheck
    no overlap RRR.timeSpec per summed type}
    invariant
```

```
    {share must be specified, unless this is meaningless for
    the specific type (indicated by shareCheck=false; in
    this case constraint "sum (LA_RRR.share) = 1 per type can
    not be applied)}
END LA_BAUnit;
```

For the LA_BAUnit class, the constraint {sum (RRR.share) = 1 per type if RRR.shareCheck} applies. The share must be specified, unless this is meaningless for the specific type (indicated by shareCheck = false). In this case, the constraint "sum (LA_RRR.share) = 1 per type" cannot be applied. This is a kind of hard constraint, but the rule is only intended for cases with RRR.shareCheck = true. In such cases, the constraint has to be valid for all of these instances, without exception.

A sample code fragment for modelling constraints in INTERLIS is displayed below:

```
STRUCTURE ExtArchive EXTENDS LADM_Base.External.ExtArchive =
    data: CharacterString;
    extraction: XMLDate;
    recordation: DateTime;
    sID: MANDATORY Oid;
    acceptance: XMLDate;
    submission: MANDATORY XMLDate;
    MANDATORY CONSTRAINT
    submission>=acceptance;
END ExtArchive;
......
CLASS GR_MarineParcel EXTENDS GR_Level =
    activity: GR_MarineActivityType;
    layerType: GR_MarineLayerType;
    resourceType: GR_MarineResourceType;
    zone: GR_MarineLayerType;
    MANDATORY CONSTRAINT
    GR_SpatialUnit.surfaceRelation = "below" OR
    GR_SpatialUnit.surfaceRelation = "mixed";"
END  GR_MarineParcel;
```

More complex constraints may include spatial data, which is useful for evaluating topological relationships (topological constraint type). The following example shows some topological rules of the COL_LADM model, along with its validation results. The "no_overlaps" function was implemented using Java plugins for iliValidator (spatial data types are explained in Section 5.3):

```
......
FUNCTION no_overlaps(
        Objects: OBJECTS OF ANYCLASS;
        SurfaceAttr: ATTRIBUTE OF @ Objects RESTRICTION (SURFACE)
        ): BOOLEAN;
......
CLASS CO_Terrain EXTENDS LA_SpatialUnit =
        ......
    geometry: MANDATORY GM_Surface2D;
    SET CONSTRAINT no_overlaps(ALL,>> geometry);
END LA_BAUnit;
```

The validation log generated by the iliValidator tool is showed below:

```
Info: ilifile <C:\tmp\uploads\ilivalidator_8040\ISO19107_V1.ili>
Info: ilifile
<C:\tmp\uploads\ilivalidator_8040\Catastro_COL_ES.ili>
Info: validate data...
Info: first validation pass...
Info: second validation pass...
...
Info: evaluate: no_overlaps Over: LADM_COL.CO_Terrain tid: 17
Error: line 20: LADM_COL.CO_Terrain: tid 17: Set Constraint
LADM_COL.CO_Terrain.no_overlaps is not true.
Info: evaluate: no_overlaps Over: LADM_COL.CO_Terrain tid: 13
Error: line 22: LADM_COL.CO_Terrain: tid 18: Set Constraint
LADM_COL.CO_Terrain.no_overlaps is not true.
...
Info: ...validation failed
```

Figure 12 illustrates a set of test data, including an overlapping area between two polygons.

Another way to define complex constraints is to create a TYPE MODEL that contains all defined functions. These functions must be implemented externally using Java code to create plugins for the iliValidator tool. After that, the model with the function definitions can be imported into any INTERLIS-based model, and the functions can then be used to create complex constraints. Here again, the advantage of using INTERLIS is the possibility of importing one model into another, applying the "modular approach", as discussed in Section 4.3.

Figure 12. Test data set created with the QGIS project generator plugin.

5.2.2. Cross-Model Constraints

Apart from the associations and constraints that exist within a single domain (i.e., legal or physical) would be wise to define some "cross-model" associations and constraints with regard to both the legal (which could refer to LA_SpatialUnit in LADM) and the physical (which could refer to either CityGML or IFC) aspects. Such associations and constraints should both exist on the conceptual level of the two models, and also be supported at the implementation stage in order to enhance the efficiency and reliability of the model [2].

In the context of this research, a cross-model constraint was defined as follows: *"the buffer of the physical object should always be inside the legal object"*. For the implementation of this constraint, a buffer is used to examine the cross legal-physical representations, as the boundaries of the object could be subject to the resolution of the representations, data quality, and accuracy. It is evident that the physical boundary of the object, including potential inaccuracies, should be inside, or at least overlap with, the legal boundary.

However, this relation is often more complex, and the multiplicity is not always one-to-one. For instance, multiple buildings may be located inside one parcel, or multiple interior spaces located inside a complex building structure; these are related to each other, and together they form one bigger legal space. Hence, given the above, the legal (or, correspondingly, the physical) space can be defined as a collection of 3D parcels in which the multiplicity is many-to-one (* ... 1).

Figure 13a,b illustrates the ownership status of the building, and its corresponding physical model. For the physical representation, a 1 m buffer has been added. Figure 14 illustrates the integrated model, comprising both legal spaces and physical elements. As depicted below, the cross-model constraint is violated, as the physical aspect is not completely inside the boundaries of the legal; the physical part that violates the constraint is shown in red.

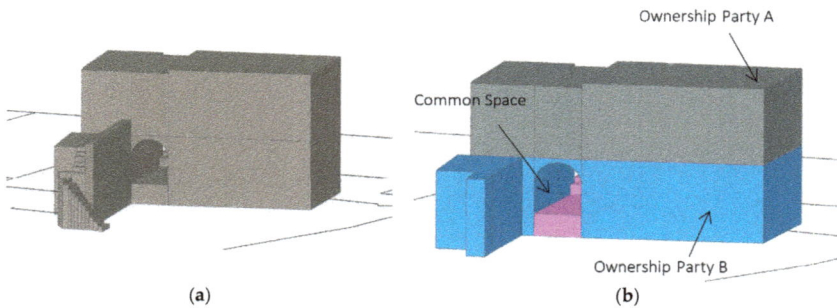

(a)   (b)

Figure 13. (a) The physical model of the building (IFC); (b) The legal model of the building.

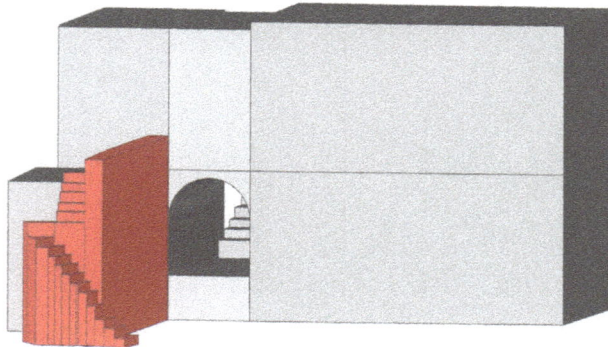

Figure 14. The integrated legal-physical model; where the cross-model constraint is violated is shown in red.

5.3. 3D Data Type in INTERLIS

LADM is based on ISO 19107, which specifies conceptual schemas for describing the spatial characteristics of geographic features and a set of spatial operations consistent with these schemas [74]. When LADM was initially translated into INTERLIS, it included the basic concepts, but was limited with respect to 3D support. The second version of LADM described in INTERLIS was released in 2016, including the second version of ISO 19107 in INTERLIS, and now supports both 2D and 3D geometries, as well as 3D structures. More precisely, the basic 3D types defined are GM_Point3D, GM_Curve3D and GM_Surface3D, and the 3D structures are GM_MultiCurve3D and GM_MultiSurface3D.

The basic 3D types are defined as follows:

```
GM_Point3D = COORD
    480000.000 .. 850000.000 [m],
    70000.000 .. 310000.000 [m],
    −1000.000 .. 9000.000 [m],
    ROTATION 2 −> 1;
GM_Curve3D = POLYLINE WITH (STRAIGHTS, ARCS) VERTEX GM_Point3D WITHOUT
OVERLAPS > 0.001;
GM_Surface3D = SURFACE WITH (STRAIGHTS, ARCS) VERTEX GM_Point3D WITHOUT
OVERLAPS > 0.001;
```

The 3D structures are defined as follows:

```
 STRUCTURE GM_Curve3DListValue =
     value: MANDATORY GM_Curve3D;
     END GM_Curve3DListValue;

STRUCTURE GM_Surface3DListValue =
     value: MANDATORY GM_Curve3D;
     END GM_Surface3DListValue;

STRUCTURE GM_MultiCurve3D =
     geometry: LIST {1..*} OF GM_Curve3DListValue;
     END GM_MultiCurve3D;

STRUCTURE GM_MultiSurface3D =
     geometry: LIST {1..*} OF GM_Surface3DListValue;
     END GM_MultiSurface3D;
```

There is no single definition for a solid, and in general, it is defined as a bounded 3D manifold, without distinguishing exterior boundaries. As stated in [74] *"A GM_Solid is the basis for 3-dimensional geometry. The extent of a solid is defined by the boundary surfaces. The boundaries of GM_Solids shall be represented as GM_SolidBoundar"*.

The authors suggested the following definition of GM_Solid as the basic 3D primitive in their previous work [2]:

```
STRUCTURE GM_Object =
END GM_Object;

STRUCTURE GM_Solid EXTENDS GM_Object =
     geometry: LIST {1..*} OF GM_Surface3DListValue;
```

END GM_Solid;

FUNCTION validateSolidGeometry(solid:ISO19107.GM_Solid):BOOLEAN;

It is noted that this structure is only a first step towards the definition of a 3D primitive in INTERLIS. A validation function (ILIFunctions.validSolid) as presented above that will ensure that the solid does not intersect with other geometries needs to be developed and added as a constraint in the GM_Solid proposed structure. If the constraint is not met, then a message should be given to the user. Finally, as presented above, in the new version of ISO 19107, the GM_MultiSurface3D structure is defined similarly to the definition proposed by the authors.

6. Conclusions

In the context of this paper, recent LADM implementation activities are briefly presented, while the need for more explicit relationships with physical models (e.g., BIM, IFC, LandXML, etc.) is underlined and the advantage of using LADM for this purpose (e.g., the external classes in LADM, such as ExtPhysicalBuildingUnit and ExtPhysicalUtilityNetwork) is highlighted. More specifically, the paper addresses an approach for enabling the representation of the relationships between physical and legal boundaries, when this is relevant in a particular country. Thus, by using the integrated LADM/INTERLIS approach, those relationships can be represented in the information model, while defining (and implementing) constraints to ensure that the actual data is valid (i.e., that there is consistency between physical and legal boundaries). Accordingly, from the research that has been carried out up until now on the link between legal and physical notions using international standards, it has been proven that the development of CityGML ADEs is becoming one of the possible solutions for the integration of LADM and CityGML.

As presented with regard to the project in Colombia (Section 4.3) being LADM-compliant will seldom be the reason behind establishing new Land Administration Systems or modernizing the existing ones. However, when a system is being modernized, then compliance with LADM is often required, while systems are benefitted. For this reason, several projects in recent years have suggested an integrated LADM/INTERLIS approach. The recent experience of implementing INTERLIS in three LADM-based country profiles has resulted in important knowledge on how to integrate those standards.

Hence, in this paper, focus is placed on the INTERLIS conceptual language and its corresponding software packages, as well as the implementation of INTERLIS formalism in the LADM standard, specifically in terms of the standard's country profiles of Switzerland, Greece and Colombia. The MDA-based system proposed by the project in Colombia could even be considered to be a basic technological solution for a Spatial Data Infrastructure related to land administration, being interesting even for implementation in the context of smaller administrations (e.g., at the level of a region or a large municipality). Thus, the integrated LADM/INTERLIS approach can be implemented in any LADM-based model to get a platform-independent exchange format linked to the conceptual model. From this integration and the projects that have adopted it thus far, it has been proved that the INTERLIS concept, supported by the development of supplementary tools, can be used as an external validating mechanism for LADM-based models. Considering all the above-mentioned factors, INTERLIS can be characterized as a promising solution for obtaining computer-processable model descriptions, and transferring conceptual model level LADM classes to technical, implementable model in XML [3].The LADM/INTERLIS approach represents a solution for the integration of legal and physical representations in a controlled manner by specifying constraints. The modularity of the LADM/INTERLIS approach (Figure 8), and hence its ability to structure country profiles in a way that consists of a core model that can be extended by several thematic or regional models of the land administration realm, is introduced in this paper.

Moreover, attention is also drawn to the system's development cycle, from conceptual model to the implementation of a working prototype within the proposed Multipurpose Land Administration

System for Greece (Section 4.2). The process followed during this development was cyclic and repetitive, providing feedback on the initial model, and improving it in terms of efficiency and technical implementation. Considering the INTERLIS tool chain, several challenges faced during this research include, among others, the formal definition of various LADM constraints (OCL and others) in INTERLIS, including cross-model constraints; the definition of extensible hierarchical and versioning code lists in INTERLIS models achieving semantic meaningful content; discussion of 3D volumetric primitives in INTERLIS; and the introduction of a holistic LADM/INTERLIS approach for developing and implementing country profiles.

Considering semantic interoperability within the domain of land administration, the need for more formal semantics (ontologies, shared terminology and concepts, semantically enriched code lists, thesaurus, etc.) is underlined in this paper. Specifically, the formalization of code list values—and especially versioned code list values (allowing the possibility of changing over time; e.g., refined definition) and hierarchically structured code lists—is proposed and modeled in INTERLIS language. With the proposed structure, the values take on a hierarchical structure, which provides some semantics, as terms higher in the hierarchy have been internationally defined and agreed upon, facilitating communication between different countries and organizations. Adding more content, meaning and structure to the current code lists for the proposed MLAS is another step in the development of the LADM.

Initially, defined LADM code lists may be used as super-classes for a number of specific code lists, whose values may be used to specify the attribute value for each country profile. Additionally, adding a unique identification code for each code list is considered necessary for the establishment of a single management system. It is noted that the unique identifier should not be changed during the life-cycle of the code list. Additionally, the first results of case studies based on the generic LADM/INTERLIS approach have been presented, showing the power of the formalized constraints.

Finally, "Legal Independency" (as introduced by Kaufmann et al. [41]) is a principle of fundamental importance in the context of LADM, ensuring that the layering of data topics is defined under a common framework of spatial reference. Along with this, the "level" concept of LADM outlines a common denominator that can be observed in land administration systems, and represents coherence in different fields (Section 4.2). Hence, it can be concluded that areas and concepts such as "level" would benefit from further international standardization and explicit modelling, as they can be shared among jurisdictions.

7. Discussion and Future Work

3D modelling techniques are rapidly being developed, providing new challenges and roles in land administration, while the demand for 3D information from stakeholders and decision makers in this field is also growing. As also stated by [75], standards like the Land Administration Domain Model are crucial to jump-starting new initiatives, and are connecting top-down and bottom-up projects, enabling involved parties to communicate based on the shared ontology implied by the model. LADM constitutes a generic expandable domain model, designed to be connected in SDI-setting to data from other domain models and other standards [61]. Additionally, LADM provides a reference model for development and refinement of efficient and effective land administration systems, and as an enabler for communication based on the shared vocabulary [76].

While standardization in land administration has been expanded to 3D and even 4D representations, adopting a multipurpose character, effective and efficient system development and maintenance of flexible systems demands further standardization and support with appropriate tools. Given this background, and as the LADM is now being studied, used and implemented around the world, it is inevitable that further issues will arise. Hence, within ISO TC211, the development of the 2nd edition of LADM has already been scheduled, and some of the identified trends, as discussed at the 6th FIG LADM Workshop [77], organized in Delft this year, are: update of current core version of LADM, potential conceptual model extensions (further modelling on RRR's or survey and spatial representation, 3D/4D Cadastre, etc.), definition of a methodology for modelling LADM country profiles, LADM implementation through application schemas, technical models and encodings (CityGML, IndoorGML, BIM, INTERLIS, RDF, etc.), development of a process and workflow standardization (initial registrations, ISO TC307, survey workflow, etc.), and development of sample implementations (open- and closed-source).

As concluded at the same Workshop [78], recent research has focused more on the link of 3D legal spaces with 3D physical spaces; as the (invisible) legal boundaries do not always match with the corresponding physical ones, 3D legal interests may exist when there is no actual construction, and vice versa. These usually lead to complex and ambiguous situations and discrepancies between the situation as presented by the legal models and the reality (physical models), while more integrated approaches have been developed (at the moment at a conceptual level) as described in Section 2. Thus, LADM implementation becomes a matter of priority, which has also been proved by the numerous research and professional projects that have been developed up to now.

Moreover, Multipurpose Land Administration Systems serve numerous goals, such as registering various rules and legal aspects for land use and tenure, spatial planning, taxation, etc. As described in Section 4.2, a holistic approach is taken when designing and implementing such systems, as they comprise land, marine and air parcels, buildings, utilities, archaeological spaces, environmental protected areas, natural resources, spatial planning zones, etc. Most of those cadastral objects have a 3D nature, which is not reflected by traditional registration in 2D systems, and often leads to confusing and complex situations.

The current version of LADM addresses survey, geometry, topology, party, legal and administrative aspects, while, as stated before, the revision of the standard has already been scheduled, and will start this year, and is expected to explore multiple extensions. One of these is LADM implementation of possibilities using either one of the existing application schemas and encodings, or a newly-developed, sample one. In this context, the possibilities of INTERLIS have been explored in terms of its complementary use (as a first step) with LADM, and the first results have been promising. However, much more work needs to be done in this direction.

As a final point in the context of LADM implementation, the "ExtPhysicalBuildingUnit" class (as represented according to CityGML, IndoorGML, or BIM/IFC) should be further explored when linking legal to physical models/ representations, including the level of detail as introduced in for example CityGML.

Considering the next steps, various technical and compatibility problems between INTERLIS tools and other systems should be addressed in their subsequent versions. Specifically, by introducing adequate general XML tools for INTERLIS, it is possible to make the concrete INTERLIS/XML-files available for an even wider range of applications. For instance, translations of standards such as CityGML, LandXML, InfraGML IndoorGML to INTERLIS, or vice-versa, would be interesting and important topics for future research. Furthermore, as INTERLIS starts to break Swiss borders, the update and development of the necessary mappings between the tools and the proposed structures is becoming necessary (i.e., mappings for the proposed hierarchical and versioned structure of code lists).

Additionally, initially defined LADM code lists may be used as super-classes for a number of specific code lists whose values may be used to specify the attribute value for each country profile. As a next step, in order to clear separate country profiles, the principles of the ISO 3166 (the International Standard for country codes and the codes for their subdivisions [79]) could be introduced. The question arises as to who will be responsible for managing the register, as well as who will have access in order to update those values, etc.

What's more, complex and extensive constraints must be defined with views (e.g., a view that connects a certain class with itself), thus allowing the comparison of any attribute combination with all other objects of the class. Such constraints can be expressed by creating functions or triggers at the database level or, in cases where iliValidator is used, as functions that could be defined in a separate

TYPE MODEL, and which call Java plugins executed by the tool. It is evident that the next version of INTERLIS will cover some of the afore-mentioned proposals.

More complete cross-model constraint and spatial constraint modelling is required, while, at the same time, performing complicated queries in order to visualize and investigate the integration of spatial and non-spatial data will lead to more conclusions. Additionally, the temporal dimension is needed, as both current and historic versions of (legal and physical objects) should always be accessible.

As INTERLIS has the necessary functionalities to be used as an external validating mechanism for LADM models, it is expected that further tools will be developed in order to support a holistic LADM/INTERLIS approach. In this context, and because the 3D primitives have now been introduced in the latest INTERLIS version, tools for 3D volumetric primitive validation should be developed. Ledoux [80] developed a methodology (val3dity) for validating solids according to the international standards, and states that having validation tools that respect the international standards helps to foster interoperability. The automatic repair of invalid solids could be considered as a next step.

As the LADM/INTERLIS approach becomes more familiar, building up adequate know-how in order to implement the LADM by means of INTERLIS should be accompanied by specific courses and on-the-job training in the use of the language and the tools, but also in understanding the LADM as a conceptual model. Some first steps have been made, and a lot of short training courses in exactly the afore-mentioned topics have been carried out in the context of the project in Colombia.

Concerning the next steps of this project, providing continued support to Land Administration stakeholders in Colombia for developing and implementing their thematic or regional models as specializations of the defined core model (spatial planning, tenure formalization, department level) is a matter of priority. The development, maintenance and improvement of the tool chain—i.e., UML/INTERLIS-Editor, ili2db, iliValidator and Project Generator plugin in QGIS—in response to user needs, experience and knowledge gained, and in terms of efficiency, is also one of the planned next steps.

Last but not least, bearing in mind LADM use and development, further "international" modelling of some areas that are frequently used in country profiles is becoming a necessity, as LADM is being more frequently used as a reference model in many countries. Indicatively, the LADM "level" approach needs further exploration in terms of standardized modelling, possibly in terms of legal, geometrical and thematic coherence. Similarly, the modular approach, where a national or core profile is extended by regional or thematic profiles, is productive, and INTERLIS provides the means to formalize such model extensions.

Acknowledgments: The authors would like to thank Germán Carrillo and Fabian Mejia for their support, for providing significant material from the project in Colombia, and contributing to the critical review of the paper with regard to the implementation of the LADM/INTERLIS approach in Colombia, and the future scheduled steps of the project.

Author Contributions: This research is a result of the collaboration and contribution of all authors. Jenni Lorenz is responsible for the description and implementation of the LADM/INTERLIS approach in Colombia, while Eftychia Kalogianni implemented the LADM/INTERLIS approach in Greece under the supervision and guidance of Peter van Oosterom and Efi Dimopoulou. Michael Germann provided valuable context for the first steps of INTERLIS and LADM integration and the implementation of the approach in Switzerland, while Wilko Quak took care of the technical issues during implementation. The coordinator of the authors is Peter van Oosterom, and each author made a substantial contribution in the preparation of the manuscript.

Conflicts of Interest: The authors declare no conflict of interest.

References

1. Object Management Group. Model Driven Architecture (MDA) MDA Guide Rev. 2.0. 2014. Available online: http://www.omg.org/cgi-bin/doc?ormsc/14-06-01.pdf (accessed on 26 September 2017).
2. Kalogianni, E.; Dimopoulou, E.; Quak, W.; van Oosterom, P.J.M. Formalizing Implementable Constraints in the INTERLIS Language for Modelling Legal 3D RRR Spaces and 3D Physical Objects. In Proceedings of the 5th International FIG 3D Cadastre Workshop, Athens, Greece, 18–20 October 2016; pp. 137–157.

3.   Germann, M.; Kaufmann, J.; Steudler, D.; Lemmen, C.; van Oosterom, P.; de Zeeuw, K. The LADM based on INTERLIS. In Proceedings of the FIG Working Week 2015 from the Wisdom of the Ages to the Challenges of the Modern World, Sofia, Bulgaria, 17–21 May 2015.

4.   Germann, M.; Kaufmann, J.; Steudler, D.; Lemmen, C.H.J.; van Oosterom, P.; De Zeeuw, K. The LADM based on INTERLIS. In Proceedings of the World Cadastre Summit, Istanbul, Turkey, 20–24 April 2015.

5.   Jenni, L.; Guarín, L.A.; Ziegler, S.; Pérez, B.V.M. Development and Employment of a LADM Implementing Toolkit in Colombia. In Proceedings of the 2017 World Bank Conference on Land and Poverty: Responsible Land Governance–Towards an Evidence-Based Approach, The World Bank, Washington, DC, USA, 20–24 March 2017.

6.   Jenni, L.; Germann, M.; Eisenhut, C.; Guarin, L.A.; Bajo, V.M. LADM Implementation in Colombia–Process, Methodology and Tools developed and applied. In Proceedings of the FIG Working Week, Helsinki, Finland, 29 May–2 June 2017.

7.   Kalogianni, E. Design of a 3D Multipurpose Land Administrative System for Greece in the Context of Land Administration Domain Model (LADM). Master's Thesis, National Technical University of Athens, Athens, Greece, 2015.

8.   Kalogianni, E.; Dimopoulou, E.; van Oosterom, P.J.M. A 3D LADM prototype implementation in INTERLIS. In Proceedings of the 10th 3D GeoInfo Conference, Kuala Lumpur, Malaysia, 28–30 October 2015; Abdul-Rahman, A., Ed.; Springer Nature: Berlin, Germany, 2017; pp. 385–408.

9.   Kalogianni, E. Linking the Legal with the Physical Reality of 3D Objects in the Context of Land Administration Domain Model (LADM). Master's Thesis, Delft University of Technology, Delft, The Netherlands, 2016.

10.   International Organization for Standardization. *ISO 19152, Geographic Information–Land Administration Domain Model (LADM)*, 1st ed.; ISO: Geneva, Switzerland, 2012. Available online: https://www.iso.org/standard/51206.html (accessed on 1 August 2017).

11.   De Vries, T.; Zlatanova, S. 3D Intelligent Cities. *GEO Inf.* **2011**, *14*, 6–8.

12.   Berry, K.J.; Mehta, S. An Analytical Framework for GIS Modeling. 2009. Available online: http://www.innovativegis.com/basis/papers/other/gismodelingframework/GISmodelingFramework.pdf (accessed on 19 April 2017).

13.   Aien, A.; Kalantari, M.; Rajabifard, A.; Williamson, I.; Bennett, R. Utilizing data modeling to understand the structure of 3D Cadastres. *J. Spat. Sci.* **2013**, *58*, 215–234. [CrossRef]

14.   Gózdz, K.; Pachelski, W.; van Oosterom, P.J.M.; Coors, V. The possibilities of using CityGML for 3D representation of buildings in the Cadastre. In Proceedings of the 4th International FIG 3D Cadastre Workshop, Dubai, UAE, 9–11 November 2014; pp. 339–361.

15.   Ying, S.; Guo, R.; Li, L.; He, B. Application of 3D GIS to 3D Cadastre in Urban Environment. In Proceedings of the 3rd International FIG 3D Cadastre Workshop: Developments and Practices, Shenzhen, China, 25–26 October 2012; pp. 25–26.

16.   Van Oosterom, P.J.M.; Stoter, J. 5D Data Modelling: Full Integration of 2D/3D Space, Time and Scale Dimensions. *Geogr. Inf. Sci.* **2010**, *6292*, 310–324.

17.   ePlan. In *ePlan Model*; Version 1.0; Intergovernmental Committee on Surveying and Mapping: Sydney, Australia, 2010; p. 61.

18.   Aien, A.; Rajabifard, A.; Kalantari, M.; Shojaei, D. Integrating legal and physical dimensions of urban environments. *ISPRS Int. J. Geo-Inf.* **2015**, *4*, 1442–1479. [CrossRef]

19.   IHO S-121. Product Specification for Maritime Limits and Boundaries. Available online: https://www.google.com/url?sa=t&rct=j&q=&esrc=s&source=web&cd=1&ved=0ahUKEwjhwZniv9DUAhUlCcAKHbrGBvEQFggiMAA&url=https%3A%2F%2Fwww.iho.int%2Fmtg_docs%2Fcom_wg%2FS-100WG%2FS-121PT%2FS121%2520Draft%2520Product%2520Specification%2520Revised%252001%2520Dec%252016%2520v2.3.8.docx&usg=AFQjCNFv_PaTC96xHdMRr9tigRZ6s9ERLg&cad=rja (accessed on 1 August 2017).

20.   BuildingSMART. IFC Product Extension. 2013. Available online: http://www.buildingsmarttech.org/ifc/IFC4/final/html/schema/ifcproductextension/content.htm (accessed on 18 April 2017).

21.   *Industry Foundation Classes (IFC) for Data Sharing in the Construction and Facility Management Industries*; ISO16739; International Organization for Standardization (ISO): Geneva, Switzerland, 2013.

22.   Groger, G.; Kolbe, T.H.; Nagel, C.; Hafele, K.H. *OGC City Geography Markup Language (CityGML) Encoding Standard*; Open Geospatial Consortium: Wayland, MA, USA, 2012.

23. Scarponcini, P. *InfraGML Proposal (13–121), OGC Land and Infrastructure DWG/SWG*; Open Geospatial Consortium: Wayland, MA, USA, 2013.

24. Scarponcini, P.; Gruler, H.C.; Stubkjaer, E.; Axelsson, P.; Wikstrom, L. *OGC, Land and Infrastructure Conceptual Model Standard (LandInfra)*; Open Geospatial Consortium: Wayland, MA, USA, 2016.

25. LandXML. LandXML 2.0 (Working Draft) Schema Announced. 2014. Available online: http://landxml.org/ (accessed on 7 October 2016).

26. Zlatanova, S.; van Oosterom, P.J.M.; Lee, J.; Lic, K.J.; Lemmen, C.H.J. LADM and IndoorGML for support of indoor space identification. In Proceedings of the 11th 3D GeoInfo Conference, Athens, Greece, 20–21 October 2016.

27. Atazadeh, B.; Kalantari, M.; Rajabifard, A. Comparing three types of BIM-Based Models for managing 3D ownership interests in multi-level buildings. In Proceedings of the 5th International FIG 3D Cadastre Workshop, Athens, Greece, 18–20 October 2016; pp. 183–198.

28. Stadler, A.; Kolbe, T.H. Spatio-semantic Coherence in the Integration of 3D City Models. In Proceedings of the 5th International Symposium on Spatial Data Quality (ISSDQ), Enschede, The Netherlands, 13–15 June 2007.

29. Dsilva, M.G. A Feasibility Study on CityGML for Cadastral Purposes. Master's Thesis, Eindhoven University of Technology, Eindhoven, The Netherlands, 2009.

30. Çağdaş, V. An application domain extension to CityGML for immovable property taxation: A Turkish case study. *Int. J. Appl. Earth Obs. Geoinf.* **2013**, *21*, 545–555. [CrossRef]

31. Rönsdorff, C.; Wilson, D.; Stoter, J.E. Integration of Land Administration Domain Model with CityGML for 3D Cadastre. In Proceedings of the 4th International FIG 3D Cadastre Workshop, Dubai, UAE, 9–11 November 2014; pp. 313–322.

32. Li, L.; Wu, J.; Zhu, H.; Duan, X.; Luo, F. 3D Modelling of the ownership structure of condominium units. *Comput. Environ. Urban Syst.* **2016**, *59*, 50–63. [CrossRef]

33. Isikdag, U.; Horhammer, M.; Zlatanova, S.; Kathmann, R.; van Oosterom, P.J.M. Semantically rich 3D building and cadastral models for valuation. In Proceedings of the 4th International FIG 3D Cadastre Workshop, Dubai, UAE, 9–11 November 2014; pp. 35–53.

34. El-Mekawy, M.; Östman, A. Feasibility of Building Information Models for 3D Cadastre in Unified City Models. *Int. J. E-Plan. Res.* **2012**, *1*, 35–58. [CrossRef]

35. Oldfield, J.; van Oosterom, P.J.M.; Quak, W.; van der Veen, J.; Beetz, J. Can Data from BIMs be Used as Input for a 3D Cadastre? Formalizing Implementable Constraints in the INTERLIS Language for Modelling Legal 3D RRR Spaces and 3D Physical Objects. In Proceedings of the 5th International FIG 3D Cadastre Workshop, Athens, Greece, 18–20 October 2016; pp. 199–214.

36. Kim, S.; Kim, J.; Jung, J.; Heo, J. Development of a 3D Underground Cadastral System with Indoor Mapping for As-Built BIM: The Case Study of Gangnam Subway Station in Korea. *Sensors* **2015**, *15*, 30870–30893. [CrossRef] [PubMed]

37. Zlatanova, S.; Lic, K.J.; Lemmen, C.H.J.; van Oosterom, P.J.M. Indoor Abstract Spaces: Linking IndoorGML and LADM. In Proceedings of the 5th International FIG 3D Cadastre Workshop, Athens, Greece, 18–20 October 2016; pp. 317–328.

38. Soon, K.H. Representing Roles in Formalizing Domain Ontology for Land Administration. In Proceedings of the 5th Land Administration Domain Model Workshop, Kuala Lumpur, Malaysia, 24–25 September 2013; pp. 203–222.

39. Soon, K.H.; Thompson, R.; Khoo, V. Semantics-based Fusion for CityGML and 3D LandXML. In Proceedings of the 4th International FIG 3D Cadastre Workshop, Dubai, UAE, 9–11 November 2014; pp. 323–338.

40. Hespanha, J.P.; van Bennekom-Minnema, J.; van Oosterom, P.J.M.; Lemmer, C.H.J. The Model Driven Architecture Approach Applied to the Land Administration Domain Model Version 1.1-with Focus on Constraints Specified in the Object Constraint Language. In Proceedings of the Integrating Generations, FIG Working Week, Stockholm, Sweden, 14–19 June 2008.

41. Kaufmann, J.; Steudler, D. Cadastre 2014–A vision for a future cadastral system. In Proceedings of the FIG XXI International Congress, Commission 8: Spatial Planning and Development, Brighton, UK, 19–25 July 1998.

42. Hespanha, J.P. Development Methodology for an Integrated Legal Cadastre. Ph.D. Thesis, Delft University of Technology, Deft, The Netherlands, 2012.

43. Hespanha, J.P.; Paixao, S.; Ghawana, T.; Zevenbergen, J.; Carneiro, A. Application of Land Administration Domain Model to Recognition of Indigenous Community Rights in the Philippines: Laws Examined with

Spatial Dimensions (7579). In Proceedings of the FIG Working Week 2015, From the Wisdom of the Ages to the Challenges of the Modern World, Sofia, Bulgaria, 17–21 May 2015.

44. Van Bennekom-Minnema, J. The Land Administration Domain Model "Survey Package" and Model Driven Architecture, Joint Programme in Geographical Information Management and Applications. Master's Thesis, Utrecht University, Utrecht, The Netherlands, University of Twente/ITC Enschede, Enschede, The Netherlands, TU Delft, Deft, The Netherlands, Wageningen University, Wageningen, The Netherlands, 2008.

45. Social Domain Tenure Model (SDTM). Available online: http://stdm.gltn.net/ (accessed on 1 July 2017).

46. Van Bennekom-Minnema, J. Example Implementation LADM: IT System Specification, Draft 25 March 2011.

47. Zein, T.; Hartfiel, P.; Berisso, Z.A. Addis Ababa: The Road Map to Progress through Securing Property Rights with Real Property Registration System. In Proceedings of the World Bank Conference on Land and Poverty, Washington, DC, USA, 23–26 April 2012.

48. Gruber, T.R. A Translation Approach to Portable Ontology Specifications. In *Knowledge Acquisition*; Academic Press Ltd.: London, UK, 1993; Volume 5, pp. 199–220.

49. Mäs, S.; Wang, F.; Reinhardt, W. Using ontologies for integrity constraint definition. In Proceedings of the 4th International Symposium on Spatial Data Quality (ISSDQ), Beijing, China, 25–26 August 2005.

50. Object Management Group. Documents Associated With Object Constraint Language (OCL™), Version 2.4. 2014. Available online: http://www.omg.org/spec/OCL/2.4/ (accessed on 26 September 2017).

51. Louwsma, J.; Zlatanova, S.; van Lammeren, R.; van Oosterom, P.J.M. Specifying and implementing constraints in GIS–with examples from a Geo-Virtual Reality System. *GeoInformatica* **2006**, *10*, 531–550. [CrossRef]

52. Xu, D.; van Oosterom, P.J.M.; Zlatanova, S. A Methodology for modelling of 3D Spatial Constraints, Chapter. In *Advances in 3D Geoinformation*; Abdul-Rahman, A., Ed.; Springer: Basel, Switzerland, 2016; pp. 95–117.

53. Chiang, R.H.L.; Barron, T.M.; Storey, V.C. Reverse engineering of relational databases: Extraction of an EER model from a relational database. *Data Knowl. Eng.* **1994**, *12*, 107–142. [CrossRef]

54. Duboisset, M.; Pinet, F.; Kang, M.A.; Schneider, M. Precise modelling and verification of topological integrity constraints in spatial databases: From an expressive power study to code generation principles. In *Conceptual Modelling–ER*; Delcambre, L., Kop, C., Mayr, H.C., Mylopoulos, J., Pastor, O., Eds.; Springer: Berlin, Germany, 2005; pp. 465–482.

55. Van Oosterom, P.J.M. Constraints in spatial data models, in a dynamic context. In *Dynamic and Mobile GIS: Investigating Changes in Space and Time*; Drummond, J., Billen, R., Joao, E., Forrest, D., Eds.; CRC Press: Boca Raton, FL, USA, 2006; pp. 104–137.

56. Lemmen, C.H.J.; van Loenen, B.; van Oosterom, P.J.M.; Paasch, J.; Paulsson, J.; Ploeger, H.D.; Zevenbergen, J. Legal Refinement of the LADM Standard: More classes or extended code lists with better defined types of Rights, Restrictions and Responsibilities? In Proceedings of the PLPR 2014 8th Annual Conference on Planning, Law and Property Rights, Haifa, Israel, 12–14 February 2014.

57. European Land Information Service (EULIS). 2017. Available online: http://eulis.eu/about-us/ (accessed on 25 March 2017).

58. Lutz, M.; Portele, C.; Cox, S.M.J.; Muraay, K. Code Lists for Interoperability—Principles and Best Practices in INSPIRE. 2012. Available online: https://presentations.copernicus.org/EGU2012-10415_presentation.pdf (accessed on 23 October 2017).

59. International Organization for Standardization. *ISO 19135:2005 Geographic Information–Procedures for Item Registration*; ISO: Geneva, Switzerland, 2005.

60. W3C OWL Working Group. OWL 2 Web Ontology Language: Document Overview (Second Edition). Available online: http://www.w3.org/TR/owl2-overview/ (accessed on 10 July 2013).

61. Janecka, K.; Karki, S. 3D Data Management–Overview Report. In Proceedings of the 5th International FIG 3D Cadastre Workshop, Athens, Greece, 18–20 October 2016; pp. 215–260.

62. W3C. SKOS Simple Knowledge Organization System Namespace. 2004. Available online: https://www.w3.org/TR/owl-ref/ (accessed on 23 October 2017).

63. Web Ontology Language (OWL). Available online: https://www.w3.org/OWL/ (accessed on 1 August 2017).

64. Çağdaş, V.; Abdullah, K.; Işikdağ, Ü.; van Oosterom, P.J.M.; Lemmen, C.H.J.; Stubkjaer, E. A Knowledge Organization System for the Development of an ISO 19152:2012 LADM Valuation Module. In Proceedings of

the FIG Working Week 2017, Surveying the World of Tomorrow-From Digitaisation to Augmented Reality, Helsinki, Finland, 29 May–2 June 2017.

65. Çağdaş, V.; Stubkjær, E. Core immovable property vocabulary for European linked land administration. *J. Surv. Rev.* **2014**, *47*, 49–60. [CrossRef]
66. Cadastre and Land Administration Thesaurus (CaLAThe). 2017. Available online: http://cadastralvocabulary. org (accessed on 2 October 2016).
67. Paasch, M.J. Standardization of Real Property Rights and Public Regulations. The Legal Cadastral Domain Model. Ph.D. Thesis, Royal Institute of Technology (KTH), Stockholm, Sweden, 2012.
68. Paassch, M.J.; van Oosterom, P.J.M.; Lemmen, C.H.J.; Paulsson, J. Further modelling of LADM's rights, restrictions and responsibilities (RRRs). *Land Use Policy* **2015**, *49*, 680–689.
69. KOGIS. *INTERLIS 2.3 Reference Manual*; Coordination, Geo-Information and Services (COGIS), a Division of the Swiss Federal Office of Topography: Wabern, Switzerland, 2006.
70. Swiss Government. Federal Act on Geoinformation. Bern. 2007. Available online: http://www.admin.ch/ opc/en/classified-compilation/20050726/index.html (accessed on 7 October 2016).
71. Lemmen, C.H.J. A Domain Model for Land Administration. Ph.D. Thesis, Delft University of Technology, Deft, The Netherlands, 2012.
72. Acuerdo Final. Acuerdo Final Para la Terminación del Conflicto y la Construcción de una Paz Estable y Duradera from Mesa de Conversaciones. Available online: https://www.mesadeconversaciones.com.co/ sites/default/files/24_08_2016acuerdofinalfinalfinal-1472094587.pdf (accessed on 26 September 2017).
73. Zulkifli, N.; Rahman, A.; Jamil, H.; Hua, T.C.; Choon, T.L.; Seng, L.K.; Lim, C.K.; van Oosterom, P.J.M. Towards Malaysian LADM Country Profile for 2D and 3D Cadastral Registration System. In Proceedings of the 10th 3D GeoInfo Conference, Kuala Lumpur, Malaysia, 28–30 October 2015.
74. International Organization for Standardization. *ISO 19107:2003 Preview Geographic Information–Spatial Schema*; ISO: Geneva, Switzerland, 2003.
75. Lemmen, C.H.J.; van Oosterom, P.J.M.; Kalantari, M.; Unger, E.M.; Teo, C.H.; de Zeeuw, K. Further Standardization in Land Administration. In Proceedings of the 2017 World Bank Conference on Land and Poverty: Responsible Land Governance–Towards an Evidence-Based Approach, The World Bank, Washington, DC, USA, 20–24 March 2017.
76. Lemmen, C.H.J.; van Oosterom, P.J.M.; Bennett, R. The Land Administration Domain Model. *Land Use Policy* **2015**, *49*, 535–545. [CrossRef]
77. The 6th Land Administration Domain Model (LADM) Workshop. Available online: http://wiki.tudelft.nl/ bin/view/Research/ISO19152/LADM2017Workshop (accessed on 25 August 2017).
78. Summary of the 6th Land Administration Domain Model (LADM) Workshop. Available online: http://wiki. tudelft.nl/pub/Research/ISO19152/WorkshopAgenda2017/8_9_LADM_prelim_decisions.pdf (accessed on 25 August 2017).
79. International Organization for Standardization/TC46. *ISO 3166–1, Codes for the Representation of Names of Countries and Their Subdivisions–Part 1: Country Codes*; ISO: Geneva, Switzerland, 2006.
80. Ledoux, H. On the Validation of Solids Represented with the International Standards for Geographic Information. *Comput. Aided Civ. Infrastruct. Eng.* **2013**, *28*, 693–706. [CrossRef]

International Journal of
Geo-Information

isprs

MDPI

Article

Working with Open BIM Standards to Source Legal Spaces for a 3D Cadastre

Jennifer Oldfield [1,*]**, Peter van Oosterom** [2]**, Jakob Beetz** [3] **and Thomas F. Krijnen** [4]

[1]   Faculty of Geosciences, Universiteit Utrecht, Willem C. van Unnikgebouw, Heidelberglaan 2,
    3584 CS Utrecht, The Netherlands
[2]   Faculty of Architecture and the Built Environment, TU Delft, Julianalaan 134, 2628 BL Delft,
    The Netherlands; P.J.M.vanOosterom@tudelft.nl
[3]   Faculty of Architecture, RWTH Aachen University, Templergraben 55, 52062 Aachen, Germany;
    j.beetz@caad.arch.rwth-aachen.de
[4]   Department of the Built Environment, TU Eindhoven, Vertigo Building, De Wielen, P.O. Box 513,
    5600 MB Eindhoven, The Netherlands; t.f.krijnen@tue.nl
*   Correspondence: jennifer.oldfield@oldfieldlanguage.com; Tel.: +31-(0)62-753-9488

Received: 1 September 2017; Accepted: 16 October 2017; Published: 7 November 2017

Abstract: Much work has already been done on how a 3D Cadastre should best be developed. An inclusive information model, the Land Administration Model (LADM ISO 19152) has been developed to provide an international framework for how this can best be done. This conceptual model does not prescribe the technical data format. One existing source from which data could be obtained is 3D Building Information Models (BIMs), or, more specifically in this context, BIMs in the form of one of buildingSMART's open standards: the Industry Foundation Classes (IFC). The research followed a standard BIM methodology of first defining the requirements through the use of the Information Delivery Manual (IDM ISO29481) and then translating the process described in the IDM into technical requirements using a Model View Definition (MVD), a practice to coordinate upfront the multidisciplinary stakeholders of a construction project. The proposed process model illustrated how the time it takes to register 3D spatial units in a Land Registry could substantially be reduced compared to the first 3D registration in the Netherlands. The modelling of an MVD or a subset of the IFC data model helped enable the creation and exchange of boundary representations of topological objects capable of being combined into a 3D legal space overview map.

Keywords: BIM; IDM; workflow; MVD; Environs Act; land registry map; superficies; complex multi-use rights

1. Introduction

1.1. Orientation

The Twenty-first Century is the century of Smart Cities, cities which are mapped to the last detail and which are used by a broad tapestry of stakeholders for seamless communication throughout the lifecycle of buildings (or other constructions) and cities at every level. Smart Cities are mapped in 3D and have buildings designed and managed by means of Building Information Models (BIMs). The mapping of buildings into 3D BIMs needs a 3D Cadastre or Land Registry [1] to complement it with respect to the legal status of the objects, land, and space, whereas the BIM primarily deals with a building as a decomposition of physical elements from an engineering perspective. While this need has long been acknowledged in the densely-populated countries (areas), the Cadastral map remains based on two dimensions, although, in reality, property units include both height and depth. Creating a 3D system, as is being pioneered in several countries around the world, would have many advantages [2,3]. One simple, cost-effective way to achieve a 3D Cadastre could be to adapt or design a development workflow from which it is possible to (re-)use information from existing BIMs to create 3D parcels.

Section 1 of this paper further introduces the reader to BIM and 3D Cadastres, respectively, and to the various open standards, which would be used to achieve the goal of reusing data. Section 2 looks at the methodology used. In Section 3, a collaborative workflow and two use cases are detailed. Section 4 contains our main conclusions and suggestions for future work.

1.2. A Background to BIM

The Building Information Council of the Netherlands (BIR) and the Dutch BIM Gateway (BIMLoket) define BIM as follows:

- Firstly, as the Building Information Model. This is a digital representation of how a (physical) building (including its facilities) is designed, is realized, and how it ends up.
- Secondly, Building Information Modelling places more emphasis on the process, both alone and in partnership. It is about working independently and cooperatively on building projects with the help of exchanging/sharing digital information models [4].

These two aspects of BIM have been developed by the industry with the aim of 'bringing the numerous threads of different information used in construction into a single environment' [5]. In turn, the need for many—often paper-based—documents is either eliminated or reduced by exchanging digital documents. BIM is also used to improve communication between parties. When it is used properly, good quality information, which can be understood by all, is on hand when it is needed. This is something that improves the construction process overall.

The world of BIM encompasses proprietary BIM such as the products produced by Bentley or Autodesk and open BIM, represented by buildingSMART. buildingSMART's BIM standards, which are used in this research, are the buildingSMART Data Dictionary or International Framework of Dictionaries (IFD), the Information Delivery Manual (IDM), the Industry Foundation Classes (IFC) and the Model View Definition (MVD).

It should be noted that commercial products are increasingly supporting open BIM standards. The two standards used in particular in this article are the Information Delivery Manual (IDM) and the IFC (Industry Foundation Class). The IDM is a methodology used to capture and specify processes and information flow during the lifecycle of a facility [5]. The creation and maintenance of a facility, for example a complex construction project, involves many different participants. Knowing what information needs to be communicated between them and when is important. The IDM Part 1 makes use of Business Process Modelling Notation (BPMN) and templates for Exchange Requirements in order to facilitate this process.

The IFC is a set of object definitions and exchange formats that promote interoperability between different platforms and that are used to transport data [6]. They promote interoperability within the industry as well. The Lakeside Restaurant data set (see Section 3.3.2 for more details), for example, was made using Autodesk's Revit and then made available to all using the IFC standard. This standard allows it to be imported back into Revit and into many other applications as well, especially into coordination tools that can aggregate IFC models from various sources into a single comprehensive view.

Complementing these standards are the MVD, which details which objects are required within a specific context, their required attributes, and the possible values for these attributes, all defined as a subset of the IFC schema. The IFD is a data dictionary that underpins building processes by mapping multilingual terms, their attributes, and their relationships [6].

1.3. Background 3D Cadastre

Land administration involves maintaining a cadastral mapping agency and a Land Titles office or Land Registry. In some countries, these roles are maintained by a single organization; in others they are separate. The organizations involved are considered the governmental authority in each region or country where they administer land. As such, they have a pivotal role in the Smart City concept as a coordinating institution. It makes sense, therefore, that, where possible, data from these organisations should be managed and maintained through the use of semantically rich 3D Models. This will bring it in line with other developments in this field, for example BIM. Not only would it bring it in line, but reusing the rich content of BIM models based on open exchange standards would save on costs and provide an interoperable result. For instance, BIM geometry could be reused for 3D Cadastral parcels, or, the other way around, Cadastral maps could be seamlessly imported into authoring environments at project initiation. A considerable increase in property values, both public and private, means that a clearer picture with relation to their property Rights, Restrictions, and Responsibilities (RRR)—the 3D RRRs of the Cadastral world [3,7]—is needed. For example, a 3D spatial representation derived from a BIM of an apartment spread across multiple rooms but belonging to one owner could bring clarity to ensuing legal issues.

The term RRR can be further explained as follows: 'rights' usually refers to property ownership but can also refer to land use; 'restrictions' refers to situations such as a building being a listed one and that it has to be painted a certain colour or to the existence of public access across a property. 'Responsibilities' is more difficult to define but could refer to the fact that 24-h access needs to be given to a water main underneath a property.

While BIM could be used to provide the geometry for a 3D Cadastre, the spaces themselves can be different from those defined by the BIM. While BIMs work with complex physical spaces, for example the rooms, corridors, walls, and floors of a building, the legal space needs to work with only one space for a single property. This is despite this space containing a number of physical spaces (rooms) or parts of physical spaces (to the middle of the wall space). In this new legal space, the boundary surface binds and defines the size of the spatial unit and thus the right that an entity such as an owner can claim on it. This new legal space can be amply represented in open BIM exchange models.

A further example is the space that might need to be left around a pipe. Imagine a pipe situated at the bottom of someone's garden. The title deed specifies a restriction. This restriction is that nothing can be built over the pipe so that access can always be gained to it. Thus the legal space is not the pipe itself but an extended space around it.

While BIM models physical infrastructure, the Land Administration Domain Model (LADM) works from the perspective of legal spaces. Its focus is on the "rights, responsibilities and restrictions which affect land or water and that land's geometrical components" [1]. The LADM International Organization for Standardization (ISO) 19152 is an open standard, which has been adopted by the International Organisation for Standardisation (ISO). The LADM is a conceptual or information model but is not a data product specification. Thus, it does not detail how to deal with what it describes in practice, nor does it provide any region-specific solutions. For example, it does not provide any encoding during exchange using EXtensible Markup Language (XML) or data storage in a database. It is a 'descriptive standard' rather than a 'prescriptive standard' [1]. Part of the challenge of extracting data from BIM for use in a 3D Cadastre is mapping the IFC information model to the LADM.

Two parts of the LADM were relevant for this paper; the first was the spatial unit package, illustrated below in Figure 1, and the second was the surveying and representation package, which is associated with the spatial unit package. The spatial unit package covers many eventualities. A spatial unit group, for instance, is the grouping of spatial units into an administrative zone such as a municipality or canton. A spatial unit level, by contrast, is "a collection of spatial units with a geometric and/or topological and/or thematic coherence"[6].

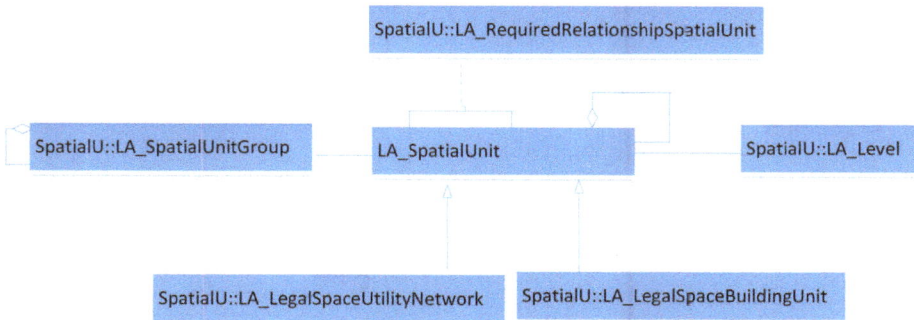

Figure 1. Land Administration Domain Model (LADM): Spatial Unit Package [1].

The LADM is a conceptual model, which has suggested geometry but no associated exchange format. This suggested geometry is found in the surveying and representation package. The most relevant entities from the LADM (with prefix LA_) and their attributes (with prefix GM_ for geometry) were LA_Point (GM_Point), LA_BoundaryFaceString (GM_Multicurve), and LA_BoundaryFace (GM_Multisurface).

1.4. Background: The Environs Act

Just like other countries around the world, the Netherlands is working towards a 3D Map of its legal spaces but has yet to realise it completely. Within the same organization—the *Kadaster*—moves are afoot to revamp the Large Scale Topographical Map (*Basisregistratie Grootschalige Topografie* (BGT)) of the Netherlands. This map is a digital map, which illustrates buildings, roads, water, railway lines, and vegetation simply and clearly. It is the result of the cooperative efforts of the municipalities, provinces, water companies, the Ministry of Industry and Trade (*Economische Zaken -EZ*), the Ministry of Defence, the organization that administers the railways (*Prorail*), and the organisation that administers infrastructure (*Rijkswaterstaat*). Each member of the group is responsible for a section of the digital map [8].

The revamp of the Large-Scale Topographical Map (*BGT*) is being completed within the legal framework of a revolutionary law; the *Omgevingswet* or 'Environs Act'. This act is replacing tens of other laws and hundreds of regulations with the aim of simplifying the system and making it more user-friendly. Concurrently, the management and development of water, air, soil, the natural environment, infrastructure, buildings, and cultural heritage will be able to be done more efficiently and effectively.

A key part of this process is the incorporation of IFC files into the system. These will be collected from building and address registrations managed by municipalities under the *Basis Registraties Adressen en Gebouwen* (BAG) or the 'Registration of Buildings and Addresses' law. The number of BIMs being created within the Netherlands is steadily increasing. At this point, buildings and other constructions such as infrastructure, which cost €10,000,000 and above, are being designed using a BIM [9]. The adoption of BIM is a process that was stimulated rather than hindered by the recession, which began in 2008 [9]. The increasing use of BIM has been as a result of market forces in an attempt to reduce costs, rather than a government mandate, and has occurred in spite of—rather than because of—European and national efforts to stimulate the use of open standards.

A further factor at play is the presence of novel integrated financing forms for building development, including options for maintenance contracts for the operation phase of the building, with the aim that integrated objectives and longer-term shared responsibilities result in a better quality building. If this DBFMO (Design Build Finance Maintain Operate) contract involves a BIM, then this could entail that the BIM is maintained and updated during the operation phase of the building. Experience in the Netherlands has shown [9] that, while the maintenance can be contracted out to a maintenance company, it is wiser if the BIM itself remains the property of the building owner or owners. A maintained BIM is of interest to a 3D Cadastre as it means that new 3D information can be obtained and the registration updated when the building or any part of it changes.

A point key to this discussion is that, while the current land registry system legally encompasses the idea that the owner of a spatial unit or land parcel is the owner of the space above it and that below it—its height and depth—it neither visualises nor fixes these boundaries. Put simply, while the Map of Legal Spaces (*Kadastrale Kaart*) works in reality with 3D spaces, it visualises and fixes them in 2D. It would be beneficial to all involved if the reality matched how it was recorded in the system.

Unfortunately, although it would provide many benefits, there could be high costs, not only with regards to setting the system, but also to maintaining it [10]. The complexity of geometrical and topological operations is obviously much higher on 3D volumes than it is on 2D parcels [11]. The data would need to be surveyed and then entered into the database. It would then need to be updated whenever any changes occurred to the property. The more detail in which a property is recorded, the larger the likelihood that changes to the property could require changes to the representation in the Cadastral system. For example, the addition of a dormer window does not change the footprint but does change the 3D volume.

1.5. First 3D Cadastral Registration in The Netherlands

One interim step, which has recently occurred and which was reported in the Dutch newspaper the *Volkskrant* [12], has been the submission of 3D visualisations to the *Kadaster*. This 3D information has been made accessible via a (direct) link to the Cadastral map [2]. These visualisations are of the newly-completed railway station in Delft. The station, which included offices, the station itself, and a tunnel, has multiple owners (NS Vastgoed, Railinfratrust, Gemeente Delft), whose ownership was illustrated by different 3D spaces delineated by colour.

This 3D interactive PDF was, however, not part of a 3D topological map of legal spaces but an addition to the current 2D registration. Its ground-breaking registration took two years to be finalised [13] because the registration was initially completed in 2D and then reworked to include 3D information [2]. If the initial registration had been accompanied from the beginning by a 3D legal space together with the design, it would have been completed in a matter of weeks rather than years [13]. Clearly, requiring that the registration be begun in 3D rather than 2D would be a more efficient workflow.

1.6. Generalising the IFC: Nagel

A BIM IFC file is semantically complex and relies largely on the use of solids formed in a different way to those found in GIS files [14,15]. One established form of research is building a conversion function from the IFC to CityGML. This began with Nagel [16] but has been extended by others, including Donkers [17]. As CityGML works from a foundation of GIS, there are parallels with what a 3D topological database would require. A key premise is to generalise, a concept taken from the world of cartography [18]. Nagel generalised IFC files in several different ways, including extruding a footprint, and then building a more complex representation, extruding from different stories of the building. The research documented in this paper worked with the first approach.

1.7. Comparing Three Types of BIM-Based Models for Managing 3D Ownership Interests in Multi-Level Buildings: Atazadeh et al.

There are several parallels between this paper and that listed above [19]. Firstly, the Australian research also looked at complex multi-level use rights with an apartment building example. The apartment had two or more single dwellings, with a single lot such as a staircase of foyer being common property. Their research explored three distinct BIM-based representations, either physical or legal or both combined. The apartment was annotated with details such as ownership.

IFC classes can encompass spaces with physical and virtual boundaries [20]. A physical boundary coincides with a physical element that delineates the space, such as a wall. A virtual boundary can exist where a section of a room has a different functional character such as an open kitchen attached to a living room [20]. Furthermore, a collection of spaces can be grouped into a zone (IfcZone, IfcSpatialZone) to define additional characteristics to a cluster of spaces. While the spaces within a zone are generally positioned adjacent to one another, they do not have to be [20]. Legal spaces can also be made up of many different volumes, which makes these spaces and zones of interest in the context of a 3D Cadastre. IFC space boundaries can be defined at different levels. First level space boundaries form simply a shell of the accessible space, excluding bounding volumes, with no references to spaces outside of it. A second level space boundary, however, does reference and aligns with the space on the other side of the bounding element [20] and, therefore, relates to the concept of a composite solid, a collection of solids glued together at overlapping faces.

The space boundaries used by [19] were presumably first level boundaries. By comparison, a key goal of our research was to find topological boundary resources within the IFC model. In further contrast to our research, the mapping between the IFC and the LADM appears to have been completed based on the .ifc files generated from the Autodesk Revit software used to build the apartment. This contrasts with our methodology, wherein we first mapped and then built objects that conformed to the mapping. As has been noted [19], there are inconsistency issues between the .ifc files generated by different forms of software. This means that working with first level space boundaries in another piece of software can result in boundaries being defined using other geometry definitions, which could hinder the data exchange process.

2. Materials and Methods

The problem which needed to be solved in this research project was how to bridge the divide between BIM and GIS and to derive GIS object from BIM files. This was done within a theoretical context from the geomatic engineering domain that Smart Cities based on interoperable urban infrastructures show greater efficiency [21]. While the aim of this theoretical context—the Dutch government's desire to find automated solutions to help govern more cost-effectively and sustainably—overlaps with that of the Internet of Things for Smart Cities, it did not involve Internet of Things [22].

The two worlds of BIM and GIS have traditionally been separate although there is overlap in their approach to spatial data. One essential difference relevant for this paper, for example, was that BIM coordinates are generally local while GIS coordinates are often geo-referenced [23]. The overlap can be seen in the fact that there is provision made in the IFC for geographical coordinates [6].

Our research combined what could be seen as a classic GIS and a classic BIM methodologies, again, however, with overlap. The GIS methodology involved creating a conversion function. Its steps included; studying other conversion functions, choosing use cases, listing model requirements, mapping one data set to another, making a database schema, implementing the mapped data and the conversion and validation function in a prototype environment, testing it using the use cases and then evaluating the process. An essential part of the process was finding suitable data to begin. This is the methodology with which this research project began but which was adapted to become an extraction function within the context of a BIM methodology due to both the complexity of the problem and its cross-disciplinary nature.

The BIM methodology began with a requirements definition which is analogous with the first three steps of the GIS methodology (studying existing functions, choosing use cases, listing model requirements). This requirements definition was made first by listing model requirements such as the objects needing to be both topological and have boundary representation but also that they should be validated, include location information, and have a 3D complex ID.

Part of the requirements definition involved putting the requirements into a wider framework. When and how the data should be obtained was detailed by designing a workflow in Business Process Modelling Notation (BPMN) from an open BIM Information Delivery Manual (IDM). While stipulating the model requirements and data modelling [24] is common GIS methodology, the design of a workflow to put the prototype into a wider context and to formally outline responsibilities by means of process modelling is, to the authors' best knowledge, new. The addition of this multi-actor workflow focused on streamlining communication for an information infrastructure currently being developed in the Netherlands and was a new scientific contribution. Part of its innovatory nature was its use of open standards such as the IFC for the encoding of BIM and the LADM for land administration semantics.

A Model View Definition (MVD) was the next step in our process. The need to first define a subset of a model before mapping can begin arises from the comparatively large size and complexity of the IFC. Not only does it have more than 800 entities, but there are many ways of associating them. This is illustrated by the three different forms of topological boundary representation suggested for use to fulfil the requirements definition; a shell-based surface model, polygonal bounded half space, and faceted boundary representation. Note that the IFC schema defines many more forms of representation items such as trees of Boolean operands, free-form double curved b-spline surfaces, and a variety of sweeps and extrusions.

A key part of our approach was to stay within the resources of the IFC's common schema and not try to extend the model further. This is the approach favoured within the world of IFC as it saves considerable time and money and promotes interoperability. This last is because there is a greater likelihood that software will recognise well-established rather than new IFC entities.

Once the MVD had been made, the resulting data model was mapped to the LADM. While the LADM is a data schema, albeit a conceptual one, drawn in Unified Modelling Language, the IFC is not. It is, rather, illustrated in the Standard for the Exchange of Product Data (STEP) [6]. STEP is not a diagram which models the design of a database. While a model view definition can be created by selecting which entities are of most use in a certain situation, what is created is not a view of a database which can then be queried but rather a set of requirements which narrows the scope of a schema. To build a database from IFC data, STEP files would need to be converted into another format.

Several use cases were then created which conformed to the model requirements and the MVD. This was done in the first instance by programming and in the second instance with proprietary BIM software. The object was first made and then exported as an IFC space object with boundary representation.

In the world of BIM, validation is often interpreted as the test performed to confirm that use cases conform to the MVD. In the world of GIS, the term validation is more often applied to testing whether objects are closed and do not overlap. Both tests were performed as part of our research.

The problem defined at the beginning of the methodology was to create GIS objects from BIM files. The model view definitions and use cases showed how this can be done. The primary methodology was a BIM one, however important foundations have now been laid for the continuation of the project based on a GIS conversion function methodology. A database schema based on the LADM could be drawn up and the data converted from IFC and further validated in a prototype environment. The objects could then, if required, be accessed via a server.

The materials included sample data obtained from the DURAARK website [25], which came in the form of .ifc files. More specifically, a data set from the British National Building Specification (for further details see Section 3.3.2) [26] was used. The architectural programs used included Autodesk's Revit and Archicad on a Microsoft Virtual Machine. Code snippets were programmatically created in the IfcOpenShell (http://ifcopenshell.org) software library. The prototype was visualized using the Solibri Model Viewer.

3. Results

The aim of the research was to build a model prototype; the procedure followed the standard BIM methodology outlined in Section 2. To begin with, a requirements definition was conducted and a workflow designed on the basis of it. A mapping between the LADM and the IFC was then made, and two use cases were made to test the MVD, which arose from the mapping.

3.1. Modelling the Collaborative Process with IDM

Just as in many other places in the world, the Dutch government is going digital. A vast information infrastructure is being built with the aim of making the system more efficient but also making it simpler to use. One part of this is the creation of an information infrastructure the Digital System (*Digitaal Stelsel*) [27]. A part of the Digital System, the Information House Buildings [28] is devoted to the automation of the process of building registrations. The subject of this workflow is leveraging off this process to obtain 3D legal spaces (if required). Please note that this workflow is modelled in the IDM to facilitate communication with the building industry, not to impose the world of BIM on the world of GIS. A BIM model is the product of many parties. At a certain point in time, there is a merging, but this involves the individual parties overlaying their own version with what they have learnt. A true unambiguous and single-source-of-reference complete model of the building may never actually exist. There is also the issue of intellectual property; the individual owners of the BIM may not like their model being generalised. This is taken into account.

A central goal of the presented research is obtaining 3D parcel data from BIMs. A collaborative workflow would support the reuse of BIM data for 3D Cadastre purposes by improving the alignment of activities. Part of the process of gaining a building permit for a building would involve providing the Land Administration System with appropriate legal spaces. This process is illustrated in Business Process Modelling Notation in Figures 2 and 3. With the aim of making this process as clear as possible, text that would usually be referred to the diagram using a number system has been left in the diagram.

Someone wanting to build or to renovate a house first seeks advice from the government concerning, for example, what can be built where and to which Cadastral regulations they should adhere. They are also given access to a digital file in IFC format showing what has already taken place. They then set to work on a 3D BIM for the construction. Part of this process is to create several space objects grouped into a zone to represent the legal spaces in the building. These are customised legal space .ifc files for inclusion in the BGT and *Kadastrale Kaart*. The resulting 3D BIM is tested digitally to see if it conforms to the advice the Land Registry gave in the original file. In addition, an assessment needs to be made as to whether the definition of the legal spaces is in accordance with the physical artefact as designed. A definitive permit is requested, and, when it has been received, the building is then constructed according to the design that has been submitted. During the building process, things change, and the BIM, which serves as the central point for the collaborative building process, changes with it. The 'as-built' BIM becomes different to the BIM that was initially submitted to the authorities. These are typically minor changes that occur out of on-site coordination works or material availability among external suppliers. The building is inspected and, where necessary, the Cadastral regulations are enforced. A further issue is that the building may be subdivided and put up for sale in a different manner than was initially communicated. Therefore, the cadastral database needs to have an 'as sold' data set for its records as well. Underpinning the entire process is the use of open BIM standards, which ensure that the data can be used by many for a variety of purposes and for a long time.

This workflow has been given in the manner of a BIM open standard (the IDM) to illustrate how different actors should collaborate. Communicating it in this manner could also be helpful when detailing the Cadastral requirements of a building project. This workflow has been modelled on a similar workflow proposed in the Netherlands for obtaining BIM data for building registrations [28,29].

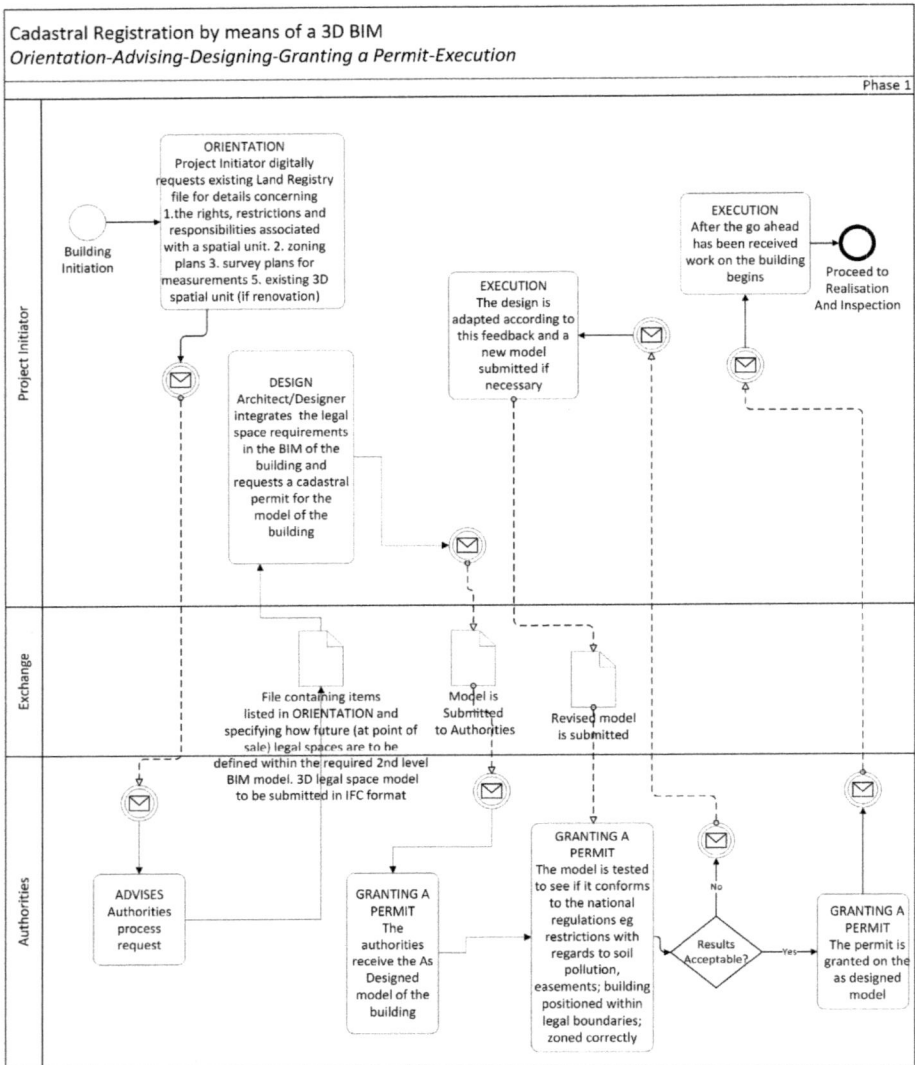

Figure 2. Information Delivery Manual (IDM) Workflow: Cadastral Registration by means of a 3D BIM Phase 1.

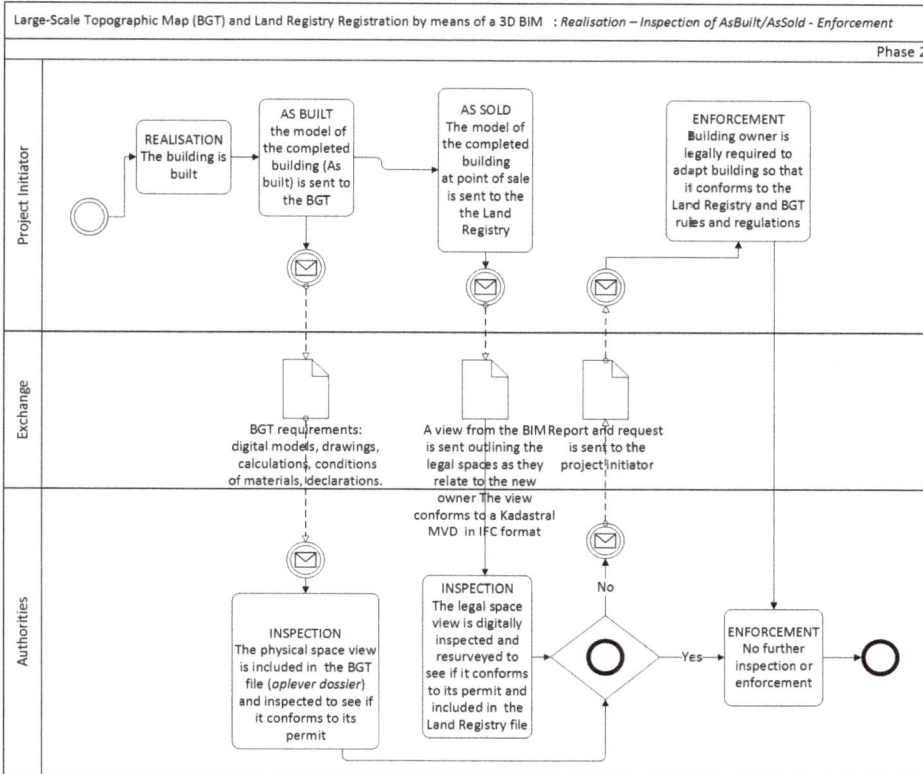

Figure 3. IDM Workflow: Cadastral Registration by means of a 3D BIM Phase 2.

3.2. Mapping the LADM to the IFC

The mapping of the LADM to the IFC has been separated into three parts for easier comprehension; these parts do, however, all connect together. The model in Figure 4 follows the typical IFC schema decomposition structure of an *IfcProject, IfcSite, IfcBuilding,* and *IfcBuildingStorey.* While there can only be one *IfcProject,* there can be multiple sites, buildings, and storeys (ground floor, first floor, etc) within a project. The *IfcSite, IfcBuilding,* and *IfcBuildingStorey* hierarchy accommodates this. It is at the *IfcSite* level that a BIM's one set of geo references can be found, something which has been mapped to an attribute of the *LA_SpatialUnit.* The LADM allows a legal space to be pinpointed by one point [1]. The *LA_SpatialUnit* has been mapped at the level of *IfcSpace.* An *LA_SpatialUnit* can be composed of many things like a utility network; for example, underground pipes but also the plumbing on a major building site. This can be in addition to a *LA_LegalSpaceBuildingUnit,* an *LA Level* or an *LA Group,* all of which are specialisations of *LA_SpatialUnit.* A space with physical space boundaries can be any of these things, which is why *LA_SpatialUnit* has been mapped to it. In common with the *LA_Spatial Unit,* an *IfcSpace* can also be made up of a number of spaces. While a number of topologically-represented spaces can be stacked in the IFC, they can also be grouped into an *IfcZone,* if required. A "spatial unit represents . . . a single or multiple volumes of space"—hence its reflexive relationship in the diagram." [1]. Because of this, both *IfcZone* and *IfcSpatialZone* have been mapped to *LA_Spatial Unit.*

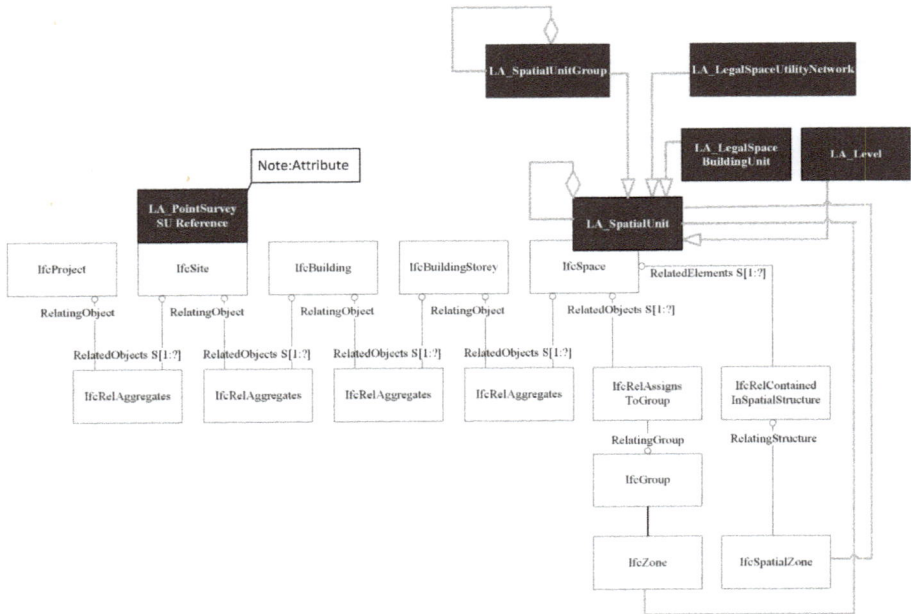

Figure 4. Mapping between the LADM and the IFC: Decomposition.

Two potential options for grouping *LA_Spatial Unit* are shown in the mapping, *IfcZone* and *IfcSpatialZone*. The advantage of an *IfcSpatialZone* is that each entity "can have its own placement and shape representation"[6]. An *IfcZone*, by contrast, is a group of *IfcSpaces*.

The IFC works primarily with solids rather than boundary representation [6,14]. There are many types of solids in the IFC, as befits a model of more than 800 entities. In general, these solids are made by extruding from a circle, rectangle, or polygon, although the IFC also allows many types of boundary representation. While the obvious place to look for these many types is in *IfcTopologyResource*, *IfcGeometricResource* can also provide options.

Our research looks at building a topological 3D Cadastre. A further constraint is that land administration databases generally work with simple geometry. This is illustrated by the geometry of the Dutch *Kadaster*'s database [29]. This database works with polyloops (*lijnketens*), consisting of straight line segments and circular arcs in 2D. Other curves are used in other countries in exceptional cases. The Dutch database is a topological database, which uses one '*kaartlijn*' to define the boundary between spatial units.

Four forms of geometric boundary representation have been chosen from the many options provided by the IFC (see Figure 5). The basic building blocks of all of them are points (nodes), edges (straight lines), and surfaces (faces). None of them involve curves, unless they are planar. They are open shell by means of infinite prisms, closed shells, polygonal bounded half-spaces, and faceted boundary representation (brep).

An open shell is a collection of surfaces. It is a topological entity but is not closed. In a Cadastral setting, it may need to be left open at the top or bottom. A closed shell is a space enclosed in a collection of connected faces or surfaces; it encloses a volume and is 'water tight'. Whereas an open shell has one or more sides missing, thus leaving it open, a closed shell does not. The closed shell is typically part of faceted Boundary Representations (BRep) in the IFC that define a closed volume by providing every bounding face explicitly. This simple form of boundary representation has planar faces and straight edges. The vertices are defined in polygonal loops (*IfcPolyLoop*), and the edges are implicitly defined as the line segments that span those vertices.

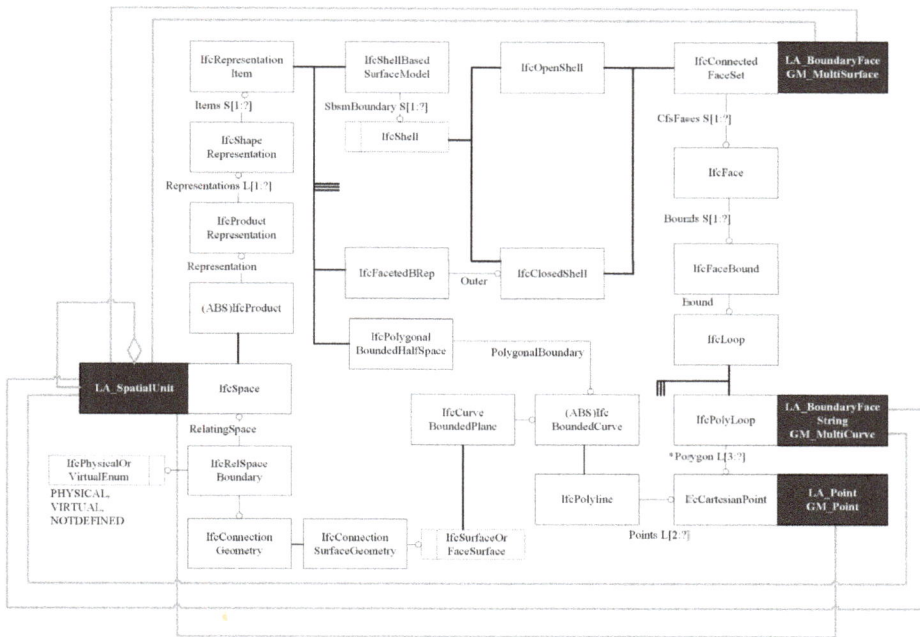

Figure 5. Mapping between the LADM and the IFC: Geometry.

Open and closed Shells and faceted brep can straightforwardly be mapped to the LADM. The cartesian points, which are their most basic building block, map to *LA_Point (GM_Point)*; the polyloops formed from the points equate with *LA_BoundaryFaceString (GM_Multicurve)* and the connected faces, which edge them (*IfcConnectedFaceSet*) to *LA_BoundaryFace (GM_Multisurface)*. When these building blocks are put together, they form the open and closed shells with faceted Breps, which represent *IfcSpace*.

The buildingSMART website describes property sets, modelled in Figure 6, as "a way to exchange alphanumeric information attached to spaces, building elements and other components" [6]. Property sets are a way to extend the IFC model without changing it at a class level. The property sets work as containers to which individual properties (key-value pairs) can be added This can work within the framework of the common or predefined property sets (*p_set*), which are included from IFC2*3 [30], a revision of the IFC, onwards. *IfcSpace*, for example, has several predefined properties (*Pset_SpaceCommon*). One of these predefined properties, an *IfcPropertySingleValue*, could be used in conjunction with *IfcIdentifier* to provide a reference ID for each spatial unit. In this instance (see Figure 6), *IfcIdentifier* from *Pset_SpaceCommon* has been mapped to an attribute of *LA_LegalSpaceBuildingUnit*, *suID*, which stands for spatial unit identity number.

Property sets have been included in the IFC model so that attribute definitions from different disciplines can be internationally standardized at a basic level. The use of *Pset_SpaceCommon* to provide an identifier is an example of this. Just as the LADM itself can be extended and adapted for regional or national use, so can the IFC. These IFC basic property sets can then be complemented by regional property sets or property sets agreed upon in projects.

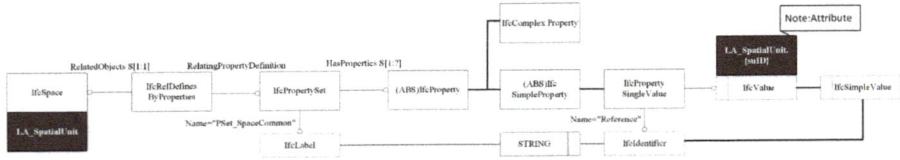

Figure 6. Mapping between the LADM and the IFC: Properties.

Once a Model View Definition (MVD) has been put together, a tool—the IfcDoc tool, which is made freely available on the buildingSMART website—can be used to document it fully. This full documentation can be exported in .mvdxml, allowing an automated interface for importing and exporting software, as well as standardized grammar to assert whether exchanged IFC models comply to the MVD [31]. Instances can then be created within this framework. The IfcDoc tool can be used to create and document, as well as to check that models comply with this framework. In the 3D use case explored in this paper, however, the object (was made in commercial software using the IFC4 Design Transfer View, which allows the inclusion of a larger variety of entity definitions.

3.3. Use Cases

The following use cases illustrate scenarios in which a 3D Land Registry System could be useful.

3.3.1. Scenario 1: Traditional Parcel Columns

The first use case is illustrated in Figure 7. A central principle of land registry systems around the world is that, although a parcel of land or a spatial unit is recorded in 2D, in reality it alludes to a 3D space. Unless there are other rights or restrictions recorded, this space reaches below the 2D representation to the earth's core and endlessly up into the sky.

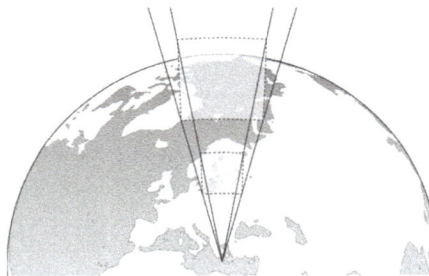

Figure 7. A central principle of 2D land registry systems: from the Earth's core up to infinity [32].

Research has been conducted into how this concept could be geometrically stored in a database. One method would be to take parcels recorded in 2D and convert them into 3D by adding a z-axis [33]. This z-axis would be added to the nodes of each parcel and would be infinite with a maximum of infinity.

Industry Foundation Classes (IFC) models typically document a building artefact by means of a decomposition of three-dimensional volumes. These volumes are referred to as the *Body* geometry. Current research [33,34] states that a reinterpretation from two-dimensional Cadastral registrations into three-dimensional entities is necessary, either for a transition period in which the Cadastre is migrated or for a hybrid stage in which two-dimensional parcels exist alongside three-dimensional volumes and need to be checked for interferences. For comparison, three approaches to unify 2D parcel data with IFC models are given below. All approaches have the fact that legal spaces are exchanged as an *IfcSpace* element in common.

In its simplest form, a two-dimensional footprint representation is provided for such a space. This footprint is illustrated first as a snippet of code (Figure 8) and, secondly, as in Figure 9a.

```
#100=IFCCARTESIANPOINT((0.,0.));
#101=IFCCARTESIANPOINT((100.,0.));
#102=IFCCARTESIANPOINT((120.,300.));
#103=IFCCARTESIANPOINT((-20.,300.));
#104=IFCPOLYLINE((#100,#101,#102,#103,#100));
#105=IFCSHAPEREPRESENTATION(#3,'FootPrint','Curve2D',(#104));
#106=IFCPRODUCTDEFINITIONSHAPE($,$,(#105));
#107=IFCSPACE('1F93ymWGCHvxyww5AXPh_y',#1,'Room 1',$,$,#2,#106,$.
.ELEMENT.,.NOTDEFINED.,$);
```

Figure 8. Snippet of IFC illustrating a section of a 2D footprint.

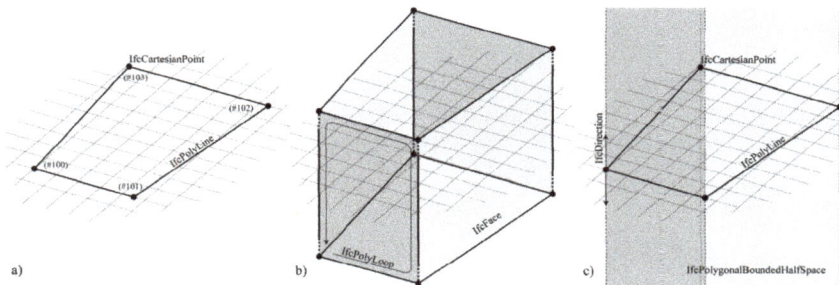

Figure 9. Three ways to represent a traditional parcel column: (**a**) footprint, (**b**) open shell with vertices near infinity, (**c**) the union of two *IfcPolygonalBoundedHalfSpaces* to represent a prism up to infinity and down to the Earth's core.

On the one hand, this is a satisfactory solution as it allows a collection of two-dimensional parcels within an IFC file. One advantage of a Cadastral database in which all parcels are recorded as three-dimensional entities is that situations of conflict are immediately apparent as the intersections of three-dimensional parcel volumes. In a database with both two- and three-dimensional parcel volumes, the two-dimensional volumes would need to be interpreted as three-dimensional prisms when conflicting situations were queried for. Therefore, it would be beneficial to look for approaches to encode such an infinite prism in IFC.

EXPRESS, the modelling language in which the IFC schema is defined, differentiates between integer and real number types of arbitrary precision, but the real number type in EXPRESS does not offer a means to encode the concept of infinity. This contrasts with other technical standards to represent floating point numbers such as IEEE 754, which is the prevalent representation of floating point numbers in hardware and software today. Therefore, from a semantic point of view, it is not possible, for example, to state that the Z-coordinate of some vertex lies at infinity. In reality, however, an arbitrary, large number can be chosen that is 'infinite enough' for all practical calculations. Geometric modelling kernels often use arbitrarily large values to represent infinity so that transformations on such infinite points are still meaningful. For instance, the geometric modelling kernel Open CASCADE, which is used by various closed and open source IFC software applications, uses 1×10^{100} as its absolute value, beyond which a length measure is considered infinite. Note that this is many orders of magnitude greater than the distance from the Earth to the sun in meters and is therefore a sufficient approximation of infinity for Cadastral purposes.

Despite there being no explicit notion of infinity present in the IFC, there are geometric representations with infinite volume or area such as infinite mathematic planes. An *IfcHalfSpaceSolid* is the portion of infinite three-dimensional Cartesian space that exists on one side of an infinite surface. This geometrical entity is frequently used to define cutting planes that trim away parts of wall segments under tilted roof slabs. A further specialization of this construct, *IfcPolygonalBoundedHalfSpace*, can be used to model an infinite extruded prism in the direction away from the surface normal. The union of two such half spaces in opposite directions can be used to model an infinite prism in both directions. The IFC standard documentation [6] includes the phrase "half space solids are only useful as operands in Boolean expressions", but no formal mechanisms are in place to exclude such items from being first-order citizens in geometric representations. Note that these body representations are distinct from the concept of space boundaries discussed earlier, but space boundaries can be used to further qualify the nature of the individual faces of the body representation; in particular, whether such a face marks a physical or virtual boundary.

3.3.2. Scenario 2: Complex Multi-Use Rights with the 3D

The Lakeside Restaurant is a data set made available as an IFC file by the United Kingdom's National Building Specification [26]. This specification is a case study project for the National BIM Library, which is used to market the National BIM Library. This library contains proprietary and non-generic objects, which are free to use and platform neutral so that they can be imported into BIM design projects [26]. The data set is a complex design that is modelled down to the last detail. The data set is used within this paper to illustrate how legal spaces can best be taken from 3D BIMs.

3.3.3. The Lakeside Restaurant in a 2D Cadastre

Buildings built above water pose a problem to a 2D Land Administration System. Imagine a situation in which a privately owned parcel is situated above water belonging to the city council (government). Only the stilts on which the building is touch the ground; the rest of the building is situated higher up and above the water. Figure 10 illustrates how it could be registered. This method was used for a similar building in Amsterdam, which was positioned above an underground carpark [34].

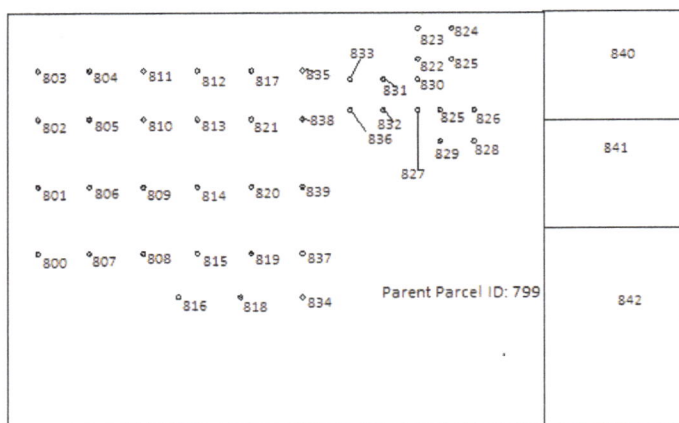

Figure 10. A traditional 2D method of registration for complex multi-use rights (image created by author).

The round circles in Figure 10 represent 41 separate spatial units or parcels, which are part of parent parcel 799, all of which are included in the database. These separate spatial units are the piles that penetrate through the water to the lake bed below. The building itself is positioned above these piles.

There are a few problems with this solution. The first is that, on the basis of these records, someone studying the records can only guess at what the building (Figure 11) looks like. The second lies in the maintenance of the records. Each one includes the rights, responsibilities, and restrictions associated with the whole building. Not only would updating them all be prone to error, but one could easily be inadvertently skipped while transferring the deed.

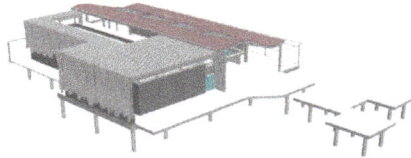

Figure 11. The Lakeside Restaurant dataset pared back and visualised in IFC [26].

3.3.4. The 3D in a 3D Cadastre

An alternate method of registering the 3D would be to do so in a 3D format. The data could be extracted from a 3D IFC model for use by a 3D Cadastre. The 3D would be a particularly interesting space to register. To begin with, it is positioned above water. Not only is it positioned above property belonging to someone else, but the boundary between the two properties is what could be described as an ambulatory natural boundary. In this context, this would refer to the fact that water levels beneath the building are not fixed but rise and fall depending on the weather, rather than the normal definition, where a river or other water boundaries expand in 2D.

A further issue is that, not only does the building encompass an open air (no top) outdoor deck, positioned above the water, but also many other spaces within the building are not fully enclosed. One such space can be seen in Figure 11, in which an outdoor eating area is shielded by a roof and partially covered walls. Another part of the building is an outdoor terrace, completely open to the sky.

One method of representation would be to represent the building as one simple surface model, as illustrated in Figure 12. In Figure 12, the whole restaurant has been redrawn as a closed volume. The terrace has also been reformed, to illustrate that it is a limited space both below and above and that the whole has become one *IfcSpace*.

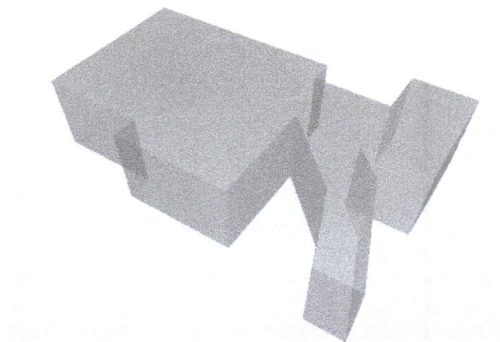

Figure 12. Conceptual Mass as an *IfcSpace* object.

A different option would be to use the notion of *IfcZones* or *IfcSpatialZones* to further decompose and specialize that nature of various sections of the building. Parts of the building could then be represented by *IfcSpaces* but with varying representations The closed part of the 3D could rely on boundary representations to accurately depict the varying heights of the roof. Partially closed sections could be bounded by virtual boundaries, i.e., ones that do not correspond to physical objects.

4. Conclusions and Future Work

This paper has looked at how data can be obtained from BIMs for input into a 3D Cadastre, first by means of a workflow illustrated using an open BIM standard, the Information Delivery Manual, and secondly from the perspective of two use cases, traditional parcel columns and 'The Lakeside Restaurant'.

The workflow, which forms the first half of the paper, is based, in part, on the difficulties experienced in creating a legal space for the Lakeside Restaurant. Communicating Cadastral requirements early in the design process would facilitate the process of obtaining legal spaces from BIM. While, in some cases, spaces could simply be extracted from BIMs at the end of the design process, legal spaces could also be defined at the beginning of the design process. Requiring that a conceptual mass of the plan of the building be sent to the Cadastre for testing early on would ensure the presence of topological legal spaces for use by the cadastre. It is also a way to check that the building is being designed and built within Cadastral regulations, that is, the rights, restrictions, and responsibilities associated with the spatial unit.

Our research suggests that a dialogue be established with the Land Registry from the construction's inception. In this way, the IFC model could be tailored to Land Registry requirements. These IFC models would not be extracted from as-built BIMs but would need to be developed concurrently. They would be manually put together by the building's designer with the reuse of the building's geometry. The workflow could also bring many other actors into the construction's lifecycle at an early stage of development. These could include various owners, developers, designers, financiers, permit providers, surveyors, notaries, and Cadastral registrars.

The use of the IFC's spaces and zones with varying topological representation could be a way for virtual Cadastral legal spaces to be defined within BIMs. The spaces can be grouped into zones and can be a number of non-adjacent, aggregate, or intersecting volumes. This capability could prove very attractive to the complex shapes of modern buildings but could also be used to encompass spatial units spread over different rooms in the same building. Additionally, the use of *IfcSpatialZone* rather than *IfcZone* could allow independent placement and shape representation.

The spaces defined within zones and thus forming a virtual legal space could be topologically bounded in three different ways: open or closed shells (*IfcOpenShell, IfcClosedShell*), polygonally bounded half-spaces (*IfcPolygonalBoundedHalfSpace*), or facetted boundary representation (*IfcFacetedBrep*). These topological volumes could, in turn, be automatically extracted, complete with referenced coordinates, and entered in a topological database and thus ultimately be used to generate an overview of legal spaces in the Netherlands or elsewhere in the world.

The Lakeside Restaurant use case demonstrates the right of superficies as "a real right to own or to acquire buildings, works or vegetation in, on or above an immovable thing owned by another" [32]. It is, therefore, an ideal volume for showing why a 3D Cadastre is needed. The fact that the Lakeside Restaurant was such a difficult space to register highlights the need for a new topic and possibly new semantics to be created within the IFC in order to define legal spaces. This means that a step further than simply achieving interoperability between the IFC and Cadastral legal spaces by using the same semantics might need to be taken. The review and expansion of the IFC is an ongoing process, of which this proposed new topic could become a part.

Before steps are taken to do this, however, a thorough investigation into the options provided by the property sets would need to be made. This could be through researching the framework of the common or predefined property sets. Property sets could also be defined on a national level [6] to reflect regional regulations. This would be in addition to the work begun in this paper on finding appropriate semantics within the current model.

Future work could be to develop a workflow in the IDM Part 1's BPMN, which brought more actors into the construction's development lifecycle, including owners, developers, designers, financiers, permit providers, surveyors, notaries, and Cadastral registrars, with the aim of improving collaboration. A further project could be to create a more complex legal space for the Lakeside Restaurant using open or closed shells (*IfcOpenShell*, *IfcClosedShell*), polygonally-bounded half-spaces (*IfcPolygonalBoundedHalfSpace*), or facetted boundary representation (*IfcFacetedBrep*); *IfcZone* or *IfcSpatialZone*; and complete, rather than partial, geo referencing.

Acknowledgments: Our thanks go to Jeroen van der Veen for providing literature about the Environs Act and for his participation in the initial paper.

Author Contributions: Jennifer Oldfield conceived the use cases, wrote the paper, and co-worked on the data modelling. Thomas Krijnen contributed sections to the work on IFC and helped finalize the mapping to LADM, in addition to reviewing the paper and providing valuable insights. Peter van Oosterom provided relevant articles and gave extended feedback on the work completed. Jakob Beetz contributed to the methodology and analysis in the early stages of research and also connected other participants to this effort for input.

Conflicts of Interest: The authors declare no conflicts of interest.

References

1. Lemmen, C.; van Oosterom, P.; Bennet, R. The Land Administration Domain Model. *Land Use Policy* **2015**, *49*, 535–545. [CrossRef]
2. Stoter, J.; Ploeger, H.; Roes, R.; van der Riet, E.; Biljecki, F.; Ledoux, H. First 3D Cadastral Registration of Multi-level Ownerships Rights in the Netherlands. In *Proceedings of the 5th International FIG Workshop on 3D Cadastres, Athens, Greece, 18–20 October 2016*; van Oosterom, P., Dimopoulou, E., Fendel, E., Eds.; International Federation of Surveyors: Copenhagen, Denmark, 2016.
3. Kitsakis, D.; Dimopoulou, E. Addressing Public Law Restrictions within a 3D Cadastral Context. *Int. J. Geo-Inf.* **2017**, *6*, 182. [CrossRef]
4. What Is BIM? Available online: http://bimloket.nl/107 (accessed on 18 May 2016).
5. BuildingSMART Norway (2007) Information Delivery Manual: Guide to Components and Development Methods. Available online: http://idm.buildingsmart.com (accessed on 18 May 2016).
6. Building Smart. Available online: www.buildingsmart.org (accessed on 18 May 2016).
7. Aien, A.; Kalantari, M.; Rajabifard, A.; Williamson, I.; Bennett, R. Advanced Principles of 3D Cadastral Data Modelling. In *Proceedings of the 2nd International Workshop on 3D Cadastres, Delft, The Netherlands, 16–18 November 2011*; van Oosterom, P., Dimopoulou, E., Fendel, E., Eds.; International Federation of Surveyors: Copenhagen, Denmark, 2011.
8. BGT. Available online: https://www.kadaster.nl/ (accessed on 31 August 2017).
9. Schapendonk, J. (van Wijnen, Rotterdam, The Netherlands). Personal communication, 2016. Interview 2016.
10. De Lathouwer, B. The 3D Standards evolution Piloting for Smart Cities and the Build Environment (GEOBIM). In Proceedings of the 5th International FIG Workshop on 3D Cadastres, Athens, Greece, 18–20 October 2016.
11. Breunig, M. An approach to the integration of spatial data and systems for a 3D geo-information system. *Comput. Geosci.* **1999**, *25*, 39–48. [CrossRef]
12. Van de Weijer, B. Kadaster Wordt Driedimensionaal. Volkskrant, 2016. Available online: www.Volkskrant.nl (accessed on 12 February 2017).
13. 3D cadastre & 3D planning Geodesign Summit EU Delft The Netherlands 2016. Available online: http://wiki.tudelft.nl/bin/view/Research/ISO19152/GeoDesignSummitEU2016 (accessed on 31 August 2017).
14. BuildingSMART IfcGeometricModelResource. Available online: http://www.buildingsmart-tech.org/ (accessed on 29 September 2017).
15. Liu, X.; Wang, X.; Wright, G.; Cheng, J.C.P.; Li, X.; Liu, R. A State-of-the-Art Review on the Integration of Building Information Modeling (BIM) and Geographic Information System (GIS). *ISPRS Int. J. Geo-Inf.* **2017**, *6*, 53. [CrossRef]
16. Nagel, C. Ableitung Verschiedener Detaillierungsstufen von IFC Gebäudemodellen. Master's Thesis, Hochschule Karlsruhe Technik und Wirtschaft, Karlsruhe, Germany, 2007.

17. Donkers, S.; Ledoux, H.; Zhao, J.; Stoter, J. Automatic conversion of IFC datasets to geometrically and semantically correct CityGML LOD3 buildings. *Trans. GIS* **2015**, *20*, 547–569. [CrossRef]

18. Geiger, A.; Benner, J.; Haefele, K.H. Generalisation of 3D IFC Building Models. In *Geoinformation and Cartography*; Breunig, M., Ed.; Springer International Publishing: Cham, Switzerland, 2015; pp. 19–35.

19. Atazadeh, B.; Kalantari, M.; Rajabifard, A. Harnessing BIM for 3D Digital Management of Stratified Ownership Rights in Buildings. In *Proceedings of the 5th International FIG Workshop on 3D Cadastres, Athens, Greece, 18–20 October 2016*; van Oosterom, P., Dimopoulou, E., Fendel, E., Eds.; International Federation of Surveyors: Copenhagen, Denmark, 2016.

20. Weise, M.; Liebich, T.; See, R.; Bazjanac, V.; Laine, T. Ifc Implementation Guide: Space Boundaries for Thermal Analysis. US Government Services Administration (GSA), 2009. Available online: buildingsmart-tech.org (accessed on 31 August 2017).

21. Roche, S. Geographic Information Science I: Why does a smart city need to be spatially enabled? *Prog. Hum. Geogr.* **2014**, *38*, 5. [CrossRef]

22. Zanella, A.; Bui, N.; Castellini, A.; Vangelista, L.; Zorzi, M. Internet of Things for Smart Cities. *IEEE Internet Things J.* **2014**, *1*, 22–32. [CrossRef]

23. Van Berlo, L.; De Laat, R. Integration of BIM and GIS: The Development of the CityGML GeoBIM Extension. Available online: www.bimserver.org (accessed on 12 November 2016).

24. Aien, A.; Kalantari, K.; Rajabifard, A.; Williamson, I.; Bennett, R. Utilising data modelling to understand the structure of 3D cadastres. *J. Spat. Sci.* **2013**, *58*, 215–234. [CrossRef]

25. Duraark. Available online: http://duraark.eu/ (accessed on 31 August 2017).

26. National Building Specification. Available online: www.thenbs.com (accessed on 31 August 2017).

27. Ondersteuningomgevingswet. Available online: www.digitaleoverheid.nl (accessed on 31 August 2017).

28. Pijpker, U.; Brouwer, K.; van Kooten Niekerk, E.; Groenendaal, W. Informatiehuis Bouw: Samen bowen aan het Digitale Stelsel Omgevingswet. Available online: www.omgevingswetportaal.nl (accessed on 31 August 2017).

29. BRK. Available online: https://www.overheid.nl/ (accessed on 31 August 2017).

30. Liebich, T. IFC 2x Edition 3 Model Implementation Guide. 2009. Available online: www.buildingsmart-tech. org (accessed on 31 August 2017).

31. Karlshøj, J. An Integrated Process for Delivering IFC Based Data Exchange. Available online: iug. buildingsmart.org (accessed on 31 August 2017).

32. Stoter, J.; van Oosterom, P. *3D Cadastre in an International Context: Legal, Organizational and Technological Aspects*; Taylor & Francis/CRC Press: Florida, FL, USA, 2006.

33. Thompson, R.; van Oosterom, P.; Huat Soon, K. Mixed 2D and 3D Survey Plans with Topological Encoding. In *Proceedings of the 5th International FIG Workshop on 3D Cadastres, Athens, Greece, 18–20 October 2016*; van Oosterom, P., Dimopoulou, E., Fendel, E., Eds.; International Federation of Surveyors: Copenhagen, Denmark, 2016.

34. Stoter, J.; van Oosterom, P.; Ploeger, H. The Phased 3D Cadastre Implementation in the Netherlands. In *Proceedings of the 3rd International Workshop on 3D Cadastres: Developments and Practices, Shenzhen, China, 25–26 October 2012*; van Oosterom, P., Fendel, E., Eds.; International Federation of Surveyors: Copenhagen, Denmark, 2016; pp. 273–288.

International Journal of
Geo-Information

isprs

MDPI

Article

A 3D Digital Cadastre for New Zealand and the International Opportunity

Trent Gulliver [1,*], Anselm Haanen [1] and Mark Goodin [2]

[1] Office of the Surveyor-General, Land Information New Zealand, Wellington 5145, New Zealand; ahaanen@linz.govt.nz
[2] Operations, Land Information New Zealand, Wellington 6145, New Zealand; mgoodin@linz.govt.nz
* Correspondence: tgulliver@linz.govt.nz; Tel.: +64-4-471-6233

Received: 31 August 2017; Accepted: 15 November 2017; Published: 21 November 2017

Abstract: New Zealand has a legal 3D cadastre, and has done since the inception of its cadastral survey and tenure systems around 150 years ago. However, the digital representation of the cadastre is 2D with 3D situations handled via static plan, section and elevation images and supporting textual information. Work is currently underway to develop a 3D digital cadastre that will enable the 3D spatial extents of property rights, restrictions and responsibilities to be captured, validated, lodged, integrated with existing data, visualised, and made available for use in other systems. This article presents the approach that is being promoted by regulators of New Zealand's cadastral survey system in discussions with suppliers of land administration systems. Previous research concluded that the most appropriate way for New Zealand to develop a 3D digital cadastre is to build upon its existing system. The 2D digital cadastre would continue to be the default layer with 3D situations incorporated as and where necessary. To enable this requires a new approach to handling parcels defined in 3D. The representation of a 3D parcel as a spatial object is being proposed to allow parcels limited in height to be integrated into the digital cadastre and subsequently maintained. While the authors discuss how New Zealand's digital cadastre may be transitioned to 3D, it is suggested that the generic nature of spatial objects could be applied to other jurisdictions. For this reason, the international appeal of the approach is considered as other jurisdictions and providers of software applications may benefit from New Zealand's efforts.

Keywords: 3D cadastral survey system; 3D cadastre; 3D parcel; rights restrictions responsibilities; spatial object

1. Introduction

This article builds on the paper, A 3D Digital Cadastre for New Zealand by 2021: leveraging the current system and modern technology [1], which was submitted by the authors to the 5th International FIG (International Federation of Surveyors) Workshop on 3D Cadastres in 2016. In that paper it was emphasised that the authors' involvement in regulating New Zealand's cadastral survey system meant a regulatory perspective was being shared with the international research community. While this perspective is continued in this article, there is a point of difference to the authors' earlier contribution. A key finding of the research at that time was the need to develop a system that best accounts for New Zealand's situation, for which there are a number of unique characteristics. Whilst this premise still stands today, the authors have since formed the view that the development of New Zealand's cadastral survey system presents opportunities at the international level. These opportunities relate to the way in which 3D parcels of rights, restrictions and responsibilities could be produced and incorporated into the digital cadastre as the survey and title system for New Zealand is further developed.

It is not the intention of the authors to offer a detailed design solution. Rather this article shares with the international community the work that New Zealand is doing to develop its current 2D digital cadastre to an operational 3D digital cadastre. To date, this work is enabling the authors and other Land Information New Zealand (LINZ) staff to have meaningful conversations with suppliers of land administration systems. The detailed design of the final solution will be the responsibility of the vendor contracted to build the new system with input from LINZ and the surveying profession of New Zealand.

The authors have been actively promoting the development of a 3D digital cadastre for New Zealand since 2013. During this time:

- the international status of 3D cadastres has been examined;
- dialogue with likeminded international jurisdictions has taken place;
- attendance and participation at national and international conferences with a 3D cadastre flavour has occurred;
- input of practising licensed cadastral surveyors has been obtained.

Outputs from these undertakings include a thesis [2], conference papers [1,3] and presentations, articles in a New Zealand survey publication [4,5] and several internal discussion documents.

The following discussion commences with an overview of New Zealand's cadastral survey system. This is an important starting point as the system provides both the environment and platform on which to develop a 3D digital cadastre. With this context the next section presents the approach being pursued to achieve a 3D solution. This section includes the documentation of requirements fundamental to a 3D digital cadastre in New Zealand. This is followed with discussion on the international opportunities that the authors' believe exist. The article is closed with concluding remarks.

2. The Present Situation in New Zealand

In this section an overview of New Zealand's cadastral survey system is presented with consideration of the institutional framework and the technical and operational processes currently employed for handling cadastral survey data. This sets the scene in which the goals for developing the system are discussed.

2.1. Property Rights and the Cadastral Survey System

New Zealand (Figure 1) benefits from a national property rights system that promotes efficiency and confidence in the transaction of property rights. The cadastral survey system is a core component of the property rights system with a prime purpose to define the location and spatial extents of land and other real property (e.g., a unit under the Unit Titles Act 2010). This information is used by managers of tenure systems to enable the registration of rights, restrictions and responsibilities and provides interested parties with confidence in the location of boundaries [5].

All cadastral surveys must be undertaken by licensed cadastral surveyors. There are currently 668 licensed cadastral surveyors [6], most of whom are employed in the private sector. Those surveyors must ensure that their surveys and the resulting cadastral survey datasets comply with rules (regulations) set by the Surveyor-General (currently the Rules for Cadastral Survey 2010 [7]) before they are lodged with LINZ. LINZ, the government agency responsible for the cadastral, geodetic and title systems, ensures each cadastral survey dataset complies with the Rules before approving it and then integrating it into the cadastre.

Figure 1. Reference [8] New Zealand—an island nation with a national property rights system.

2.2. A 3D Legal Cadastre

Associated with the cadastral survey system is the 'cadastre'. The Cadastral Survey Act 2002 defines 'cadastre' to mean "all the cadastral survey data held by or for the Crown and Crown agencies" [9] (s. 4). In practise, this describes the repository of cadastral survey datasets lodged with LINZ and integrated into its database (currently Landonline). These integrated data are referred to in the Surveyor-General's strategic document, Cadastre 2034 [10], as being fundamental to the cadastral survey system.

New Zealand has a well-established 'legal' 3D cadastre. The ability to define the extents of rights, restrictions and responsibilities in 3D has existed since the inception of New Zealand's cadastral survey and tenure systems at the time when the country was being colonised by the British in the 1800s [2]. The freehold title system under the Torrens based Land Transfer Act 1952 [11] supports rights, restrictions and responsibilities in property, regardless of whether they are restricted in height or not. The Land Transfer Act 1952 is set to be repealed and replaced by the Land Transfer Act 2017, although the fundamental principles of the system will remain [12] (s. 3).

2.3. The Cadastre and the Role of Monuments

The spatial definition of the New Zealand cadastral survey and property rights systems, and hence cadastre, is founded on physical monuments in the ground. The legal position of boundaries of a parcel of rights, restrictions and responsibilities is defined by original and undisturbed boundary monuments. These boundary monuments, which are often wooden or plastic pegs, are connected by survey observations (via bearing and distance vectors) to nearby reference marks (known as witness and permanent reference marks). The reference marks, which are often iron spikes or iron tubes set into the ground, are in turn connected to other survey marks, including geodetic control. These connections help surveyors to confirm the reliability of old boundary marks and also to relocate boundary positions if the original mark is determined to be disturbed or no longer there.

2.4. The Cadastre and the Role of 2D Coordinates

Connection of the cadastral network to the geodetic network allows 2D coordinates to be assigned to all survey and boundary marks that are integrated into Landonline. It is through these coordinates that the cadastral network can be digitally managed in 2D. Coordinates have also enabled the highly-automated capture, validation, recording and supply of cadastral survey data, thus promoting high levels of efficiency and accuracy of in-coming data [5].

While coordinates enable surveyors to readily relocate survey and boundary marks, they do not provide a legal definition of boundary location (i.e., New Zealand does not have a legal coordinated cadastre). The legal definition of boundaries continues to be provided by original and undisturbed boundary marks established by the surveyor and supported through observations connecting those marks, as documented in the certified cadastral survey dataset.

2.5. The Cadastre and the 2D Parcel Fabric

A valuable output of the cadastral survey system is the parcel fabric, being a continuous surface of connected parcels that covers the whole of New Zealand. The parcel fabric is a substantive layer of information that is used extensively in a myriad of spatial applications, well beyond cadastral surveying. Although the parcel fabric is derived from information provided through cadastral survey datasets, it should not be confused as being the legal cadastre. The parcel fabric is a digital, non-legal representation of boundaries with a positional accuracy that may not coincide with original and undisturbed boundary marks in the ground and may not represent their relative positions as documented in the legal cadastre [5].

2.6. The Cadastre and the Effects of Ground Movement

New Zealand's geographical location astride the collision zone between the Australian and Pacific tectonic plates adds to the complexity of cadastral surveying in this country. Survey observations between local marks accommodate 'general' ground movements across the country because the distortions over small areas are normally insignificant (i.e., the relativity between boundary and witness marks is preserved). In the case of significant movements, such as those resulting from a powerful earthquake (especially around fault lines), the differences might be too great to ignore. In these situations resurvey work is required, based on old marks in conjunction with other information available to the surveyor. The effect of ground movement also impacts on the accuracy of coordinates used for the digital management of the cadastral network. As New Zealand moves, discrepancies occur between the in-the-ground position of marks and their coordinated representation [5]. The magnitude and unpredictable nature of ground movement in New Zealand is the prime reason for a monument-based cadastre.

2.7. 2D Cadastral Survey Datasets

In situations where there is neither need nor desire to restrict the lower and/or upper limits of parcels, surveyors present the detailed survey information in 2D. In the past (prior to Landonline) surveyors documented this information on paper-based survey plans (e.g., Figure 2). Today, this information is spatially captured in the Landonline system via a cadastral survey dataset. Cadastral survey datasets include all the survey data (e.g., marks, measurements, boundaries, parcels) in structured digital form (Figure 3) as well as a static diagram image ('plan'—Figure 4) of the dataset. Within Landonline, the dataset components are directly linked to the same components from previous surveys. For example, it is possible to see the measurements between two marks from many different surveys.

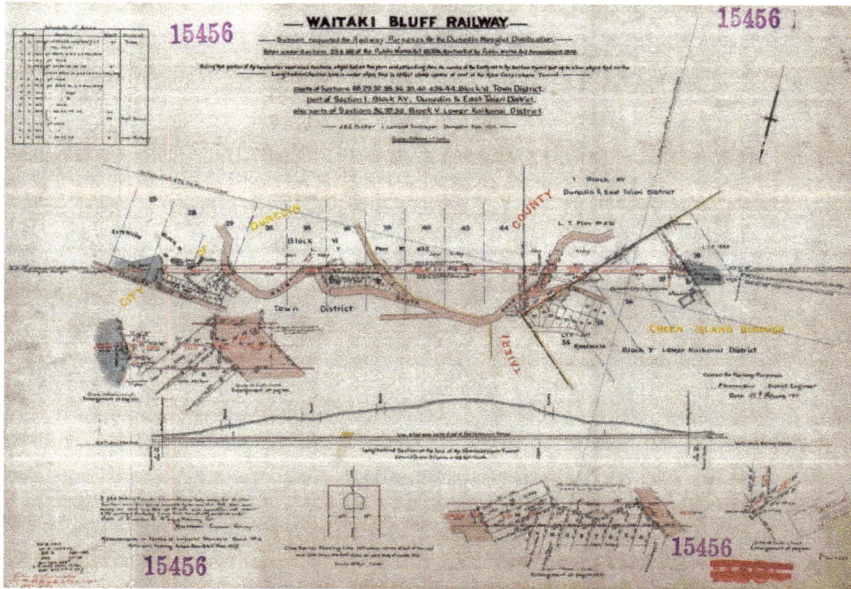

Figure 2. Reference [13] An early example of rights defined in 3D. The 1911 survey plan relates to a railway tunnel in the South Island of New Zealand. The tunnel traverses beneath multiple privately held parcels of land. The upper limit of the railway corridor is restricted in height by way of reference to an offset from a physical structure, being the ceiling of the tunnel. The lower limit of the tunnel is not defined; therefore, it is assumed that the tunnel parcel extends downward to the centre of the Earth.

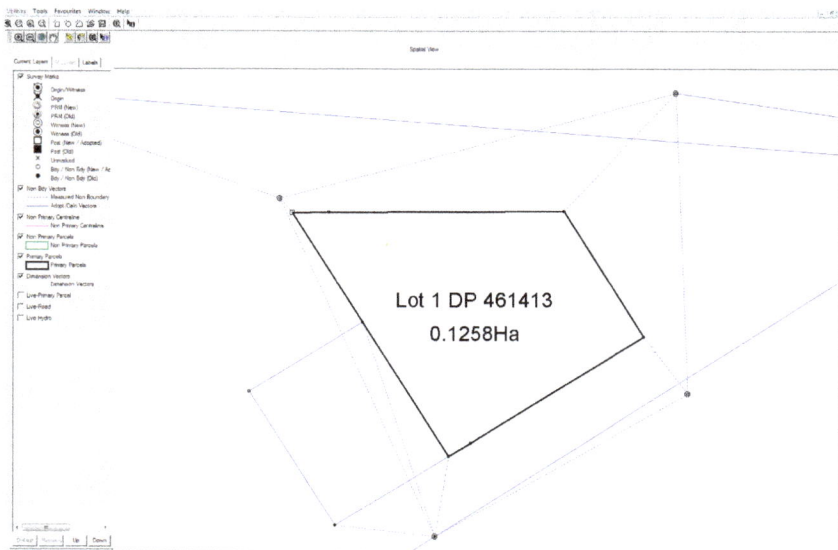

Figure 3. Landonline spatial view of captured cadastral survey dataset.

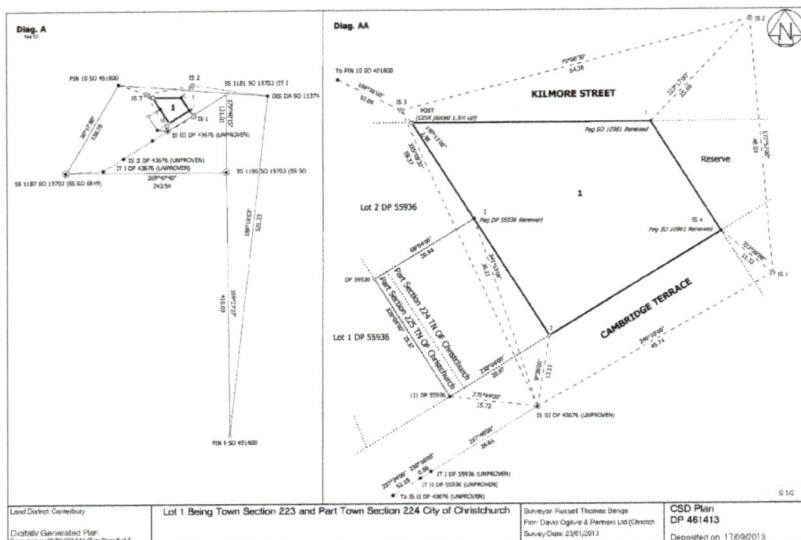

Figure 4. Static diagram image (plan) of cadastral survey dataset.

2D cadastral survey datasets are integrated into the Landonline database. Integration includes recording all the data in the cadastral survey dataset as lodged by the surveyor, as well as generating 2D coordinates for all survey and boundary points in terms of the geodetic control network using least squares adjustments. The coordinates are assigned an accuracy 'order' based on their compliance with the Rules, although, as noted in Section 2.4 they do not define the legal boundaries. Integration also includes meeting topology requirements to ensure that there are no gaps or overlaps recorded in the network of 'primary' parcels. That primary parcel network covers all of New Zealand and consists of 2,538,167 parcels as at 24 August 2017 [14].

Landonline also has a 2D network of 'secondary' parcels (e.g., for rights-of-way or other easements) which are generally related to the corresponding underlying primary parcels. In this secondary parcel network gaps and overlaps are permitted, although secondary parcels may not cross a primary parcel boundary.

2.8. 3D Cadastral Survey Datasets

In situations where a parcel is defined in 3D, the surveyor is required to draft a plan, with section and elevation graphics supported by textual descriptions. This information is provided as a TIFF file and uploaded to Landonline as a static image and not intelligent spatial data as for a 2D cadastral survey dataset (and hence 3D parcels are not visualised in the digital cadastre). These plans can be difficult to interpret by surveyors, let alone the layperson, especially where the boundaries are not uniform.

There are a variety of scenarios where a surveyor may define the vertical extents of a parcel of rights, restrictions and responsibilities, both below ground level (e.g., a tunnel as in Figure 2) and above ground (e.g., an air space covenant). In New Zealand the most common situation in which 3D parcels are created are for unit titles under the Unit Titles Act 2010 (previously the 1972 Act) [15]. Unit titles are the most widely used form of multi-unit property ownership in New Zealand. As at 21 August 2017 there are 14,263 residential and commercial unit title developments that comprise a total of 168,039 individual unit titles [14].

Figures 5 and 6 are examples of plan and elevation views of a 3D unit title development over the primary parcel (surveyed in 2D) that was discussed under Section 2.7 above.

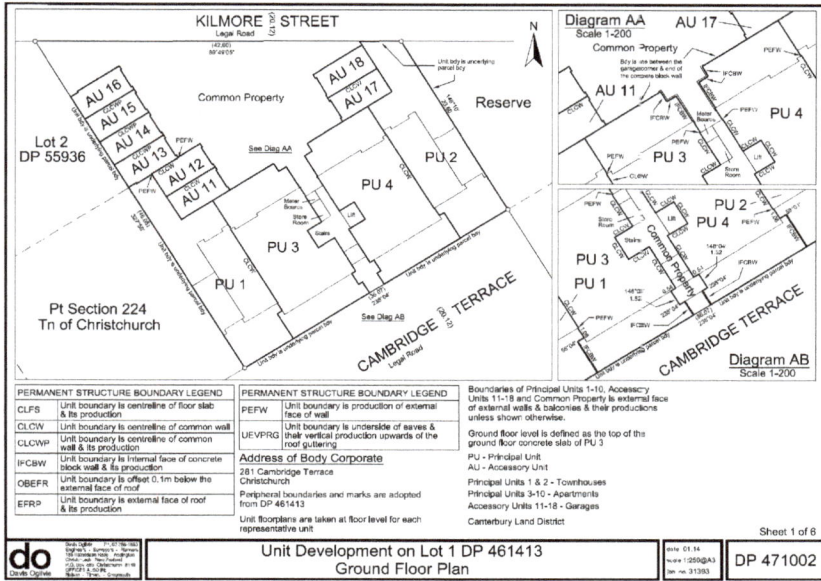

Figure 5. Plan view of unit title development.

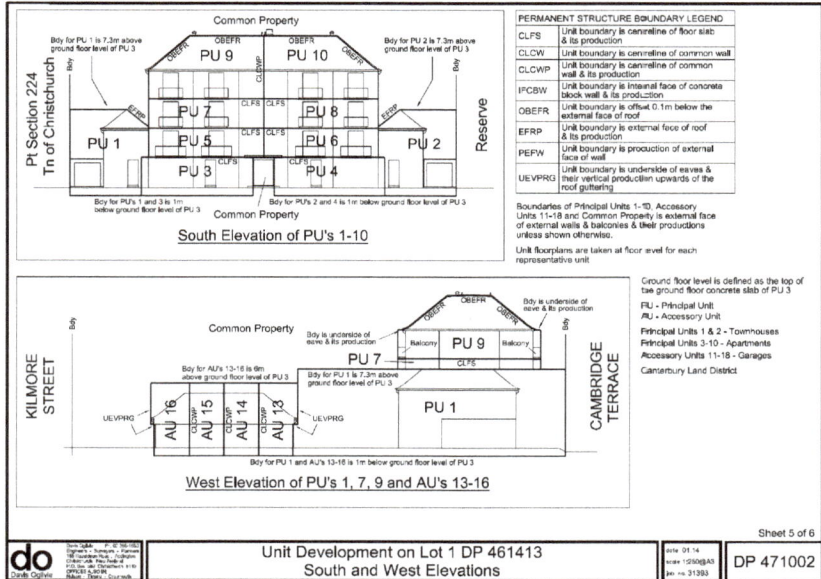

Figure 6. Elevation views of unit title development.

2.9. Future Goals for the New Zealand 3D Cadastre

The introduction of Landonline in the early 2000s represented a significant step forward into the age of digital information. It enabled an end-to-end cycle of digital cadastral survey data (discussed further in Section 3.3 below), at least in a 2D sense. However, the cycle is incomplete in situations where parcels are defined in 3D because analogue procedures are incorporated (as emphasised in Section 2.8 above). The aim is for New Zealand's current legal 3D cadastre to be represented and managed digitally.

A 3D digital cadastre would permit data associated with the real world extents of property rights, restrictions and responsibilities to be digitally captured, automatically checked against requirements, combined with existing data (and subsequently maintained), and exported for re-use in other systems. The primary goal is for the entire cadastral survey process (from 'field to finish') to move away from any remaining analogue processes and maximise the benefits associated with digital data. In this respect New Zealand is already well advanced in relation to 2D surveys (i.e., those that do not have height information), but not for 3D.

Cadastral surveyors in New Zealand already utilise digital technologies when undertaking their surveys, calculations, and verification. They are also obtaining the digital models generated by architects and engineers for new buildings (e.g., through Building Information Modelling—'BIM') and utilise them to define internal boundaries (with ground truthing). However, the current processes do not allow surveyors to submit that data to the cadastre, but instead require them to produce a plan image that cannot be interpreted by a computer and which require human interpretation.

Full digital representation of 3D survey data will enable quality to be maximised through digital verification techniques (e.g., clash detection for 3D boundaries). Visualisation tools offer the opportunity to significantly improve the quality and interpretation of 3D cadastral survey datasets and the digital models they contain. The functionality of zooming into a 3D model and changing the point of view is now readily available (e.g., 3D PDF viewers).

Availability of 3D digital cadastral data will enable it to be presented in various forms and utilised with other geospatial data for a multitude of purposes, both current and yet to be realised [3]. This is routinely undertaken for 2D cadastral data, but can also be readily achieved for 3D datasets.

3. Approach to Achieving Digital 3D

In this section an approach to achieving a 3D digital cadastre is identified with rationale provided as to why it was selected. This leads into discussion around the application of the approach at the (reasonably high) technical level.

3.1. Identifying an Approach

Solutions need to consider how the extents of 3D property rights, restrictions and responsibilities are legally defined in a digital environment and also how the related digital data is incorporated into and managed within the system. As noted earlier New Zealand's legislative framework already supports the definition of property rights, restrictions and responsibilities in 3D and does not inhibit the development of the cadastral survey system to cater for 3D digital data. It is at the technical level where modification is required.

Gulliver (2015) reviewed the literature associated with the development of 3D cadastres, particularly stemming from the research of Stoter and van Oosterom [16], where three fundamental interpretations of 3D cadastre are presented: fully 3D cadastre, hybrid solution and 3D tags. The option being pursued for New Zealand is based on a variation on the concept of 'hybrid cadastre'. Under this approach, 3D property rights, restrictions and responsibilities can be integrated into the digital cadastre and subsequently maintained. In situations where the upper and lower height limits of property rights, restrictions and responsibilities are defined, a full 3D spatial depiction would be used. Otherwise '2D' parcels would be maintained as a default.

The development of a 3D digital cadastre using a variation of Stoter's hybrid approach is deemed to be the most appropriate solution to enhance New Zealand's cadastral survey system. This approach builds on the existing robust 2D digital cadastre by allowing 3D data to be digitally captured, validated, maintained and made available for reuse as and where necessary. Importantly the approach also allows New Zealand's monument-based legal cadastre (discussed in Sections 2.3 and 2.6 above) to be preserved as the foundation of the digital cadastre.

3.2. Spatial Objects to Represent 3D Parcels

The concept of a 'spatial object' (Figure 7) is being pursued to allow parcels defined in 3D to be submitted and integrated into the digital cadastre and subsequently maintained. Spatial object modelling is a seasoned tool used in GIS (Geographic Information System) applications [17] and its relevance to advancing cadastral systems was foreseen in Cadastre 2014: A vision for a future cadastre system [18]. In the context of this article, a spatial object describes (within specified accuracy standards) the size, shape and extent of property rights, restrictions and responsibilities as a 'watertight' 3D volume. In a GIS context, the spatial object is a coordinated 3D volume—defined in terms of x, y and z. Modern spatial technologies, including GIS and BIM, have functionality for creating, manipulating, viewing, and managing such spatial objects.

Figure 7. A perspective view of a unit title development, being a set of spatial objects which represent the 3D extents of parcels of rights, restrictions and responsibilities associated with a multilevel apartment complex (a single 3D parcel is highlighted in magenta). The digital model is based on the paper-based 3D cadastral survey dataset discussed in Section 2.8 above.

3.3. Establishing the Fundamental Requirements

The requirements considered to be fundamental for handling 3D cadastral survey datasets in a 3D digital cadastre are considered in terms of the Cycle of Digital Cadastral Survey Data, as presented in Figure 8.

In the Cycle of Digital Cadastral Survey Data a cadastral survey typically commences with a surveyor searching the cadastral survey database for records that relate to the area of interest. Relevant information is obtained and spatial data is extracted and uploaded into the surveyor's survey software where it is combined with new data from their field survey. A dataset is prepared in the surveyor's software and checked for initial compliance against Rules and system requirements via the validation service. The surveyor then sends their dataset to the LINZ staging environment.

Here the surveyor finalises the cadastral survey dataset and checks it for accuracy and completeness, again through the validation service. The surveyor then certifies the cadastral survey dataset and submits it to LINZ for approval and integration into the cadastral survey database.

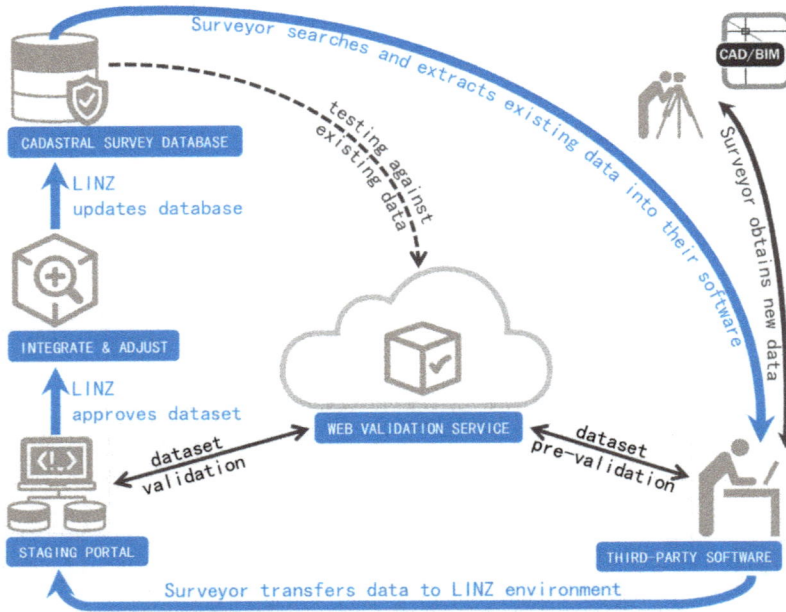

Figure 8. Cycle of Digital Cadastral Survey Data. Note: LINZ denotes Land Information New Zealand.

From the Cycle of Digital Cadastral Survey Data, the fundamental requirements of a 3D digital cadastre are identified (refer to Table 1) and then discussed in further detail in the following subsections.

Table 1. List of fundamental requirements of 3D digital cadastre.

Requirement
Search, visualise and retrieve existing 3D parcels stored in the digital cadastre
Create new 3D parcels via 3D cadastral survey datasets
Lodge new 3D cadastral survey datasets
Validate 3D cadastral survey datasets against regulatory and system requirements
Integrate 3D cadastral survey datasets into the digital cadastre
Maintain spatial alignment of 3D parcels in digital cadastre

3.3.1. Search, Visualise & Retrieve

A search function would need to have the ability to visualise, interrogate and extract digital 3D survey and boundary information. Specifically, users of the system should be able to:

a. see 3D parcels as 2D birds-eye (plan) view against the 2D cadastral parcel network (Figure 9);
b. see 3D parcels in 3D against a 3D projection of the 2D parcel fabric in an area (Figure 10), which also enable different 3D cadastral survey datasets to be related to each other spatially;
c. interrogate 3D parcels defined in a cadastral survey dataset by visualising and interacting with them in 3D;
d. retrieve all data in the 3D cadastral survey dataset as provided by the surveyor.

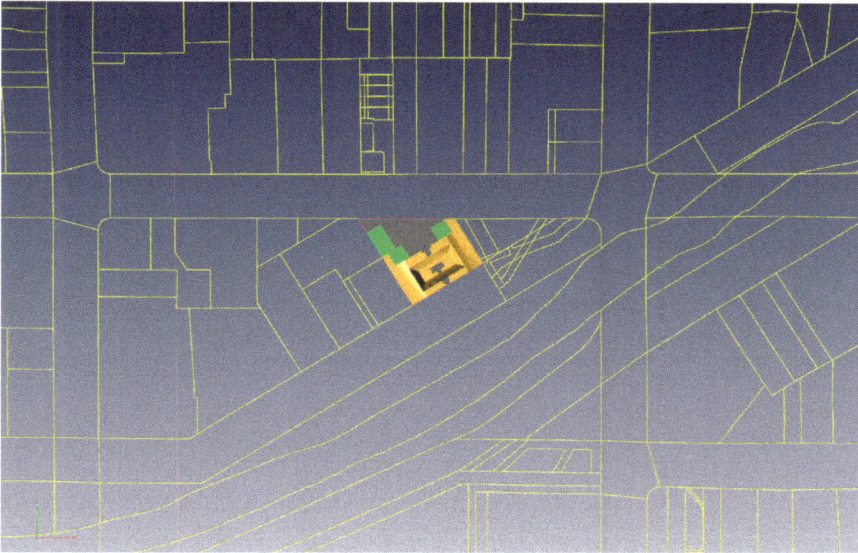

Figure 9. Plan view of 3D parcels against the 2D parcel fabric.

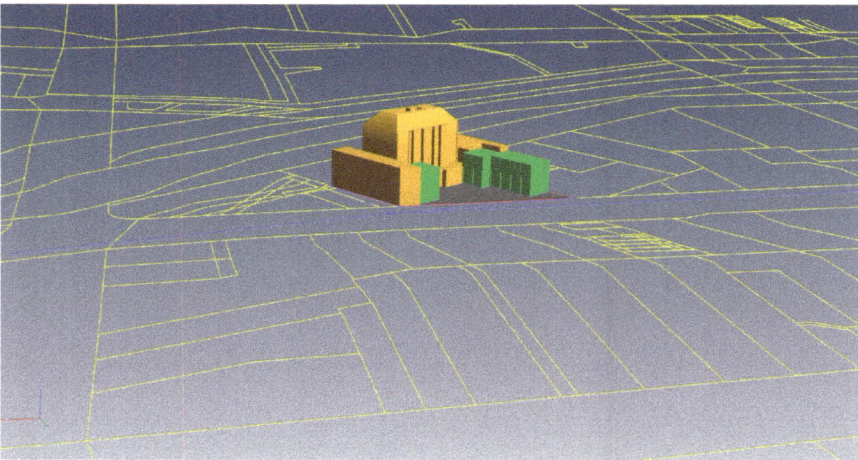

Figure 10. Perspective view of 3D parcels against the 2D parcel fabric.

Existing 3D parcels can be changed (e.g., subdivided, boundaries shifted) just as ordinary '2D' parcels can be altered. In this case, surveyors would obtain existing 3D survey and parcel data from the system and then upload it into their third-party software. The surveyor would then combine new survey work with the existing data as they create a new cadastral survey dataset.

Exports of 3D parcels for use in other software would be dependent on the user's requirement of the data. Two different options for extracting a 3D parcel would be available:

1. as it was lodged, certified and approved in the cadastral survey dataset, as that is the authoritative record of the legal position of the boundary; and;

2. as transformed to fit the digital cadastre, recognising that positions change over time due to improved data and geodetic shifts.

In the former case the future surveyor could then transform that data to fit marks found on the new survey to accurately determine the location on the ground at that time.

3D parcels as spatial objects would be able to be integrated into GIS or other spatial information systems and overlaid with other datasets, whether in 2D or 3D. Being coordinated in terms of the official geodetic datum would enable the boundaries of the primary and secondary parcels (position of the spatial object) to be readily determined on the ground, especially through the use of positioning technologies.

It will be important for users to have the ability to become informed of the location and spatial extents of all property rights, restrictions and responsibilities through visual interrogation and analysis of the 3D digital cadastre. 3D parcels need to be suitably displayed in spatial views in terms of the underlying primary parcel fabric. In addition to 3D perspective views, 2D plan views could permit a quick assessment of the 'footprint' of all property rights, restrictions and responsibilities in relation to underlying primary parcels (refer to Figures 9 and 10). The user would be alerted to situations where a 'footprint' represents a right defined in 3D. Views in 3D could then be explored if further understanding was required by the user.

3.3.2. Creation of 3D Parcels

A 3D parcel would be based on data collected and verified as correct by the surveyor responsible for certifying and lodging the cadastral survey dataset. These data could be obtained through a variety of sources, including digital architectural and engineering designs, BIM data, and via direct survey measurements made in the field. It is expected that the latter would also be used to ground-truth each of the preceding scenarios.

As discussed in Section 3.2 above, every 3D parcel represented as a spatial object is by definition coordinated in terms of x, y, and z coordinates. In order to have a defined relationship between a 3D parcel and its underlying primary parcel, there would be a need to link the 3D parcel to the underlying parcel and nearby permanent reference marks (refer to Figure 11).

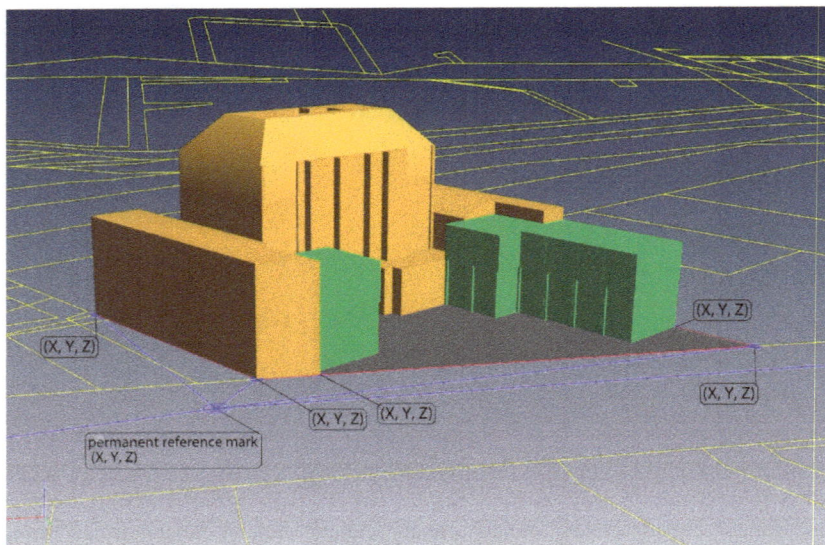

Figure 11. 3D parcels coordinated in terms of the underlying parcel and permanent reference mark.

From a survey definition perspective, the 3D parcel (a parcel with defined height limits) must be defined in relation to its underlying primary parcel. This ensures that the boundaries of the 3D parcel could be identified by firstly relocating boundary points on the underlying primary parcel which will have been fully defined by monuments.

Similarly, in the cadastral survey dataset, the relationship between the 3D parcel and its underlying primary parcel needs to be explicitly defined. This can be achieved by including in the dataset, horizontal coordinates that correspond to boundary points previously defined on its underlying primary parcel. Wherever the boundaries of the 3D parcel and the underlying parcel are coincident, the horizontal coordinates of both ends of the underlying boundary would also need to be included in the dataset. The accuracy between those coordinated points on the underlying parcel and the vertices of the spatial object would have to meet the relevant standard in the Rules. These requirements would also ensure that the spatial object for the 3D parcel could be maintained in alignment with the primary parcel network.

The vertical position of the 3D parcel needs to be capable of being re-established in the future, and also be reflected spatially in the digital cadastre. Therefore, the 'z' vertices inherent in the 3D parcel would need to be related to 'z' coordinates on existing boundary corners of the underlying primary parcel and/or permanent reference marks, all in terms of the official vertical datum.

3D Parcel Represented by Permanent Structure Boundaries

The Rules for Cadastral Survey 2010 allow certain types of secondary parcels to be defined by permanent structure boundaries. These boundaries are described in relation to a physical feature (e.g., the outer face of a wall, or an offset to the feature). The licensed cadastral surveyor who certifies the cadastral survey dataset is responsible for defining the position and accuracy of the permanent structure boundary in relation to the permanent structure.

Two options have been identified for recording this relationship:

1. Three-dimensional parcel representation of the permanent structure boundary only (as presented in Figure 7). The parcel and its boundaries would be defined by a 3D spatial object, along with a description of the physical structure to which it is related and the relationship (e.g., 'boundary through centre of wall' or 'boundary follows centre of concrete floor'). The description of the relationship between the permanent structure boundaries and the permanent structure is of great importance as it defines the legal position of the boundary.
2. Spatial object representation of the permanent structure boundary and the permanent structure (e.g., the physical structure of the apartment complex associated with the 3D parcels depicted in Figure 7). The 3D parcel and its boundaries would be defined by a spatial object, as would the permanent structure itself (i.e., two layers of data would be provided). A description of the relationship between the two would not necessarily be required as this would be able to be determined from the spatial objects using a measurement tool in spatial software. This approach, in which legal spaces are associated with the physical elements to which they relate, was raised by Aien et al. [19] and is being further promoted in recent research, by Atazadeh, Rajabifard and Kalantari [20].

Common property, being land or a building that is for the use of all the property owners, would not necessarily be defined by a 3D parcel. It could be that part of the space remaining after the 3D parcels have been excluded, and would be viewable using suitable software that extruded the 2D underlying parcel boundaries as appropriate.

3.3.3. Lodgement

The system should be capable of receiving 3D cadastral survey datasets (that may include both 2D and 3D information) as digitally certified by the surveyor (i.e., without change). Cadastral survey datasets would be prepared by surveyors in third-party software. The contents of a 3D cadastral survey dataset would include information that describes the size, shape and spatial position of each of its 3D parcels in relation to:

- the '2D' underlying primary parcel, or the 3D parcel where it is being subdivided or redefined;
- any permanent structure to which it is referenced (e.g., wall, floor and ceiling of an apartment complex).

A 3D cadastral survey dataset may also include ordinary ('2D') parcels and related survey information. For example, in a unit title development the surveyor may choose (or be required) to re-survey the '2D' primary parcel to which new 3D parcels of an apartment block are referenced.

As is the situation for ordinary parcels, each 3D parcel will need to be identified through a unique appellation (label). This is required for the management of the cadastre and to enable tenure system managers to register (create) the associated right, restriction or responsibility.

3.3.4. Validation

Data being submitted into the cadastre needs to be validated. The primary purpose of validation is to ensure that the data complies with the Rules and that it is able to be integrated into the digital cadastre. Automated business rules should check that:

a. new 3D parcel boundaries do not overlap underlying parcel boundaries to which they relate;
b. new 3D parcels in a 3D cadastral survey dataset do not illegitimately overlap (some types of overlaps are permitted) other 3D parcels (new and existing);
c. 3D cadastral survey datasets are in terms of the official national vertical and horizontal datums;
d. the 3D volume of any existing 3D parcel that is being subdivided (e.g., a unit redevelopment of an apartment complex) is completely taken into account.

Validation through automated business rules is a key part of the system, and this is emphasised in research by Thompson and van Oosterom [21] and Karki et al. [22]. It is likely that surveyors will be required to obtain and carry through links to existing data (e.g., underlying parcels, points) held in the system to enable automated validation. For maximum effect, automated validation should be available to surveyors as a cadastral survey dataset is being prepared in third-party software prior to certification and lodgement. LINZ staff are also required to perform validation prior to the approval of the certified cadastral survey dataset.

Three-dimensional clash detection routines found in current surveying, engineering, GIS and BIM software indicate that the validation of 3D parcels ought to be readily achieved. Despite any validation procedures, the responsibility for the correctness of certified data will continue to be the obligation of the surveyor responsible for the cadastral survey dataset.

3.3.5. Integration

The vision is for a digital cadastre where 3D property rights, restrictions and responsibilities are represented digitally in 2D or 3D as appropriate in a single integrated and seamless system. To achieve this, the data contained in 3D cadastral survey datasets, as provided by the surveyor, will need to be integrated into the digital cadastre (i.e., combined with existing data). The integration process will store the components of the cadastral survey dataset in the database in a similar manner to other cadastral survey datasets and will be completed by LINZ. To enable automated integration processes, there is likely to be a need for surveyors to obtain and carry through links to existing data held in the system (as for validation, discussed above). Once a cadastral survey dataset is approved, it would be adjusted into the digital parcel network and the representation of 3D parcels would be repositioned relative to their underlying parcel and the permanent reference marks in terms of the official coordinate datum and projection.

Currently primary and secondary parcels defined by nodes and lines are fully integrated into the boundary network. This means they are managed topologically and coordinates are generated through least-squares adjustment of the vector (bearing and distance) data. However, under this proposal 3D parcels would be managed through a different process. Topology would not be directly managed, and alignment would be maintained by applying a transformation to the 3D parcels, using the connection points to the cadastral network as the 'control' (i.e., the boundary points on the underlying parcel and the permanent reference marks, as recorded in the cadastral survey dataset).

3.3.6. Maintaining Spatial Alignment

LINZ needs to be able to maintain and update the spatial location of 3D parcels in the network over time. Any movement of geodetic marks (such as that caused by tectonic movements) that affect cadastral marks may have an effect on the underlying parcels which in turn has an effect on any associated 3D parcel. The processes for adjusting the 2D parcel network will need to be extended to also adjust any 3D parcel to keep them in the correct relationship. The requirement for connections that meet the accuracy standards in the Rules is intended to ensure that the transformation would result in negligible distortion of the 3D parcel during future alignment processes. Similarly, the height values of 3D parcels will need to be adjustable as the height values of geodetic control marks change over time.

4. New Zealand's Progress to a 3D Digital Cadastre and the International Opportunity

In this section New Zealand's progress to a 3D digital cadastre is presented in light of work currently being undertaken. This work couples with the approach to realise a 3D digital cadastre to create an opportunity at the international level.

4.1. Advanced Survey and Title Services to Replace Landonline

Landonline is built on technology that was considered to be leading edge in the late 1990s. There have since been substantial advancements in technology, knowledge and expertise in developing land administration systems. There are also changing expectations from an increasingly diverse range of customers and consumers that include land professionals, such as surveyors and conveyancers, along with experts and non-experts in spatial science, systems and information. These expectations relate to the functionality and performance of the system and the applications for which its data can be used. In particular there is increasing demand from the spatial community, and indeed the general public, for 3D property information in a digital and readily consumable format.

Advanced Survey and Title Services (ASaTS) is a current project of work that aims to deliver next generation technology and significantly improve the quality and range of survey and title services that LINZ provides to its customers. ASaTS will increase the availability and quality of property information to support high quality decision-making, while moving the survey and title service to a responsive and sustainable technology platform. Importantly, ASaTS is supported by the New Zealand government, and in the context of this article, it provides a pathway to realising a 3D digital cadastre.

LINZ is considering an 'as a Service' (aaS) approach to developing ASaTS. Rather than owning the system, as is currently the case for Landonline, LINZ would select a provider that could deliver a suitable system that LINZ would pay to access. It is important to note that while the new system would be owned by another party, LINZ would retain control of the data and its use. LINZ and its staff would also continue to process and assure the quality of survey and title transactions.

At this point in time, LINZ is working with a preferred supplier through a discovery and definition phase. LINZ is hopeful of completing contractual arrangements in 2018 with the development of and transition to the new system scheduled to be completed by 2021 (see Figure 12).

4.2. Applicability of New Zealand's Approach to other Jurisdictions

New Zealand's pursuit of advanced survey and title services provides an opportunity for other jurisdictions to benefit from the work resulting from the ASaTS Project. In particular, the work being undertaken could be of interest to other jurisdictions looking to develop digital 3D capabilities of their systems. This is especially due to the coordinate-based approach to create and manage 3D parcels.

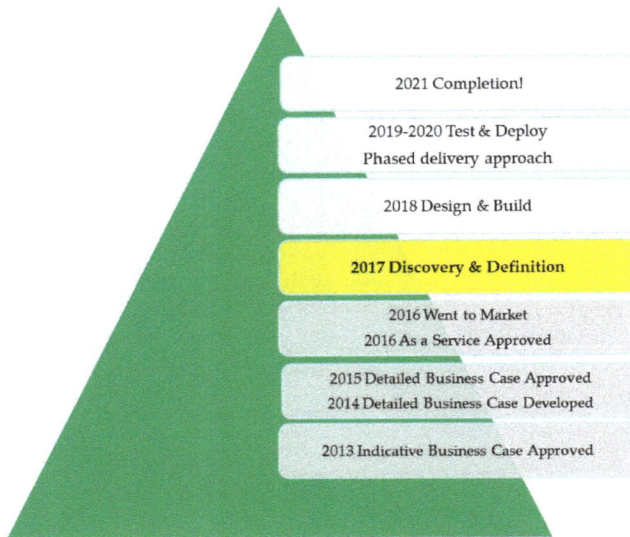

Figure 12. Advanced Survey and Title Services (ASaTS) timeline.

Other jurisdictions would not necessarily need to share a similar base system to that of New Zealand. The GIS-centric approach of 3D spatial objects should lend itself to any existing cadastre, whether it be advanced or developing, and irrespective of whether the legal definition of underlying parcels is based on monuments (supported by observations—e.g., New Zealand) or legal coordinates (e.g., Singapore [23]). While the authors have described how 3D parcels defined by spatial objects will be integrated into a base system that is monument-based, the establishment of a relationship between 3D coordinated objects and underlying or abutting parcels would seem universal.

4.3. Development of Software Applications

A key feature being explored is the potential for the ASaTS system to work in concert with third-party software applications via a set of secure standards-based application programming interfaces (APIs). The role played by these third parties is likely to include functionality to create cadastral survey datasets, including any diagrams, for both 2D and 3D surveys. Where possible, data formats and schemas will leverage existing LINZ/international standards. Where gaps exist, LINZ will endeavour to work with professional and industry bodies and communities to ensure that data standards are consistent, fit-for-purpose, and easily consumable by providers of third-party software solutions.

5. Conclusions

New Zealand is in the process of transitioning to a national 3D digital cadastre. In New Zealand's favour is a mature property rights system that already accounts for 3D situations and a robust 2D digital cadastre. These combine to provide a sound platform on which to develop 3D capabilities.

Despite this, the approach presented in this article to digitally manage 3D parcels of rights, restrictions and responsibilities would seem generic and hence agnostic of a jurisdiction's base cadastre. For this reason, the authors are of the view that the work being undertaken by New Zealand has

ISPRS Int. J. Geo-Inf. **2017**, *6*, 375

international appeal with opportunities for other jurisdictions and providers of software applications to benefit from this work and thus better realise the efficiencies and value of 3D digital cadastral survey data.

Author Contributions: This content of this article is the result of collaboration and contribution of all three authors.

Conflicts of Interest: The authors declare no conflict of interest.

References and Notes

1. Gulliver, T.; Haanen, A.; Goodin, M. A 3D Digital Cadastre for New Zealand by 2021: Leveraging the Current System and Modern Technology. In Proceedings of the 5th International FIG 3D Cadastre Workshop, Athens, Greece, 18–20 October 2016.
2. Gulliver, T. Developing a 3D Digital Cadastral System for New Zealand. Master's Thesis, Department of Geography, University of Canterbury, Christchurch, New Zealand, 2015.
3. Gulliver, T.; Haanen, A. Developing a Three-Dimensional Digital Cadastral System for New Zealand. In Proceedings of the 25th FIG Congress, Kuala Lumpur, Malaysia, 16–21 June 2014.
4. Gulliver, T. ASaTS and Opportunities for the Cadastral Survey System. In *Surveying + Spatial*; New Zealand Institute of Surveyors: Wellington, New Zealand, 2016; pp. 35–37.
5. Gulliver, T. A Coordinate Cadastre for New Zealand? Yeah, nah! In *Surveying + Spatial*; New Zealand Institute of Surveyors: Wellington, New Zealand, 2017; pp. 31–33.
6. Cadastral Surveyors Licensing Board of New Zealand. Register of Licensed Cadastral Surveyors. 2017. Available online: https://www.cslb.org.nz/ (accessed on 17 August 2017).
7. Rules for Cadastral Survey 2010. New Zealand.
8. Land Information New Zealand. Map of New Zealand. 2016.
9. Cadastral Survey Act 2010. New Zealand.
10. Land Information New Zealand. Cadastre 2034: A 10–20 Year Strategy for Developing the Cadastral System. 2014.
11. Land Transfer Act 1952. New Zealand.
12. Land Transfer Act 2017. New Zealand.
13. Land Information New Zealand. Survey Plan SO 15456, Otago Land District. 1911.
14. Land Information New Zealand. Landonline Interrogation. August 2017.
15. Unit Titles Act 2010. New Zealand.
16. Stoter, J.; van Oosterom, P. *3D Cadastre in an International Context: Legal, Organizational, and Technological Aspects*; CRC Press: Boca Raton, FL, USA, 2006.
17. Molenaar, M. *An Introduction to the Theory of Spatial Object Modelling for GIS*; CRC Press: London, UK, 1998.
18. Kaufmann, J.; Steudler, D. Cadastre 2014: A Vision for a Future Cadastral System. Available online: http://www.eurocadastre.org/pdf/cadastre_2014.pdf (accessed on 31 August 2017).
19. Aien, A.; Kalantari, M.; Rajabifard, A.; Williamson, I.; Wallace, J. Towards integration of 3D legal and physical objects in cadastral data models. *Land Use Policy* **2013**, *35*, 140–154. [CrossRef]
20. Atazadeh, B.; Rajabifard, A.; Kalantari, M. Assessing Performance of Three BIM-Based Views of Buildings for Communication and Management of Vertically Stratified Legal Interests. *ISPRS Int. J. Geo-Inf.* **2017**, *6*, 198. [CrossRef]
21. Thompson, R.; van Oosterom, P. Validity of Mixed 2D and 3D Cadastral Parcels in the Land Administration Domain Model. In Proceedings of the 3rd International FIG Workshop on 3D Cadastres: Developments and Practices, Shenzhen, China, 25–26 October 2012.
22. Karki, S.; Thompson, R.; McDougall, K. Development of Validation Rules to Support Digital Lodgement of 3D Cadastral Plans. *Comput. Environ. Urban Syst.* **2013**, *40*, 34–45. [CrossRef]
23. Boundaries and Survey Maps Act 1998. Singapore.

International Journal of
Geo-Information

MDPI

Article

Supporting Indoor Navigation Using Access Rights to Spaces Based on Combined Use of IndoorGML and LADM Models

Abdullah Alattas [1,*], Sisi Zlatanova [1], Peter Van Oosterom [1], Efstathia Chatzinikolaou [2], Christiaan Lemmen [3] and Ki-Joune Li [4]

[1] Faculty of Architecture and the Built Environment, Delft University of Technology, Julianalaan 134, 2628 BL Delft, The Netherlands; s.zlatanova@tudelft.nl (S.Z.); p.j.m.vanOosterom@tudelft.nl (P.V.O.)
[2] Faculty of Environment, Geography and Applied Economics, Department of Geography, Harokopio University of Athens, Eleftheriou Venizelou 70, 17671 Kallithéa, Greece; e.chatz175@gmail.com
[3] Netherlands Cadastre, Land Registry and Mapping Agency (Kadaster), Hofstraat 110, 7311 KZ Apeldoorn, The Netherlands; Chrit.Lemmen@kadaster.nl
[4] Pusan National University, Kumjeong-Gu, Pusan 46241, Korea, lik@pnu.edu
* Correspondence: a.f.m.alattas@tudelft.nl; Tel.: +31-639-898-691

Received: 31 August 2017; Accepted: 22 November 2017; Published: 24 November 2017

Abstract: The aim of this research is to investigate the combined use of IndoorGML and the Land Administration Domain Model (LADM) to define the accessibility of the indoor spaces based on the ownership and/or the functional right for use. The users of the indoor spaces create a relationship with the space depending on the type of the building and the function of the spaces. The indoor spaces of each building have different usage functions and associated users. By defining the user types of the indoor spaces, LADM makes it possible to establish a relationship between the indoor spaces and the users. LADM assigns rights, restrictions, and responsibilities to each indoor space, which indicates the accessible spaces for each type of user. The three-dimensional (3D) geometry of the building will be impacted by assigning such functional rights, and will provide additional knowledge to path computation for an individual or a group of users. As a result, the navigation process will be more appropriate and simpler because the navigation path will avoid all of the non-accessible spaces based on the rights of the party. The combined use of IndoorGML and LADM covers a broad range of information classes: (indoor 3D) cell spaces, connectivity, spatial units/boundaries, (access/use) rights and restrictions, parties/persons/actors, and groups of them. The new specialized classes for individual students, individual staff members, groups of students, groups of staff members are able to represent cohorts of education programmes and the organizational structure (organogram: faculty, department, group). The model is capable to represent the access times to lecture rooms (based on education/teaching schedules), use rights of meeting rooms, opening hours of offices, etc. The two original standard models remain independent in our approach, we do not propose yet another model, but applications can fully benefit of the potential of the combined use, which is an important contribution of this paper. The main purpose of the combined use model is to support the indoor navigation, but could also support different applications, such as the maintenance and facility management work, by computing the cleaning cost based on the space floor area. The main contributions of this paper are: a solution for the combined use of IndoorGML-LADM model, a conceptual enhancement of LADM by the refinement of the LA_Party package with specialization for staff and student (groups), and the assessment of the model by converting sample data (from two complex university buildings) into the model, and conducting actual access-rights aware navigation, based on the populated model.

Keywords: spatial unit; private rights; common rights; indoor; navigation; party

1. Introduction

Buildings in our cities have become very large and complex, which affects the movement of the users within such an environment. In some buildings, users require guidance to reach their destinations, and, thus, a variety of navigation models have been developed. The navigation process is a primary daily activity, which allows the user to explore the indoor space of a building [1]. The main goals of the navigation process are to define the location of the users and to distinguish the best path for the navigation process [2].

IndoorGML is an Open Geospatial Consortium (OGC) standard that provides a framework for an indoor navigation system that presents an elaboration of the indoor space and Geography Markup Language (GML) syntax for encoding geoinformation [3]. There are two components of the IndoorGML standard. The first component is the core data model that is responsible for describing geometry and topology connectivity. The second component is the data model that is used in the navigation process [4]. There are many types of buildings, such as educational, shopping centers, public transportation, residential, and hospitals, where each has different types of users. The indoor environment of each type of building is unique and has many relationships between spaces and users. The indoor space for each building is defined according to physical boundaries, such as floors, walls, and ceilings. However, some indoor spaces do not have well-defined physical boundaries, therefore, they require geometric and semantic information to define the space according to its function or the right of use [5]. Thus, by using the semantic information of the subdivided indoor spaces (NavigableSpaces), an accurate network and indoor navigation path could be generated [2]. Each space has different usability characteristics, which leads to different relations between users and spaces. According to these relationships, the NavigableSpace has different rights, restrictions, and responsibilities, based on an individual or group of users. User-space relationships are defined by the Land Administration Domain Model (LADM), which is an ISO standard (The International Organization for Standardization) [6] that focuses on rights, responsibilities, and restrictions (RRR) influencing lands and space/construction elements. LADM is considered as the foundation for countries or organizations to provide proper development on the basis of security of tenure, valuation, taxation, spatial planning, and land management.

Currently, navigation inside buildings is performed to reach different destinations based on the user's desire. However, each indoor environment has indicators that are based on the rights, restrictions, and responsibilities of the user and space. Therefore, to provide an effective navigation for different types of users, the indoor environment, and the space-user RRRs must be considered. Thus, when space subdivision process according to the effect of RRR, the result will enhance the navigation based on the actual user access to spaces. LADM is a conceptual model that covers a wide scope of information that is related to our daily life activity in educational or medical institutions, transportation hubs, shopping malls, etc. This includes a great deal of information that is relevant for navigation access rights for specific types of buildings. For example, for educational buildings, such as universities, it can be related to: student groups/cohorts, staff groups/departments, on individuals (staff, students, visitors), on lecture schedules, on room/desk assignments, on opening/access hours for various parties, on meeting room reservation, etc. Thus, this is very broad, but, in reality, it is already used in practice (perhaps not per se in one system, but it is good that there is a conceptual model including all of these aspects). We assessed the conceptual model, by adding instance level (real) data to the model for a number of real world cases. We further showed how data from this model can be used to do routing, which takes into account the access rights of a specific user. We are not aware of any other research of actual system having the same capabilities. We therefore think this is a significant breakthrough. Due to the potential data richness, additional applications of this combined use of IndoorGML and the LADM model are possible; e.g., computing cleaning costs based on square meters of floor, designing/maintaining cables, etc. In this work, the contribution can be stated as follows:

- Developing an approach for the combined use of IndoorGML and LADM (and refine/adapt/extend the combined model when needed) to support user navigation.
- Assess the proposed combined model, by testing with actual data from two university buildings, including the involved RRRs and parties (i.e., populate the model with instances/data).
- Demonstrating the use of the populated combined model to perform actual navigation for different parties, while considering their RRRs (i.e., only use spaces allowed for access).

This paper is organized as follows: the next section introduces the LADM standard, the IndoorGML standard, and the combined use approach. Section 3 describes the methodology to define/develop the combined use model, including refined modeling of parties. Section 4 represents actual data from the two cases studies that are included in the combined use model (based on the rights, restrictions, and responsibilities). The case studies are on purpose two examples from the same class (University buildings), because this is already a very challenging conceptual modelling exercise. After mastering this type of building with its users for navigation purpose, our further work will include modelling other types of building and their users, which may indeed be quite different (and also other applications than routing; e.g. maintenance, valuation/taxation, disaster management, etc.). Section 5 provides some actual navigation examples based on RRRs. The paper ends with the discussion in Section 6.

2. Background

2.1. LADM

Land administrations systems (land registry, cadastre) have different origins in different countries. The information was sometimes collected for taxation purposes [7], and in other cases, for legal security. Over the years, in many countries, the land administration systems more and more served both applications; e.g., in the area of spatial development or spatial planning. In this context, the term multi-purpose cadastre is used. Based on the initiative of the FIG (International Federation of Surveyors), ISO has developed the standard Land Administration Domain Model (LADM) [8]. In the standard, land administration is described as the process of determining, recording, and disseminating information about the relationship between people and land (or rather 'space'). These recognized rights are, in principle, eligible for registration, with the purpose being to assign a certain legal meaning to the registered right (e.g., a title). Therefore, land administration systems are not just 'handling geographic information', as they represent a lawfully meaningful relationship among people, and between people and land. The model has been used by several institutions in different counties [9].

2.1.1. LADM Content and Model

The LADM standard defines a basic administrative unit ('basic property unit') as an administrative entity, which is subject to registration (by law), or recordation, consisting of zero or more spatial units ('parcels') against which (one or more) unique and homogeneous RRRs (rights, e.g., ownership rights or land use rights, responsibilities, or restrictions) are associated to the whole entity, as included in a land administration system. A 'spatial unit' [10] is a single area (or multiple areas) of land and/or water, or a single volume (or multiple volumes) of space. A spatial unit can be described by two-dimensional (2D) or three-dimensional (3D) geometry, or even by textual descriptions [5]. 'Homogeneous' means that a right, restriction, or responsibility affects the whole basic administrative unit. 'Unique' means that a right, restriction, or responsibility is held by one or more parties (e.g., owners or users) for the whole basic administrative unit. Making the unit any larger would result in the combination of rights not being homogenous. Making the unit smaller would result in at least two neighbor parcels with the same combination of rights. The spatial units are called legal or virtual objects because they do not need to be visible in the real world.

However, it should be noted that, quite often, the boundary of a parcel coincides with a physical real-world object; e.g., a fence, wall, or the edge of a road. In the case of 3D spatial units, this is even more true; e.g., the geometries of physical objects, such as tunnels, buildings (parts), or other constructions often correspond also to legal spaces with unique and homogeneous RRRs that are attached [10,11]. Perhaps indoor navigation itself is not directly a 3D cadastre topic, but is it strongly related because there is significant overlap between the indoor spaces as used in navigation and the 3D spatial units for registering the RRRs (in the environment of an apartment and other buildings and constructions).

The main characteristic of the LADM can be summarized as follows:

- The model is object-oriented: Unified Modeling Language (UML) class diagrams, supporting the model-driven architecture (MDA).
- The basic classes of the model are (1) parties (people and organizations); (2) basic administrative units, rights, responsibilities, and restrictions (e.g., ownership rights); (3) spatial units (parcels, and the legal space of buildings and utility networks); and, (4) spatial sources (surveying) and spatial representations (geometry and topology) (Figure 1).
- The spatial units are abstract spaces. They are geometric/topological representations of rights and administrative units. Spatial units can coincide with topographic features. LA_LegalSpaceNetwork is used to define rights for utility networks (not to be confused with navigation network as in IndoorGML).
- 3D Spatial units can be unbounded; e.g., op column of space with no top.

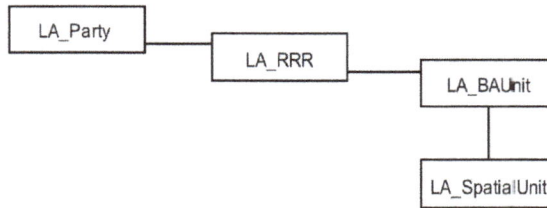

Figure 1. Core classes of the Land Administration Domain Model (LADM) taken from ISO 19152:2012.

2.1.2. Relationship between Physical and Virtual Objects

A (3D) building registration is something other than a (3D) cadastre. Cadastre concerns the legal spaces. That is, spaces that are described by geometry (and topology), to which certain RRRs are attached. Thus, all kinds of building details, such as different rooms/spaces, may not always be relevant from a legal perspective (when the same RRRs apply), but are often defined (when designing a building). Only when the RRRs are different, a separate geometry is needed. Thus, most likely only a part of the indoor building model information may be relevant in the 3D cadastre context (and perhaps that geometry is even implicit; e.g., a 3D boundary face, defined by the 'middle of the wall'). The geometries of the real world (physical) objects and the geometries of the legal objects should be consistent and we should design rules for this, which could be specified via constraints in OCL.

Annex K from ISO 19152 Figure 1 is a UML diagram showing the core classes of the LADM standard: LA_Party (person), LA_RRR (right, etc. such as ownership), LA_BAUnit, and LA_SpatialObject (parcel). LA_SpatialObject has several specializations, such as LA_LegalSpaceNetwork and LA_LegalSpaceBuildingUnit (not shown in the diagram, but could be linked to physical building registration). LADM is more a conceptual framework, defining concepts and terminology, than a prescriptive standard. A country should first develop an LADM country profile that is supporting the legislation of the country (and is described in terms of the international standard), before transforming this into a land administration implementation.

2.2. IndoorGML

The requirements for indoor spatial information are commonly specified according to the types of applications: management of building structures, analysis of human activities, air quality and control, facility management, registration of properties, or indoor navigation. Amongst all, indoor navigation and indoor LBS are the most prominent application [12,13]. The goal of IndoorGML is to reflect the requirements for indoor navigation by providing a general framework for specifying spatial models for indoor navigation, which present the properties of the indoor space and provides a formalism to derive a network for the purpose of pathfinding Figure 2. Presently, diverse systems are available for indoor navigation following a range of approaches. Each application applies its own semantic and geometric description of space, which complicates the re-use of indoor spatial information. Therefore, it is critical to uniquely specify the type of objects and their properties, and identify the spaces that are available for navigation for a specific user.

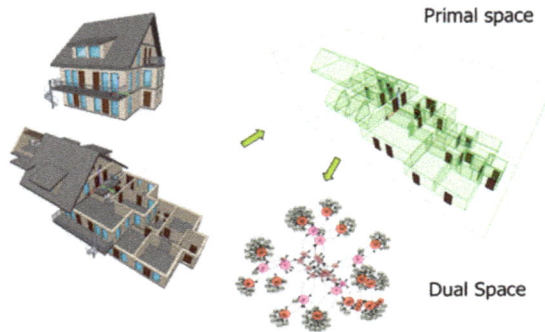

Figure 2. The general concept of IndoorGML: three-dimensional (3D) model of a building (**left**); room (green) and door (brown) spaces used for navigation (**upper right**); navigation network (**lower right**).

2.2.1. Primal Space

Space subdivisions can be defined on the basis of different properties of a building (construction, functional use, security, sensor coverage) or a user profile (e.g., walking, driving). Each specific space subdivision can be organized in a space layer, i.e., a multi-layered space model, in which spaces from different layers may overlap and spaces from the same layer do not overlap. A specific theme space partitioning can either subdivide or unite several topographic units (rooms), e.g., 'security area', 'check-in area' or 'dangerous area'. Such subdivision may also change dynamically (Figure 3).

Figure 3. Union of topographic spaces to represent dangerous areas.

The primal spaces of the current version of IndoorGML have the following characteristics:

- IndoorGML provides a UML class diagram and eXtensible Markup Language (XML) implementation.
- Spaces include all architectural components (entrances, corridors, rooms, doors, and stairs), which are of importance for moving through the building. Construction elements are seen as obstacles and are clearly indicated as non-navigable spaces.
- Components irrelevant to the description of spaces, such as furniture, are not included in the scope of the current version.
- Spaces are closed objects, represented by areas in 2D and volumes in 3D. They may touch but may not overlap within the same layer.
- Spaces can be bordered by topographic or fictional boundaries, or combinations of them.
- Spaces are uniquely identified.
- IndoorGML offers the concept of 'thin' objects (e.g., walls and/or doors), which may be represented as surfaces (in 3D or lines in 2D).
- All of the spaces are semantically classified. The current version of IndoorGML is focused on topographic representation. The top-level semantic consists of navigable (rooms, corridors, doors) and non-navigable (walls, obstacles) spaces.
- Navigable spaces are those used to derive the navigation network. They are further specialized into general (e.g., rooms), connection (e.g., doors), and transfer (e.g., corridor) spaces.

General spaces are those in which people stay. Transition spaces are those that people use to move from one general space to another. Doors and other openings are classified as connection spaces. Doors that provide connection to outdoor navigation networks have special meaning and are noted as 'anchor spaces' (Figure 4).

Figure 4. Semantics of the primal space in IndoorGML.

2.2.2. Dual Space

The dual model follows straightforward from the primal model applying the Poincaré duality [14]. According to Poincaré duality, a k-dimensional object in N-dimensional primal space is mapped to (N−k) dimensional object in dual space. Thus, solid 3D spaces in 3D primal space, e.g., rooms within a building, are mapped to nodes (0D object) in dual space. 2D surfaces shared by two solid objects are transformed into an edge (1D) linking two nodes in dual space [14]. While the complete network is derived from the primary space using only adjacency relationships, the connectivity between the spaces relies on semantics (i.e., the notion of doors). As mentioned previously, the type of the obtained graph depends on the spaces (navigable or non-navigable). Figure 5 shows examples of dual spaces derived from the same primary space. In the Figure: W stands for walls, D for doors, and S for spaces, and red lines represent edges in dual space.

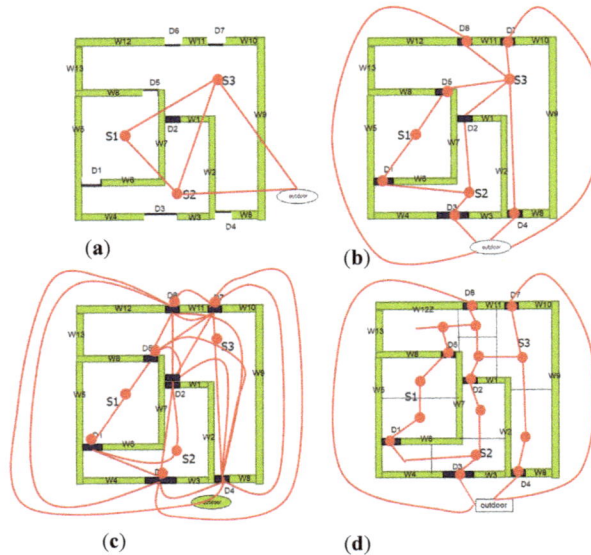

Figure 5. Examples of networks: (**a**) 'thin' doors which is a symbolically used; (**b**) 'thick' doors which are considered spaces itself; (**c**) only door spaces which are all the possible door-to-door connections; and, (**d**) the subdivision of rooms in which each new space has only one door.

The nodes in IndoorGML represent the state of a user and the edges of the transition between two states, as shown in Figure 6. IndoorGML can be used to exchange information in three forms: provide primal and dual space, provide only dual space and integrate it with another semantically rich standard (CityGML, IFC, KML), or provide only dual spaces (or path of dual space, i.e., a navigation path).

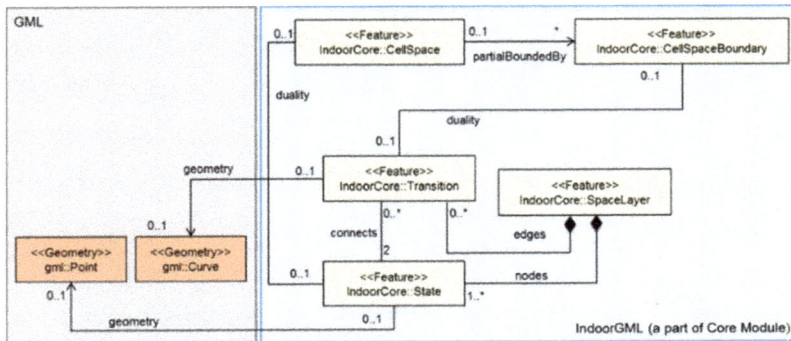

Figure 6. A part of the core IndoorGML.

IndoorGML is a relatively new standard, and, therefore, some concepts are not strictly defined or are presented in a simplistic way. They are left for further investigations and clarifications. For example, IndoorGML does not provide a strict definition for indoor and outdoor. Construction units, such as balconies, terraces, or sheds, are, therefore, not discussed in the standard. Furniture is also excluded from the scope. However, some researchers have shown that furniture can be an important indicator for identifying functional areas, such as 'at front of the coffee machine', 'at the bench', etc. [15], which can be used either for localization or for a better estimation of space that is free of obstacles for more detailed navigation. The standard does not have any strict rules on the creation of space layers, yet. The space subdivisions for a visitor and an employee are generally the same but the visitor has access to a subset of the spaces that the employee can visit. It is an interesting topic whether these two subdivisions be organized as two individual space layers ('visitor' and 'employee') or one single layer. The semantics needs to be tested in different use cases, whether it is complete or insufficient.

2.3. IndoorGML and LADM: Synergies

The two standards have been developed for different purposes (navigation vs. land administration) and have different scope (indoor vs. indoor/outdoor, above/below surface). The two standards have many differences and similarities [3]. The similarities between the two models are:

- Both models (can) deal with semantically annotated 3D spaces, which have properties.
- Both models operate with abstract spaces. Abstract spaces in IndoorGML can be defined on the basis of user or environment properties. Abstract spaces in LADM are based on legal regulations. Similarly, IndoorGML allows subdivision and aggregations of spaces, such as accessibility, security, etc. The same is true in LADM: legal spaces can be grouped in LA_BAUnit or LA_SpatialUnit and organized in a hierarchy.
- Both models have a notion of primal space with geometry and topology. The 3D partitioning of LADM can be seen as primal space. LADM maintains links to external classes, of which some are mentioned in annex K of the standard: building units, utility networks. IndoorGML provides links to CityGML, IFC, and KML.
- Both models can support several subdivisions of space. The mechanism in IndoorGML is by defining specific space layers. LADM abstract subdivisions are embedded in the conceptual schema (and called LA_Level).
- Both models maintain relationships between objects. LADM supports extensive set of relationships and constrains. Spatial relationships can be based on topology, but could be also without topology (just geometry or even textual descriptions). IndoorGML does not have specific notions of constraints between objects, but rather topological relationships (i.e., adjacency and connectivity) are used to derive the dual space.

There are also a number of significant differences:

- LADM is only a conceptual schema, while IndoorGML has XML implementation.
- IndoorGML requires non-overlapping subdivision of spaces, LADM may have overlapping abstract spaces, but spatial units that are related to full ownership may not overlap with each other (but these might overlap with a spatial unit related to a certain restriction; e.g., because of an environmental protection zone).
- IndoorGML maintains primal and dual space, while LADM has only primal space.
- LADM models legal and administrative concepts, such as ownership rights of spaces that are related to certain (group) parties. IndoorGML might use such rights to specify subdivision, but no explicit space layer has been developed so far.

According to the similarities and differences of the standards, LADM could be applied to determine a framework for space subdivision. As defined in the scope of ISO 19152, LADM is not only about ownership rights, but also about all kinds of rights to cover various relationships between spaces and parties, including those that cover the function rights of indoor spaces to determine the accessibility for the parties based on RRRs for the IndoorGML. Thus, the topological spatial units do not have gaps or overlaps in the partition in LADM. The rights, restrictions, and responsibilities, and the administrative unit play a critical part during this process. We explore the combined use IndoorGML and the LADM by creating a link that connects each navigable space of IndoorGML to the corresponding LA_SpatialUnit of the LADM without adjusting IndoorGML and LADM. As a navigable space in IndoorGML can correspond to various spatial units of the LADM (and vice versa), a many-to-many association is needed. In this way, it is possible to model or to subdivide the spatial units in the LADM. Via LA_BAUnit, the associated rights and parties can be obtained (for navigable spaces linked with a LA_SpatialUnit). Note that in order to be able to use one-to-one correspondence, each space of IndoorGML would needed to be defined based on the constraint terms of the spatial unit of the LADM (and vice versa), which is considered as less convenient. The combined use model will determine indoor space rights, and that will perform efficient navigation.

3. Methodology to Define/Develop the Combined Use Model

In this section, we investigate and develop in more detail the combined use model, which enables the link between the two conceptual schemas, LADM and IndoorGML.

This paper concentrates on an educational building that has different types of users, such as students, employees, and visitors, and they have functional rights to use the spaces of the building. These rights do not have a legal foundation, but they are still based on the classes of the LADM, representing rights, restrictions, and responsibilities.

The rights, restrictions, and responsibilities affect the motion of users (use, manage, transfer, add, receive) in indoor spaces by regulating the access and use of space. Figure 7 represent a general overview of the combined use model of the LADM, IndoorGML, and an external party database. IndoorGML associates spatial data that contains information about primal space and the external database associates information about users. The LADM associates the subdivision of the indoor space to IndoorGML, based on the rights, restrictions, and responsibilities.

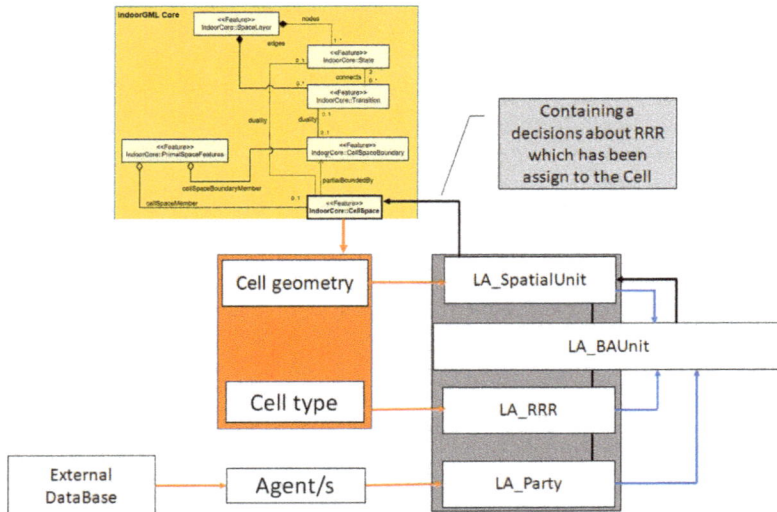

Figure 7. The combined use process for IndoorGML and the LADM.

The major link between the spatial features of indoor space from the cell space in the IndoorGML to the Spatial Unit Package in the LADM is modelled as an association. The association provides the identification (cell number) and the function of the cell. The spatial information of the cell collected by LA_SpatialUnit and the cell function information gathered by the LA_RRR, which is a class of the Administrative Package. The user's information is associated from the external database to the Party Package. The LA_BAUnit, which is a class in the Administrative Package, collects the information that is to be registered based on the LA_Right class information and LA_SpatialUnit. Based on the registration of information, the LA_BAUnit and LA_SpatialUnit associate the subdivision of indoor spaces (NavigableSpaces) to the cell space in IndoorGML.

Each type of building has a unique indoor space, which is represented by a variety of cells in IndoorGML. Each cell has a unique ID and a type of function that is used during the navigation process. IndoorGML provides this information with the cell geometry to the LADM to create a subdivision of the indoor space (NavigableSpaces), according to the RRRs. To apply a subdivision of the indoor spaces, an analysis has to be performed of the indoor space (NavigableSpaces), based on the type of building (hospital, educational, shopping mall, institutional building, etc.) and the type of users (doctors, nurse, students, visitors, customers, etc.). The relationship between the indoor space and the users defines the functional use rights.

The LADM classifies the GeneralSpaces and the TransferSpaces into categories that are based on the RRR of each cell. LADM uses the LA_RightType (with possible values in a code list) and the associated parties to classify the indoor space to a different type of cells, each type having different functional rights, as shown in Figure 8.

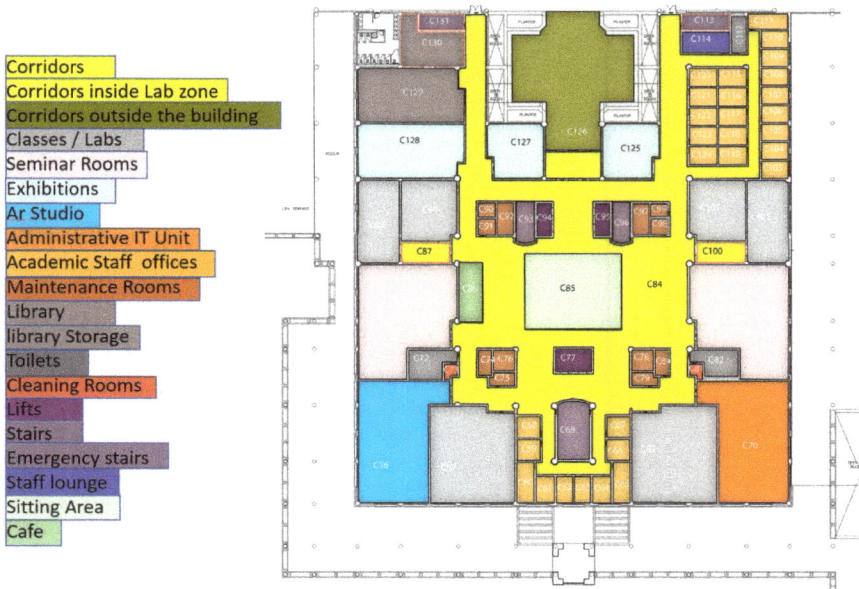

Figure 8. Classification of different types of cells based on rights, restrictions, and responsibilities and the usage function of the space.

According to the parties' rights, there are two type of cells that are categorized as (1) cells that have private rights for individual, or groups of, parties, and (2) cells that have common rights for individual, or groups of, parties. The first category means that a specific individual or group of a party has the right to use the cell, while the second category shows a common right may exist between the users. The LADM uses the same method to subdivide each cell into smaller cells that contain private rights for the individual party. For example, LADM divides an administrative department into offices, and each office held some certain rights for individual, or groups of, parties, and then divide the office into smaller cells representing each desk space and their rights based on the party, as shown in Figure 9. The result of the classification represents the relation between private rights cells and common rights cells, based on the user's activity. The LADM requires the information about the party, which is stored in an external database. The user's type is determined according to the kind of the building which consists of different kinds of groups of parties.

Figure 9. Division of the cell into smaller sub-cells based on functional use rights.

In order to start the combination, the cell space associates the primal space to the LADM. As shown in Figure 10, the cell space in IndoorGML provides the spatial information about the indoor space type to LA_SpatialSource, which is a class of surveying and representation that is responsible for gathering/documenting spatial source information for the LADM. It has a relation with LA_Point, LA_BoundaryFaceString, and LA_Boundary Face to cover all kinds of spatial data. LA_SpatialSource associates the spatial information to LA_SpatialUnit. Based on the information, the LA_SpatialSource carries the role of surveyor or architect of the spatial source to LA_Party and a description of the extent of the property via LA_SpatialUnit to LA_BAUnit.

On the other hand, LA_AdministativeSource is a subclass of LA_Source that provides the party information to LA_Party, and specifies the party that plays the role of the conveyancer. The LA_AdministativeSource documents the origin/source of right, restrictions, and responsibilities for LA_BAUnit, and a description of the rights, restrictions, and responsibilities that are held by a party and affect the LA_BAUnit for LA_RRR.

The Party Package is responsible for the representation of party information and consists of the class LA_Party and the subclass LA_GroupParty, and the LA_PartyMember, which is an optional association class that is located between the LA_Party and LA_GroupParty. The current LA_Party class has several attributes, such as the ID number of the party, the type of party, and the party role type. For situations where the party is not an individual, the LA_GroupParty defines the group ID and the group party type. Additionally, the Party Package contains code lists that are used to represent lists of values for Party Package attributes. Based on the case studies, the party information was divided into three categories, students, employees, and visitors (either at individual or group level), see Table 1. Each category had several necessary attributes, such as for students, the educational programs, and educational levels. For the employee category, there are different attributes that are based on party function/career; for example, there are academic staffs, administrative staffs, and security staffs. The Party Package could not represent the needed information correctly by using the existing attributes and requiring the extension of the code lists of the LA_Party class or the LA_GroupParty subclass. Thus, to properly cover all information of the three categories (student, staff, visitor), we explored different modelling options to extend the expressiveness of the Party Package. Two of the options will be presented and discussed: *Option A* (just adding Attributes to LA_Party) and *Option S* (adding Subclasses to LA_Party and LA_GroupParty).

Table 1. Represents different types of users and their type of rights.

Party	Description
Individual student	Has a unique ID number and a private right to use the space
Individual employee	Has a unique ID number and a private right to use the space
Individual visitor	Has a unique ID number and a private right to use the space
Group of students	Has a unique ID number and share a common right to use the space
Group of employees	Has a unique ID number and share a common right to use the space
Group of visitors	Has a unique ID number and share a common right to use the space

The *Option A* model includes additional attributes in LA_Party class, which are: *party employee type*, *party department type*, and *party educational level*, as shown in Figure 11a. Each attribute has a code list that represents the allowed values. For the attribute *party employee type*, the code list includes academic staff, administrative staff, security team, cleaning team, and maintenances team. The code list of the attribute *party department type* contains architecture, landscape, geomatics, urban (and for certain students also the values orientation year and student mix departments), and not related to faculty. The last additional attribute is *party educational level*, which includes the following code list values: bachelor, master, and PhD. The subclass LA_GroupParty is kept without any modification to represent the group ID and the group role type. By including the additional attributes to the LA_Party class, this option supports the representation of the individual party. However, this simple extension could lead to the misuse/confusion of the model in certain cases. For example, the representation of the employee's types, educational program, and educational level information in the LA_Party class cannot be used for all the categories of parties (staff, student, visitors). This disorder is caused as a result of trying to represent all types with only one class. Moreover, the same information will be repeated to all parties that are sharing the same information (e.g., staff members of same department of students of the same cohort), and that will increase the size of the data when access rights are stated for every individual party. The advantage of option A is that it is relatively simple.

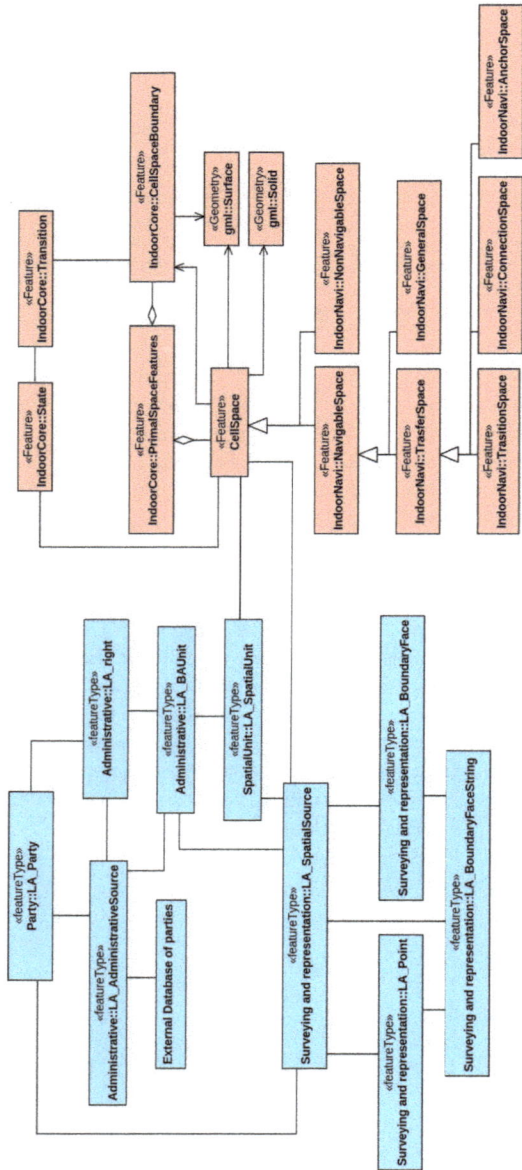

Figure 10. LADM-IndoorGML combined use model, the LADM classes are in blue and IndoorGML classes are in coral.

Figure 11. *Cont.*

(a)

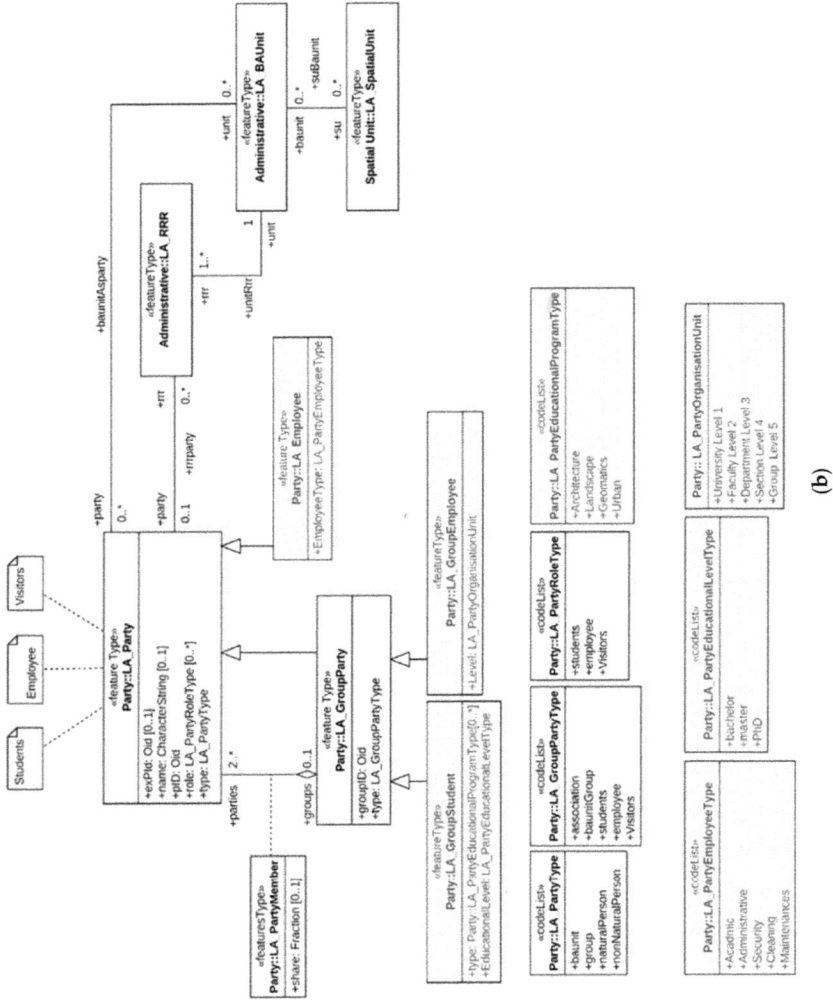

Figure 11. (**a**) *Option A*: Content of the Party Package of the first option model and the code lists to represent the party of indoor space; and (**b**) *Option S*: Content of the Party Package of the second option model and the code lists to represent the party of the indoor space.

The *Option S* model tries to avoid the problems associated with *Option A*. It first concentrates on developing the LA_GroupParty to represent each user category in the different subclass to avoid mixing the representation of the categories. The LA_Party class holds the same attributes that are similar to the original class with subclass LA_Employee, which has one optional attributes to represent the *party employee type* as shown in Figure 11b. The rest of information will be included in the subclasses of LA_GroupParty based on the *group party type*. For the description of the 'students' category, a new subclass called LA_GroupStudent is created to represent the *educational program type* and *educational level*. For the attribute *educational program type*, the code list includes values, such as architecture program, landscape program, geomatics program, and urban program. For the attribute *educational level*, the code list contains values, such as bachelor, master, and PhD. The second subclass is created to represent the employee category. The LA_GroupEmployee contains one attribute to represent the *party origination unit* which contains the hierarchy organizational levels of the organization: Level 1 represents the university, level 2 represents the faculty, and level 3 represents the departments. Note that there can be groups of groups; e.g., a number of departments to together form a faculty. Model *Option S* shows the ability to cover several types of parties by using LA_GroupParty. The Party Package is flexible and straightforward based on this second option, which we further used in our research. The only drawback is that the model may look more complex due to the increased number of (sub)classes. Note that we did not add a subclass at individual level for students as there were no new attributes (all of the student attributes could be included at group level), and we want to keep the model as simple as possible. The same reasoning for not introducing subclasses for visitors (at individual or at group level). Table 1 represents the types of users with a description of the functional rights of each type, which is also shown in Figure 11b. The class LA_Party carries the party information for instances which can be either individual or group of party associated to LA_RRR.

The LA_RRR is an abstract class that contains three subclasses, LA_Right, LA_Restrictions, and LA_Responsibilities. The right associates between LA_Party and LA_BAUnit. LA_RRR describes the rights, restrictions, and responsibilities that affected the LA_BAUnit. LA_BAUnit collects the spatial unit that has the same rights, restrictions, and responsibilities, and the party that is attached. More than one spatial unit can be registered as part of a LA_BAUnit if all of them share the same rights, restrictions, and responsibilities. It defines the type of basic administration unit type and defines the ID number of the unit. LA_BAUnit could play the role of a party.

LA_SpatialUnit contains information about the spatial cell from the IndoorGML cell space. It defines the type and dimensions of the spatial unit. Additionally, it can provide a textual description of the spatial unit. The spatial unit can be a part of another spatial unit. The spatial unit associates the cell to the LA_BAUnit. Since the LA_BAUnit could cover more than one spatial unit in the same instance object, the LA_SpatialUnit carries the subdivision of the primal space to the CellSpace in IndoorGML.

4. Case Study of Two Different University Buildings

In this section, we represent two cases study with a real data for the users and different information from the facility management for the Faculty of Environmental Designs at King Abdulaziz University, Saudi Arabia, and the Faculty of Architecture and the Built Environment at TU Delft, Netherlands (both shown in Figure 12) to explore the combined use model between the LADM and IndoorGML with real world data.

Figure 12. (**a**) Faculty of Environmental Designs at King Abdulaziz University; and, (**b**) Faculty of Architecture and the Built Environment at TU Delft

4.1. Faculty of Environmental Designs at King Abdulaziz University, Saudi Arabia

The building of the faculty consists of five floors and four departments, namely, the Architecture Department, Geomatics Department, Landscape Department, and Urban Department. Each department includes some classrooms and labs that are located on different floors. The study starts by associates the cell information from IndoorGML (the cell space) to the spatial source subclass in the LADM. The information contains the geometry information, and each cell has an ID number with the function of the cell. To determine the functional rights for each cell, the LADM has to be aware of the parties of the building. Based on the type of the building, the parties are divided into three categories; students, employees, and visitors, as shown in Figure 13.

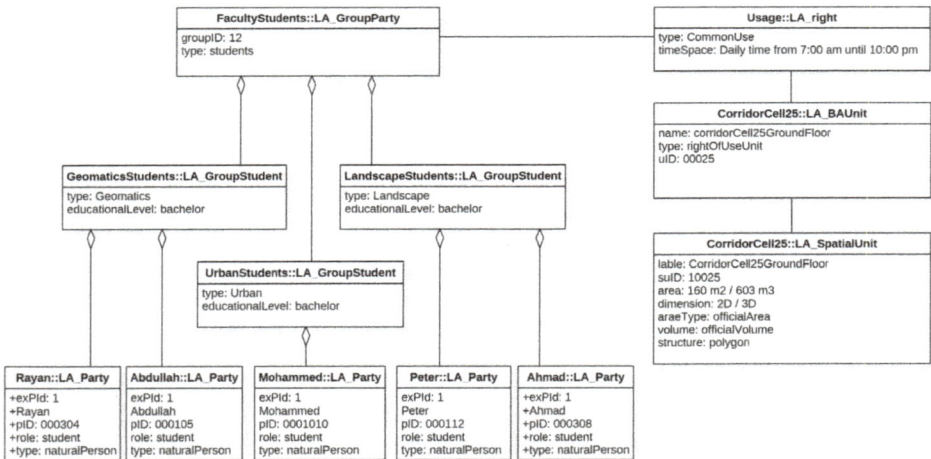

Figure 13. Instance level diagram showing the party common rights based on the building type.

Based on the relationship between the users and the function of the indoor space of the building, the LADM divides the NavigableSpaces into several types of cells that have various rights. Each color represents specific rights of use for a specific party or groups of parties, as shown in Figure 14. To build the 3D model of the Faculty of Environmental Designs at King Abdulaziz University, we had to start by collecting information about each indoor space of the building, such as architectural 2D plans, and the function of use for each room. The next step is to use software called Revit (Autodesk software) to build the 3D model and to define the 3D space for each room. The software allows creating a schedule to store all the information about the space, such as the name of the space, the 3D volume, usage function, space number, and the space level. By exporting the 3D model from Revit to the IFC extension, the 3D space of each room is connected to the schedule and is ready to be classified based on the usage function of the space. Solibri Model Checker is a BIM software that can classify the 3D space model based on the property set. By using this feature of the software, we can connect the 3D with the schedule and represent the 3D model, as shown in Figure 14.

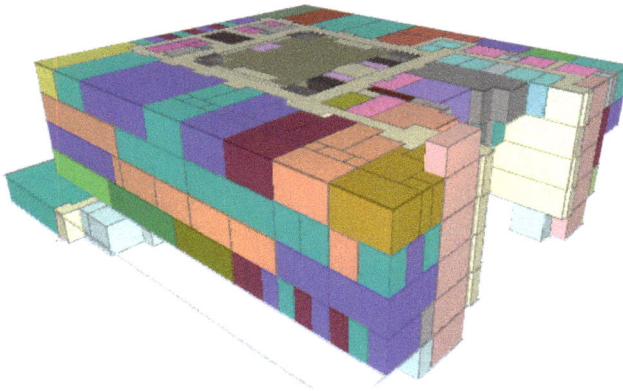

Figure 14. Classification different type of cells based on functional use rights.

The next step is to extract common areas (public areas), where common rights may exist between the parties from all of the departments. Based on the relationship between the parties and the function of the cell, we could determine the common area, as shown in Figure 15. The sitting areas, stairs, lifts, corridors, and toilets are the common areas that all of the parties have the same right to use and no users could be excluded. The common rights areas cover around 20% of the floor area. The next step is to include one private rights areas of one specific group (department) to the common areas to represent the relationship between them, as shown in Figure 16.

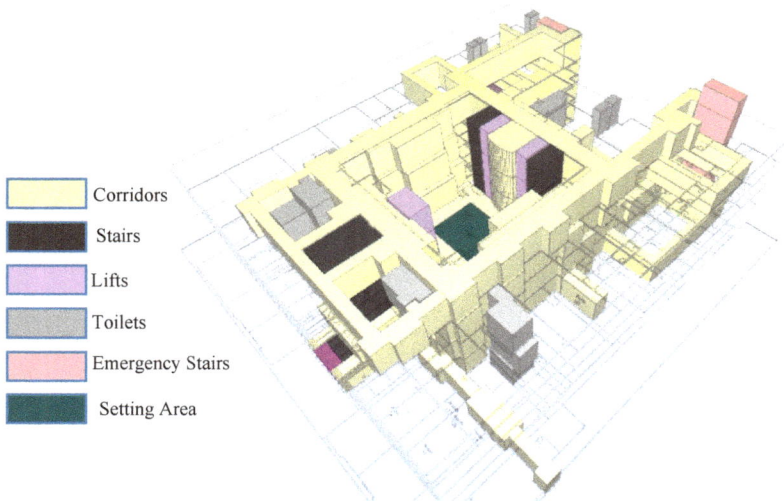

Corridors
Stairs
Lifts
Toilets
Emergency Stairs
Setting Area

Figure 15. The common rights areas based on functional use rights.

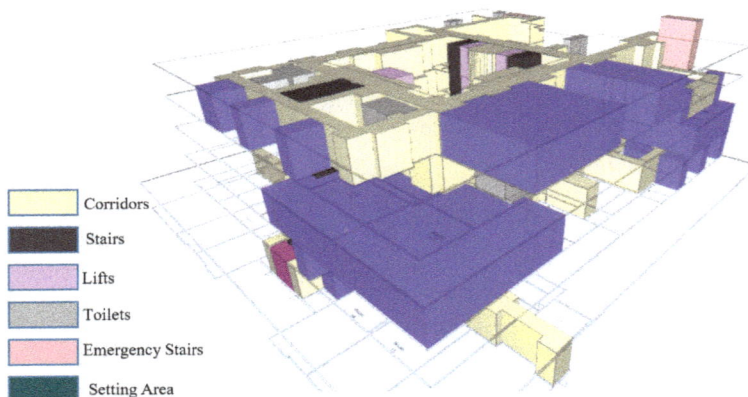

Figure 16. The common rights and private rights areas of the Urban Department based on functional use rights.

The parties of the Urban Department have access to the blue cells as private rights areas with the common areas in their daily life activities. However, the rest of the floor is considered as a non-accessible area for the Urban Department. In some of the other floors parties of the Urban Department have accessibility to use the common classrooms and labs. As we mentioned before in Figure 9, the LADM can divide each cell into smaller cells to represent the rights of the space, based on rights, restrictions, and responsibilities. The LADM attaches the right of each cell based on the party and register it to the LA_BAUnit, as shown in Figure 17. In this Figure, the LADM splits cell 16 into many smaller cells. Each sub-cell connected with a main cell is to be registered in the LA_BAUnit, and then the private rights with a description will be created to be assigned to the party.

4.2. Faculty of Architecture and the Built Environment at TU Delft, Netherlands

The second case study is the Faculty of Architecture and the Built Environment at TU Delft, which consists of four floors and two sub-floors. There are five departments, including Architecture (AR), Architecture Engineering and Technology (AET), Management in the Build Environment (MBE), Research for the Built Environment (OTB), and Urbanism (URB). Each department has supporting spaces, such as staff offices, secretary offices, meeting rooms, and printing space. According to the facility management information about the usage function of each space of the faculty, we have created a room schedule for the faculty, as shown in Figure 18. Based on the room schedule, the IndoorGML cell space associates the usage function and the geometry information to LA_SpatialSource in the LADM to classify each space based on the relationship with the parties of the building to assign the rights, restrictions, and responsibilities for each space. Based on the building type, which is an educational building, we define the parties into three different categories, which are students, employees, and visitors.

Based on the relationship between the party and the function of the indoor space of the building, the LADM divides the indoor space into several types of cells that have various rights. Each color represents specific rights of use for specific parties or groups of parties, as shown in Figure 19.

According to the relationship between the parties and the function of the cells in the LADM, we could extract the common rights areas, as illustrated in Figure 20. The common rights areas consist of corridors, stairs, lifts, toilets, restaurants, and copy/printing areas. All types of parties could use the areas without any restrictions. However, the building has other types of cells with different types of rights based on the rights, restrictions, and responsibilities, such as educational and administrative areas and departments.

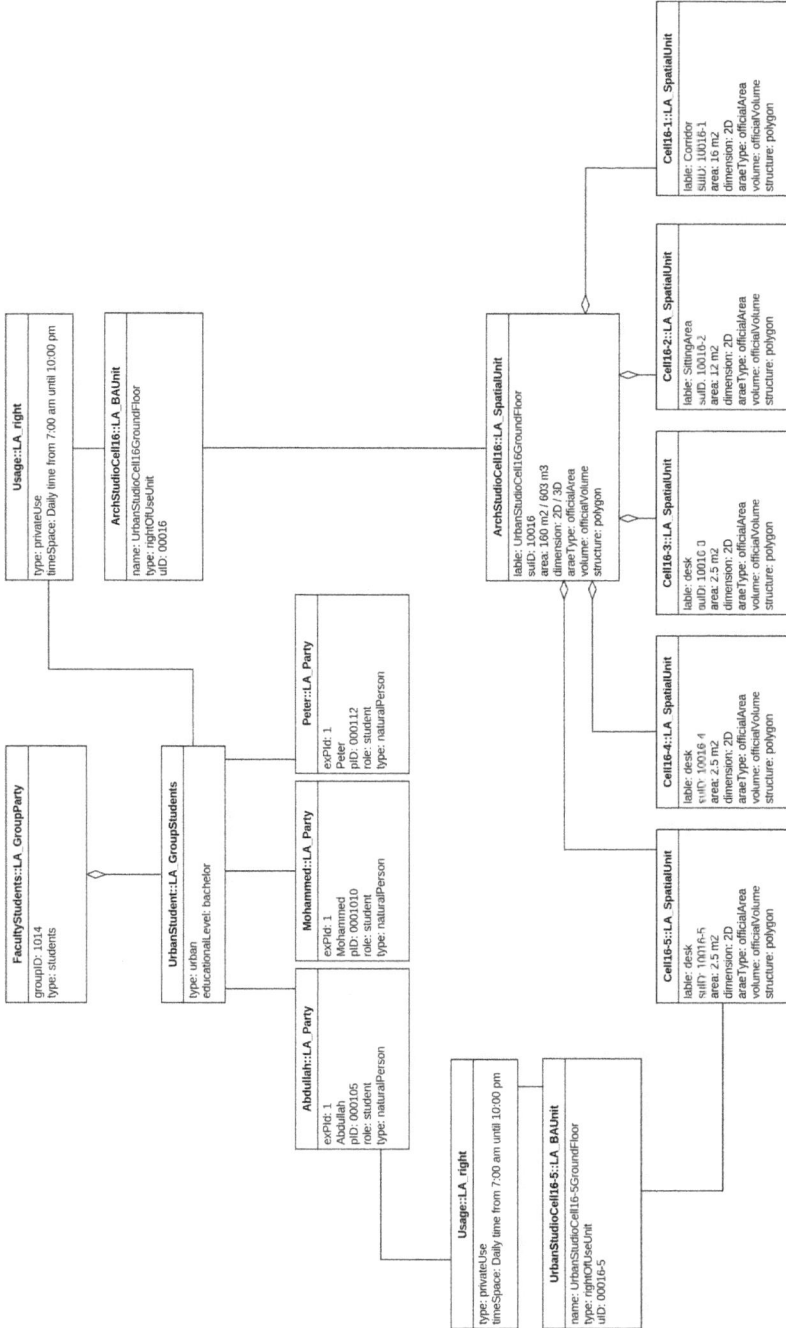

Figure 17. Instance level diagram shows the division of the cell into smaller sub-cells based on functional use rights.

Number	Room Number	Name	Space Name	Usage function	Type	unit/program	Capacity	Area	Level	Volume	Upper Limit
				Room Schedule BK							
1	08.01.01.010	01.Mid.010	BIBLIOTHEEK	A Education	A4 Study room / area	no	0	190 m²	1st floor	583.43 m³	1st floor +
2	08.01.01.050	01.Mid.050		H Horizontal traffic	H2 Hall	no	0	10 m²	1st floor	10.49 m³	1st floor +
3	08.01.01.100	01.Mid.100	BIBLIOTHEEK	A Education	A4 Study room / area	no	0	196 m²	1st floor	1124.69 m³	2nd
4	08.01.01.801	01.Mid.801		H Horizontal traffic	H1 Times	no	0	74 m²	1st floor	424.32 m³	2nd
5	08.01.01.802	01.Mid.802		H Horizontal traffic	H1 Times	no	0	63 m²	1st floor	192.20 m³	1st floor +
6	08.01.01.803	01.Mid.803	BIBLIOTHEEK	A Education	A4 Study room / area	no	0	162 m²	1st floor	706.29 m³	2nd
7	08.01.01.804	01.Mid.804		H Horizontal traffic	H1 Times	no	0	74 m²	1st floor	424.39 m³	2nd
8	08.01.01.851	01.Mid.851		V Vertical traffic	V1 Stairs	no	0	6 m²	1st floor	19.98 m³	1st floor +
9	08.01.01.853	01.Mid.853		V Vertical traffic	V2 Lift	no	0	2 m²	1st floor	6.65 m³	1st floor +
10	08.01.01.854	01.Mid.854		V Vertical traffic	V1 Stairs	no	0	11 m²	1st floor	33.94 m³	1st floor +
11	08.01.01.855	01.Mid.855		V Vertical traffic	V1 Stairs	no	0	2 m²	1st floor	6.22 m³	1st floor +
12	08.02.01.010	01.Oost.010	Zaal H	A Education	A2-4 drawing room	no	0	103 m²	1st floor	594.54 m³	2nd
13	08.02.01.050	01.Oost.050	Zaal M	A Education	A4 Study room / area	no	20	83 m²	1st floor	477.41 m³	2nd
14	08.02.01.110	01.Oost.110	01.Oost.110	A Education	A4 Study room / area	no	0	81 m²	1st floor	468.33 m³	2nd
15	08.02.01.150	01.Oost.150		S Sanitary	S1 Toilet room	no	0	4 m²	1st floor	25.05 m³	2nd
16	08.02.01.170	01.Oost.170		S Sanitary	S1 Toilet room	no	0	33 m²	1st floor	192.10 m³	2nd
17	08.02.01.180	01.Oost.180		S Sanitary	S1 Toilet room	no	0	23 m²	1st floor	69.55 m³	1st floor +
18	08.02.01.190	01.Oost.190		S Sanitary	S3 Workplace	no	0	4 m²	1st floor	12.08 m³	1st floor +
19	08.02.01.240	01.Oost.240	COLLEGEZAAL A	A Education	A1 lecture rooms	no	300	321 m²	1st floor	1863.56 m³	2nd
20	08.02.01.310	01.Oost.310		E Supporting space	E2-2 SER / MORE	no	0	19 m²	1st floor	107.21 m³	2nd
21	08.02.01.320	01.Oost.320		E Supporting space	E3-2 Copy / Print Space	no	0	31 m²	1st floor	175.81 m³	2nd
22	08.02.01.321	01.Oost.321		T Installations		no	0	1 m²	1st floor	6.29 m³	2nd
23	08.02.01.330	01.Oost.330		D Office	A4 Study room / area	no	0	80 m²	1st floor	462.71 m³	2nd

Figure 18. The usage function information for each space of the faculty.

Each department holds private rights on specific spaces of the building based on the cell's information that is provided by the IndoorGML cell space to the LA_SpatialUnit. The Architecture Department, which is located on the east side of the building has private rights to three offices, two meeting rooms, and a copy/printing space. The rest of the departments are located on the west side of the building and hold private rights for the same types of spaces, as shown in Figure 21. Each office space of the Architecture Department are divided into several spaces to determine the rights of the parties that are sharing the space, as shown in Figure 22.

The educational areas include several kinds of spaces, such as lecture rooms, drawing room, project room, studios, workshop, and study areas. All of the educational spaces have common rights for departments and students. The educational spaces cover a major area of the building and are located on different floors, as shown in Figure 23, with a concentration on the top floor. However, each department has the right to hold private rights to certain educational spaces based on the lecture schedule. The LA_Right class in the LADM provides a TimeSpace that is attribute for defining the period of the type of rights that the party can use, as illustrated in Figure 24.

Alongside the departments and the educational spaces, there are support spaces that are used by the parties, and they have common rights, such as meetings rooms and copy/printing areas, and they are distributed among the floors. The faculty includes support administrative departments consisting of the Dean's Office, Finance and Control, Facilities Management, Human Resources Management, Information and Communication Technology, Marketing and Communication, and Education and Student Affairs. They are located on the east side of the building with private rights for the spaces that are used by each department. Each office space is divided into smaller spaces to assign private rights for each employee as shown in Figure 9. Figure 19 represents the academic departments and their relationship with the spaces that have common rights, such as the meeting rooms, restaurants, and copy/printing areas, and the private areas, such as the support administrative departments.

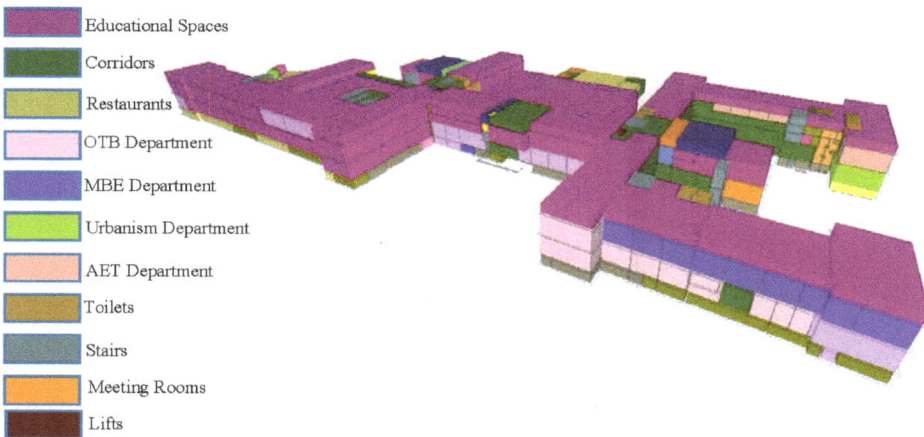

Figure 19. The classification of different types of cells based on functional rights.

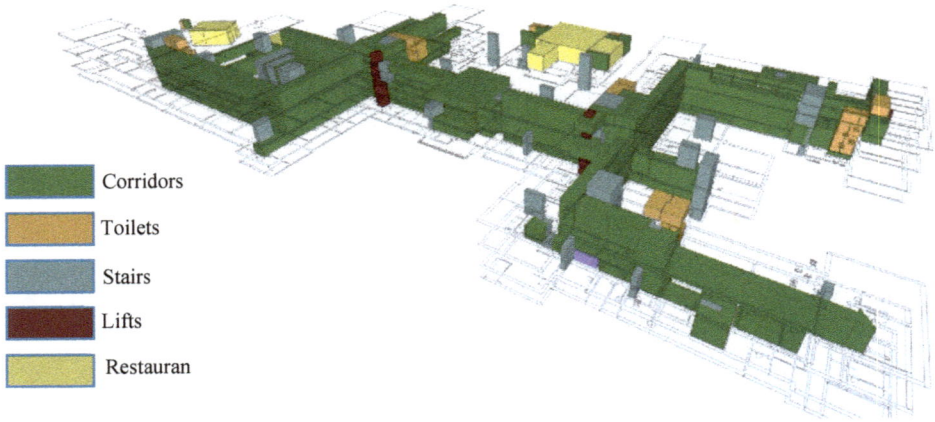

Figure 20. The common areas of the Faculty of Architecture and the Built Environment based on the rights, restrictions, and responsibilities.

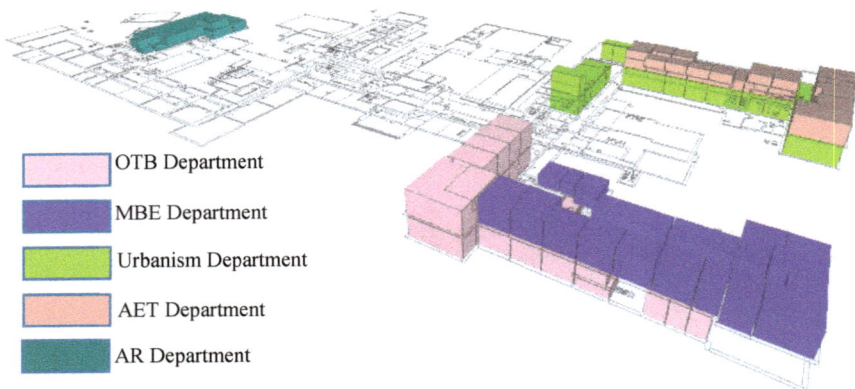

Figure 21. The private spaces of the five departments of the Faculty of Architecture and the Built Environment based on the rights, restrictions, and responsibilities of the LADM.

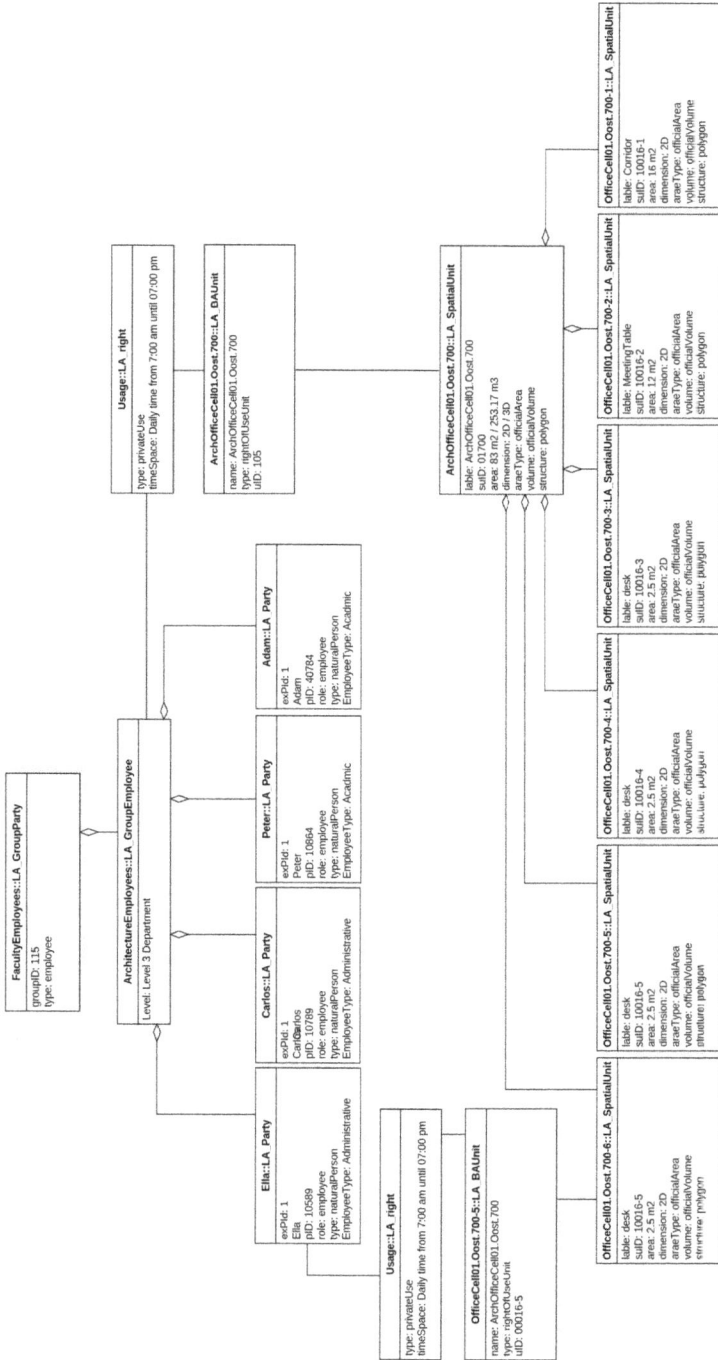

Figure 22. Instance level diagram showing the division of the office cell into smaller sub-cells based on functional rights.

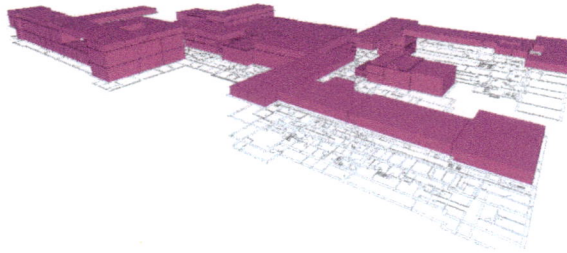

Figure 23. The educational spaces of the Faculty of Architecture and the Built Environment.

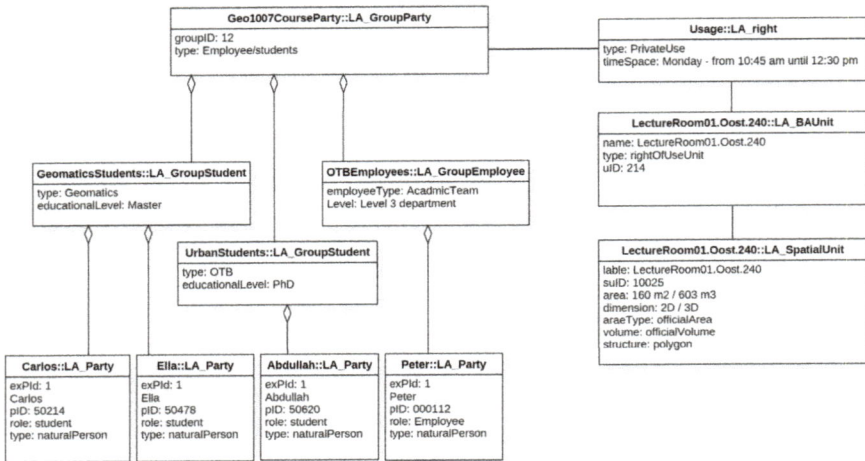

Figure 24. Instance level diagram shows converting a common rights lecture space into private rights for a specific period.

4.3. Comparison

The two case studies have shown the impact of the RRR of the LADM on the space accessibility in the buildings by categorizing the spaces based on the relationship between the party and the function of the space. Both of the buildings have the same type of use and parties. However, they have different indoor components, such as the number of floors, departments, educational spaces, and support spaces, as shown in Table 2.

Through determination of the functional right for each space based on the rights, restrictions, and responsibilities, the navigation areas will be influenced according to the party type. For example, when the party is a student, the accessibility areas will be different in both of the cases. All of the departments of the Faculty of Environmental Designs at King Abdulaziz University have private rights to educational areas. Thus, the students will have rights to use common rights spaces that belong to the faculty and the private rights areas that belong to their departments. On the other hand, the student of the Faculty of Architecture and the Built Environment at TU Delft will have the rights to use all of the educational spaces because they have common rights for all of the parties and they do not belong to any department. In both cases, when the party is an employee, the right of access will depend on the employee type and department. The academic employee will have rights to use the spaces that belong to the department and the common right spaces of the building. The spaces that belong to other departments are considered as non-accessible spaces because the party does not hold the rights to use the space. Therefore, by considering the effect of the rights, restrictions, and responsibilities on the indoor spaces, the indoor navigation becomes more efficient and more straightforward for the parties.

Table 2. The components of each faculty.

Usage Function	Space Type	Faculty of Architecture and the Built Environment at TU Delft	Faculty of Environmental Designs at King Abdulaziz University
Departments	-	Architecture Engineering and Technology. Architecture. Management in the Build Environment. Research for the built environment. Urbanism.	Architecture. Urban. Landscape. Geomatics.
Education	Lecture rooms	yes	yes
	Study area	yes	no
	Practicum	yes	no
	Practice room/Labs	yes	yes
	Drawing room	yes	no
	Project room	yes	no
	Studio	yes	yes
	Workshop	yes	yes
	Library	yes	yes
Supporting space	Central Facilities	yes	yes
	First Aid/BHV space	yes	no
	SER	yes	yes
	Meeting room	yes	yes
	Copy/print space	yes	yes
	Pantry	yes	no
Restaurant/Cafe	Restaurant	yes	yes
Storage space	Storage space	yes	yes
Horizontal traffic	Corridors	yes	yes
Vertical traffic	Stairs Lifts	yes	yes
Sanitary	Toilets		yes
	Shower room	yes	no
	Workplace		yes
Installations	Accessible Shafts Technical Workshop Installations	yes	yes
Administrative offices	-	yes	yes

5. Navigation Example Using Rights, Restrictions, and Responsibilities

This section shows the use of the populated combined IndoorGML–LADM model for three navigation examples, which the use of rights, restrictions, and responsibilities on the 3D geometry of the building based on the party type. We have used ifcSpace of IFC which is an object-oriented data model that represents the building components and spaces (primal space) and is commonly used to derive dual space (navigation network and path) for IndoorGML. We can generate a network of the indoor spaces through the node relation graph (NRG) to represent the topological associations, such as adjacency and connectivity between indoor spaces, which are the basis to generate a connectivity graph of all the spaces. Based on our approach, we do not generate a network for the whole building. First, we select the available spaces based on the RRRs of the specific user or group of users, and then we create a network and generate a navigation path using only the subset of spaces available for the party.

The NavigableSpace consists of two subclasses, and they are the TransferSpace and the GeneralSpace. The TransferSpace is used as a passage between the GeneralSpaces. The GeneralSpace is defined as any NavigableSpace not considered as a TransferSpace, for example, a lecture room, office room, toilet, and labs. To link between TransferSpace and the GeneralSpace, the navigation network has to be derived from the primal space by generating the state (node) and the transition (edge). The state is representing the space object in the primal space, and the transition is representing the connectivity relationship between the states.

The first example shows a case study for a student that has the right to use the lecture room (ArchitectureCell181) on the third floor, as shown in Figure 25 (above). The student is located at the entrance space on the first floor. Based on the user rights, the network has been created and is used to generate the navigation path that is a series of states (nodes) that are linked by transitions (edges) to connect the CellSpaces. The user (student) started the journey from the entrance to the lecture space, passing several TransitionSpaces to reach the GeneralSpaces. The CellSpaceBoundary described the boundary between the two spaces by surface geometric objects to show the relationship between the spaces, as shown in Figure 25. Additionally, the user has different rights between the TransferSpaces and the GeneralSpace, which are described in the LA_Right classes. Figure 26 represents the NavigableSpaces that are available based on the user rights and depicts the shortest navigation path between their current location and the lecture space.

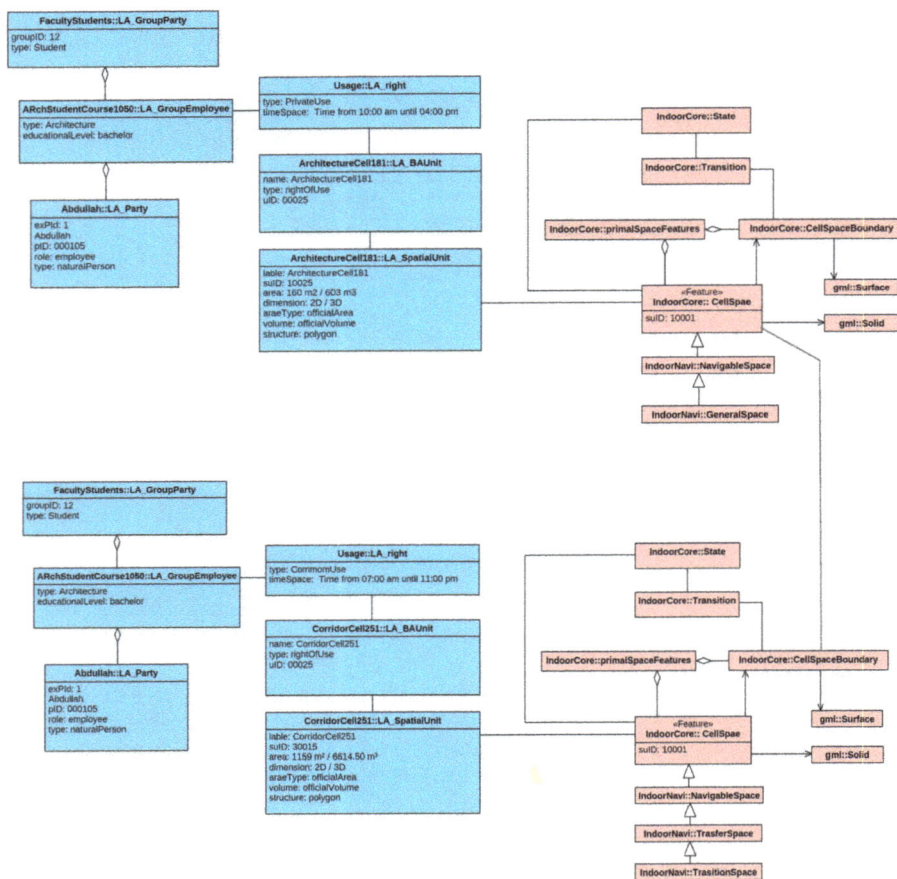

Figure 25. (**top**) Instance level diagram shows private rights of use a lecture space (GeneralSpace) held by a student from the architecture department; (**bottom**) and the instance level diagram shows common rights of the use of a lecture space (TransferSpace) held by a student from the architecture department.

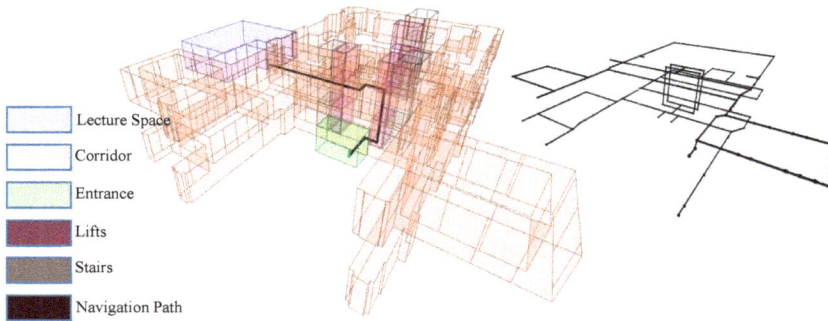

Figure 26. The primal space and a navigation path (**left**) and the accessibility network (**right**) derived from the primal spaces, both for student's rights.

The second navigation example, using the populated combined IndoorGML–LADM model, represents an academic team member that has the right to use a lecture space (GeomaticsCell15) on the ground floor, as shown in Figure 27. The navigation path starts from the party office (GeneralSpace), which is located on the third floor through the corridor (TransitionSpace) passing all of the non-accessible spaces to reach the lift (TransitionSpace), which is located on the other side of the building, and arrive to the ground floor corridor (TransitionSpace) to the final destination (GeneralSpace). As shown in Figure 28, the party has hardly any rights to use the first, second, and fourth floors (just the spaces of the stairs and elevators). Again, it is clear that the 3D geometry of the navigation path is heavily impacted by the party rights.

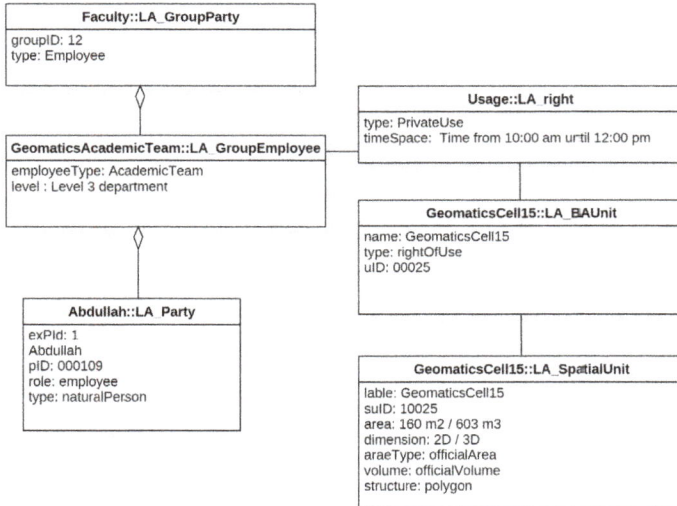

Figure 27. Instance level diagram shows private rights of the use of a lecture space held by an academic team member from geomatics department.

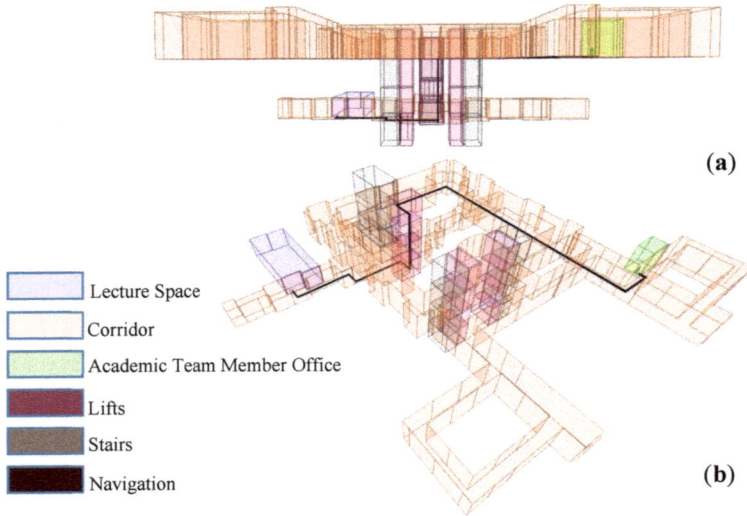

Lecture Space
Corridor
Academic Team Member Office
Lifts
Stairs
Navigation

Figure 28. The primal spaces (**a**) and a navigation path (**b**) based on the academic team membership rights.

The last example is about a cleaning team member that has a private right to enter and clean the dean's office (DeanOfficeCell308), as shown in Figure 29. Both of the spaces are located on the fourth floor. Figure 30 illustrates the navigation path between the cleaning room and the dean's office, which are both GeneralSpaces, connected via a series of TransitionSpaces. The cleaning team member has the right to clean the office between 07:00 am and 08:00 am, as indicated in the timeSpec attribute of LA_Right.

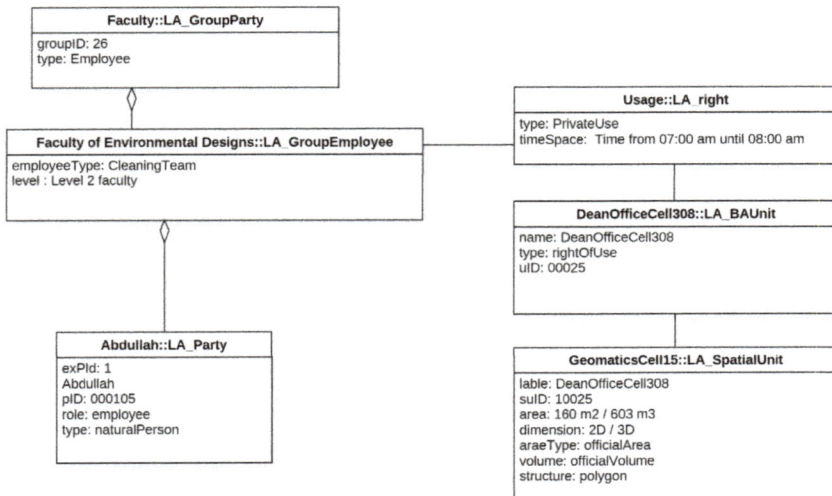

Figure 29. Instance level diagram shows private rights held by a cleaning team member for cleaning the office of the dean.

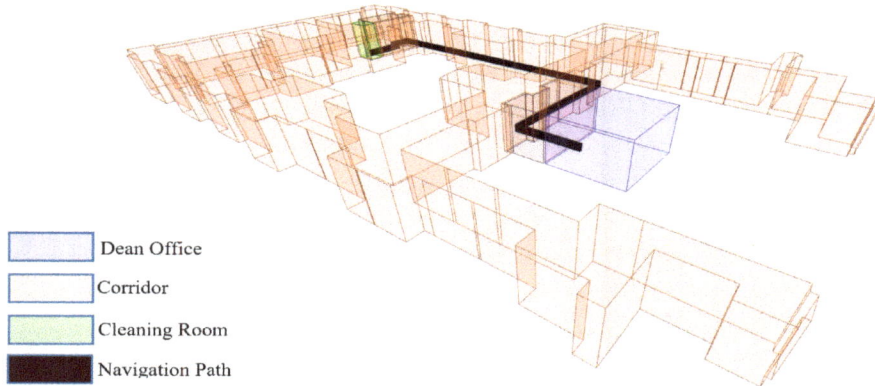

Figure 30. The primal spaces and the navigation path between the cleaning room and the dean's office based on the rights of the cleaning team member.

6. Discussion

This paper has presented the first combined use of IndoorGML and LADM models for supporting the accessibility of the indoor spaces based on the user rights. The combination is realized by linking the abstract level of LADM with the XML implementation level of the IndoorGML to reflect the effect of rights, restrictions, and responsibilities on the indoor navigation process based on the party rights. Each type of building has unique indoor spaces that are used by several types of parties. The Party Package has to represent the different types of parties based on the function of the indoor spaces. However, the standard ISO 19152 version of the Party Package could not represent the variety of parties information. For the refinement of the Party Package, two options have been presented for the refined model; the first option (*Option A*) has better support for the individual's representation by adding several attributes to the LA_Party class. However, this option has some severe drawbacks, such as unused attributes for certain party types and the repetition of the same information for staff of students at the individual level. Therefore, the second option (*Option S*) focused on the representation of the group parties by developing the LA_GroupParty to include several subclasses that are based on the type of party. Each subclass has unique attributes and code lists to cover the party information. This option provides a flexible and straightforward representation of several kinds of parties. By classifying the indoor space based on rights, restrictions, and responsibilities, there are two kinds of spaces: common rights spaces and private rights spaces. The IndoorGML CellSpace associates the geometry and function of the spaces to LADM to assign the functional rights of use for each NavigableSpace based on the party right type.

The case studies which are based on real data about the users, demonstrate the impact of rights, restrictions, and responsibilities on the NavigableSpaces (GeneralSpaces and TransferSpaces), and how that will change the 3D geometry of the building. The cases studies are the Faculty of Environmental Designs at King Abdulaziz University and the Faculty of Architecture and the Built Environment at TU Delft. The two cases have the same building type function and parties. The rights, restrictions, and responsibilities have impacted the 3D geometry of the building based on the relation between the spaces and the parties. Each party has certain rights to use the spaces of the building, and according to these rights, the navigation paths can be more precisely computed. In future work, we will extend our work in several directions as follows:

- Check the integration model on different types of buildings and parties, such as hospitals, hotels, train stations, and airports, to ensure the representation of all types of parties during their navigation activity in the indoor environment.

- Investigate how to determine the accessible and non-accessible spaces in an automated way.
- Study the contribution of the automated subdivision space in large buildings.
- Devise methods to increase the granularity of spaces inside rooms, e.g., around furniture.
- Investigate the effect of RRRs on the navigation path during a crisis.

Acknowledgments: We would like to thank the ISO TC211, OGC SWG IndoorGML and STW/M4S Sims3D teams for the fruitful discussions and resulting standardized information models. The PhD thesis research of Abdullah Alattas is funded by King Abdulaziz University. The Erasmus+ programme provided the support for the internship of Efstathia Chatzinikolaou at TU Delft. The test data sets were provided by respectively the Faculty of Environmental Designs at King Abdulaziz University, and Adriaan Jung, Facility Management office of TU Delft.

Author Contributions: Abdullah Alattas, Peter Van Oosterom, and Sisi Zlatanova contributed to organizing the structure of the article. Peter Van Oosterom and Christiaan Lemmen wrote the background information of LADM. Sisi Zlatanova and Ki-Joune Li contributed to writing the IndoorGML part. Peter Van Oosterom, Christiaan Lemmen, Sisi Zlatanova, and Ki-Joune Li provide some possible approaches for the integration of IndoorGML and LADM. Efstathia Chatzinikolaou has created the 3D model of the Faculty of Architecture and the Built Environment at TU Delft, Netherlands. Abdullah Alattas created the 3D model of the Faculty of Environmental Designs at King Abdulaziz University, Saudi Arabia, and classified the space of the two models based on the rights, restrictions, and responsibilities.

Conflicts of Interest: The authors declare no conflict of interest.

References

1. Timpf, S.; Volta, G.S.; Pollock, D.W.; Egenhofer, M.J. A conceptual model of wayfinding using multiple levels of abstraction. In *Theories and Methods of Spatio-Temporal Reasoning in Geographic Space*; Springer: Berlin/Heidelberg, Germany, 1992.
2. Becker, T.; Nagel, C.; Kolbe, T.H. A Multilayered Space-Event Model for Navigation in Indoor Spaces. In *3D Geoinformation Sciences*; Lee, J., Zlatanova, S., Eds.; Springer: Berlin, Germany, 2009.
3. Zlatanova, S.; Van Oosterom, P.J.M.; Lee, J.; Li, K.-J.; Lemmen, C.H.J. LADM and IndoorGML for Support of Indoor Space Identification. In Proceedings of the 11th 3D Geoinfo Conference on ISPRS Annals of the photogrammetry, Remote Sensing and Spatial Information Science, Athens, Greece, 20–21 October 2016.
4. Lee, J.; Li, K.-J.; Zlatanova, S.; Kolbe, T.H.; Nagel, C.; Becker, T. OGC IndoorGML, OGC 14-0051r1. 2014. Available online: http://www.opengeospatial.org/standards/indoorgml#downloads (accessed on 15 May 2016).
5. Zlatanova, S.; Li, K.J.; Lemmen, C.; Oosterom, P. Indoor Abstract Spaces: Linking IndoorGML and LADM. In Proceedings of the 5th International FIG 3D Cadastre Workshop, Athens, Greece, 18–20 October 2016; pp. 317–328.
6. Lemmen, C.H.J.; van Oosterom, P.J.M.; Bennett, R. The Land Administration Domain Model. *Land Use Policy* **2015**, *49*, 535–545. [CrossRef]
7. Simpson, S.R. *Land Law and Registration*; Cambridge University Press: Cambridge, UK, 1976.
8. ISO. *ISO 19152:2012, Geographic information-Land Administration Domain Model*, 1st ed.; ISO: Geneva, Switzerland; 118p.
9. LADM, ISO19152, LADM WIKI. 2017. Available online: http://isoladm.org (accessed on 10 August 2017).
10. Lemmen, C.H.J.; van Oosterom, P.J.M.; Thompson, R.; Hespanha, J.P.; Uitermark, H. The Modelling of Spatial Units (Parcels) in the Land Administration Domain Model (LADM). In Proceedings of the XXIV FIG International Congress, Sydney, Australia, 11–16 April 2010.
11. Zulkifli, N.A.; Rahman, A.A.; Van Oosterom, P.J.M. Developing 2D and 3D Cadastral Registration System based on LADM: Illustrated with Malaysian cases. In Proceedings of the LADM Workshop, Kuala Lumpur, Malaysia, 24–25 September 2013.
12. Brown, G.; Nagel, C.; Zlatanova, S.; Kolbe, T.H. Modelling 3D Topographic Space against Indoor Navigation Requirements. In *Progress and New Trends in 3D Geoinformation Science*; Springer: Heidelberg/Berlin, Germany, 2013.
13. Zlatanova, S.; Liu, L.; Sithole, G.; Zhao, J.; Mortari, F. *Space Subdivision for Indoor Applications*; GISt Report No 66; TU Delft: Delft, The Netherlands, 2014.

14. Munkres, J.R. *Elements of Algebraic Topology*; Addison-Wesley: Menlo Park, CA, USA, 1984.
15. Kruminaite, M.; Zlatanova, S. Indoor Space Subdivision for Indoor Navigation, ISA'14. In Proceedings of the Six ACM SIGSPATIAL International Workshop on Indoor Spatial Awareness, Dallas/Fort Worth, TX, USA, 4–7 November 2014.

International Journal of
Geo-Information

MDPI

Article

Spatial Data Structure and Functionalities for 3D Land Management System Implementation: Israel Case Study

Ruba Jaljolie [1], Peter van Oosterom [2] and Sagi Dalyot [1,*]

[1]   Mapping and Geo-Information Engineering, The Technion, Haifa 3200003, Israel;
      sruba93@campus.technion.ac.il
[2]   GIS Technology, OTB—Research for the Built Environment, Faculty of Architecture and the Built
      Environment, Delft University of Technology, Julianalaan 134, 2628 BL Delft, The Netherlands;
      P.J.M.vanOosterom@tudelft.nl
*   Correspondence: dalyot@technion.ac.il; Tel.: +972-4-829-5991; Fax: +972-4-829-5708

Received: 31 August 2017; Accepted: 5 December 2017; Published: 1 January 2018

Abstract: With the existence of mature technologies and modern urban planning necessities, there is a growing public demand to improve the efficiency and transparency of government administrations. This includes the formation of a comprehensive modern spatial land management (cadastre) system having the capacity to handle various types of data in a uniform way—above-terrain and below-terrain—enabling the utilization of land and space for various complex entities. To utilize existing knowledge and systems, an adaptive approach suggests extending and augmenting the existing 2D cadastre systems to facilitate 3D land management capabilities. Following a comprehensive examination of the Survey of Israel's operative cadastral system that supports 2D land administration, it turned out that it is crucial to outline new concepts, modify existing terms and define specification guidelines. That is, to augment and provide full 3D support to the current operative cadastral system, and to create a common and uniform language for the various parties involved in the preparation of 2D and 3D mutation plans required for modern urban planning needs. This study refers to the legal and technical aspects of Survey of Israel's CHANIT, which is the legal set of cadastral work processes specifications, focusing on database, data structure, functionality, and regulation gaps while emphasizing on 3D cadastral processes. The outcome is recommendations concerning data structure and functionalities needed to be addressed for the facilitation and implementation of an operative 3D land management system in Israel.

Keywords: 3D cadastre; land management; spatial data and functionalities; 3D volumetric parcel

1. Introduction

A 3D cadastral information system is the framework for defining and understanding the spatial restrictions, responsibilities, and rights related to spatial land arrangements. According to Aien et al. [1], the information model should allow the understanding of the various parts in the three-dimensional cadastre (geometries, classes, and attributes), explain how they are arranged, organized and conserved using computerized systems (instances of classes and constraints), and simplify data understanding required by all parties involved. 3D cadastral systems should support the implementation of spatial cadastre processes, including (1) The promotion of spatial standards used by the involved parties; (2) The establishment of 3D databases and facilitation of processes and functionalities concerned with data dissemination; (3) The interoperable exchange, combine and share of datasets; (4) The provision of management functionalities required for the establishment of 3D cadastre data, ensuring the integrity and legitimacy of geometry, topology, and semantics.

Establishing a 3D cadastral information system is a direct response to the increasing public demand aimed at improving the existing efficiency and transparency of government administrations, required today to manage the ever-growing complex modern urban planning necessities. To date, cadastre systems rely mostly on 2D registration and management procedures, such that they are limited in managing multi-layered environments. 3D cadastre, on the other hand, should handle various types of data in a uniform way—both spatially and temporally, with emphasis on infrastructure development that must be addressed and registered with respect to the third dimension—above-terrain and below-terrain. That is, establishing a conventional and well-organized series of conditions and functionalities, which will enable utilization of land and space for various complex entities and projects, which are individually or publically owned. This paper addresses and suggests the definition of an appropriate topology for spatial parcels and the implementation of 3D cadastral workflows, ensuring that the data-structure and datasets defined for both 2D and 3D representations in the land management system are compatible with the existing spatial reality.

In Israel, the Survey Of Israel's (SOI) operative cadastre system requires the submission of 2D mutation plans according to the CHANIT specifications (http://mapi.gov.il/professionalinfo/chanit/documents/mifratlehagashatkvatsim.pdf (in Hebrew)), defining a unified set of instructions for data-transfer required for Cadastral work processes. This ensures that the submitted digital form and data of the mutation plan passes properly through the detailed automatic checklist, which validates a rigorous set of rules that are concerned with the existing 2D geometric, topologic and public law restrictions existing in Israel. CHANIT specifications are defined as the data package for mutation plan and border documentation scheme and offer the technical, legal and cadastral guidelines for submission and validation of 2D mutation plans. Since SOI's approach is to adapt and augment the operative 2D cadastre system to facilitate the management of 3D cadastre [2], investigation and assessment of the CHANIT specifications are made. These, in turn, derive (among others) geometric and topologic modifications and updates required to be made in the existing 2D cadastre system database, which are addressed in this paper, focusing on the data structure, database, and functionalities.

Current studies mainly suggest new data models and systems for the representation and management of 3D cadastral data. Only several studies have addressed the issue by mapping the existing gaps to allow the transition from 2D to 3D land management systems. Our study identifies and categorizes the existing data structure, database and functionality gaps that are required to be resolved for the implementation of a 3D land management (cadastre) system in Israel. This paper will outline the recommendations and required new processes (actions) needed for advancing the capacity of validation workflow and the establishment of comprehensive 3D mutation plans. This includes, among others, aspects related to guidelines and regulations, data structure and database, data integrity and legality, cadastral processes and functionalities, visualization and presentation. Current requirements of all aspects are discussed and detailed, together with the necessary recommended amendments required for the future 3D extension, with the description of several examples.

2. Related Research

Recommendations made by the SOI related to 3D cadastre consisted mainly of two key aspects: preparing appropriate legislation and regulation, and establishing the technological base required for implementing the solution [3]. The idea of a united 3D volumetric parcel is made, representing the volume of a 3D spatial parcel, which can be part of (subtracted from) a number of 2D parcels. Already in 2009, an in-depth legal analysis made in Israel argued that the use of existing legal tools (especially leases and concessions), with no change made to the nature of existing features, might create a huge gap between the factual and legal reality [4]. Documents were drafted, stating four possible legal paths to reach this goal, where the one discussing the structuring and implementation of specific legislation of spatial 3D volumetric parcels was favored by the Israeli Ministry of Justice. These issues were reinforced when planning with an emphasis on complex urban clearly moved towards multi-layered spatial planning: above and under-ground [5]. Assessments made in Felus et al. [6], proved that these should lead to better use of land, protecting its rights and treatment, while preventing illegal use or misuse.

To define a data model for storing 3D objects, Kazar et al. [7] suggested using Oracle's data model for storing 3D geometries. Stoter and van Oosterom [8] implemented a polyhedron data type in the DBMS as an extension to the geometry model of Oracle Spatial [9], while Groger et al. [10] proposed using existing OGC (Open Geospatial Consortium) standard CityGML. A method for recovering the geometry of a 3D polyhedron depicted in a single parallel projection presented in Lee and Fen [11] uses two sets of information: the list of faces in the object, obtained automatically from the drawing, and a user-identified cubic corner, to compute the coordinates of the vertices in the drawing. Using these two sets of information establishes the 3D geometry of the whole polyhedron. The algorithm exploits the topological structure of the polyhedron, implicit in the connectivity between the faces, resulting in a complexity that is linear in the number of faces. The method is extended to objects with no cubic corners as well. A similar approach is suggested by Ying et al. [12], which can "efficiently recognize and construct solids and represent the topological relationships of valid volumetric solids in 3D space in a straightforward way, whether there is an isolated solid, or a multi-solid/solid set". Kazar et al. [7] presented different types and rules for storage, validation, and querying of 3D models, arguing that the GM_Solid representation consists of an unsophisticated qualitative model—in comparison to more topological models—for describing 3D geometry. In the same context, validation rules are addressed together with examples of valid and invalid geometries. Authors noted that actual validation rules are domain dependent, as in Cadastre. For example, it is unclear if dangling faces (patches) or self-intersection are allowed. Currently, both Oracle and ESRI geodatabases do not yet fully support 3D topology structure [6]. In conformity with the jurisdiction of Queensland, Australia, a specific set of digital data validation rules in realizing a 3D cadastre is proposed, in which 2D parcels are treated as infinite 3D columns containing the volume above and below ground [13]. Cadastral processes are designed to check and verify different aspects of 3D cadastre, such as verifying 3D encroachments using a cadastral database, disjointing of 3D rights, supporting 3D common property and curved surfaces. Shojaei et al. [14] presented a set of visualization requirements for 3D cadastral visualization systems, identifying three categories: (1) features that are specific to the visualization of the 3D cadastre, such as underground view and cross-section view; (2) features that pertain to visualization systems, such as interactivity and visual representation; (3) additional features that define how a 3D visualization system must behave, such as usability and interoperability.

3D cadastre studies are made, both on a national and international level (e.g., [15–23]). These studies conducted a detailed analysis of various 3D spatial configurations in an attempt to examine and finally evaluate the ability to provide a unified and proper configuration of a 3D cadastral system prototype. On a national level, Ho and Rajabifard [24] illustrated a case study on the mandatory introduction of BIM for compliance checking in Singapore, and developed a cyclic framework that could be used to inform the choice of strategic activities to support change. Drobež et al. [25] presented the existing cadastral registration system in Slovenia, discussing the required data for representing 3D cadastre, providing an example of the transition from 2D to 3D. Kim and Heo [26] proposed a data model that is based on LADM for expanding the current 2D cadastral system in Korea into a 3D system. Janečka and Souček [27] presented the Czech Republic LADM-based country profile, creating a conceptual model of the cadastre and tables (classes). Kitsakis and Dimopoulou [28] addressed the public law restrictions of the legal framework in applying a 3D cadaster management system in Greece, featuring the importance of 3D modelling based on a case study. Atazadeh et al. [29] provided a methodology for extending a BIM-based data model to support 3D digital management system of complex ownership spaces in Melbourne, Australia, presenting the data elements that are required for this purpose.

Most studies focused on specific aspects of such systems, such as legal and technical issues concerning 3D cadastre, with an intention to provide an optimal solution for defining and solving certain aspects, including the multiplicity of theoretical alternatives for spatial land registry standards of multi-level property. Van Oosterom et al. [30] and Van Oosterom [31], for example, conclude that no complete 3D Cadastre system, covering all aspects, is yet operational, where in most cases 3D cadastral parcels represent only housing units. Still, a number of countries (jurisdictions) investigated the transition to full spatial registration (e.g., [32]). Accordingly, it seems that in terms of conceptual and technological maturity obtained in Israel, now is the right time to reconsider the required technological and operative processes in accordance with the preliminary productive steps made.

3. Methodology

This research aims at setting an approach for expanding current 2D cadastral systems into 3D systems. The methodology to formalize this approach is set in accordance with the recommendations made by SOI, which structure the implementation of legislation related to 3D volumetric parcels, together with the review of SOI's "CHANIT" specification; accordingly, recommendations that will set new methods are made, related to processes, functions, and data.

The first recommendation is to define 3D cadastral processes that should be handled by the management system. Since no 3D procedures are involved in the current 2D system, specifications should identify and detail the orders and sequences of 3D cadastral operations, as well as the possible cadastral procedures that might be applied to the 3D system. Main procedures are subdivision, consolidation/union and transfer (between different lots), in addition to other processes that are detailed later. The system is expected to enable automatic validations and verifications of the propriety of performing cadastral procedures with respect to geometric, topological and public law restrictions, on both the vertical and/or horizontal domains. For effectively managing 3D cadastral information, our research suggests the formation of three main processes: (1) Insertion of a new 3D object (2) Visualization of 3D objects (via search criteria entered by the user) (3) Creating a mutation plan. The processes are later detailed in the article, and implemented in python, with the presentation of a conceptual prototype.

The second recommendation focuses on defining the functions needed for performing the described processes. This research concluded that there is a need for carrying out the functions of: spatial intersection, spatial overlap/overlay, spatial buffer/extrusion, spatial union/merge, spatial clip/extract/select, spatial split, spatial delete/erase, distance calculation, area/projection calculation, and volume calculation. These are crucial to enable the system to perform the cadastral procedures effectively. For determining the relationship between two simple spatial parcels, the study uses an algorithm that is based on detecting the relation between the parcels' nodes [33].

The third recommendation deals with creating a fit-for-purpose data structure and database for archiving the 3D objects. The database is created using SQL in python, where the next step will include the structuring of a small-scale prototype, and implementing a pilot for checking the reliability of the functions (this is not presented here, and is planned for future research).

The recommendations made here will serve as guidelines for constructing 3D cadastral management system. CHANIT specification will be also expanded accordingly. For the sake of expanding the existing 2D management system, database and functionality gaps in CHANIT specification were identified and categorized. Figure 1 illustrates the legal and technical aspects of CHANIT specifications that were reviewed, and that are needed to be modified. Recommendations relevant to the database, the data structure, and the cadastral processes are as follows.

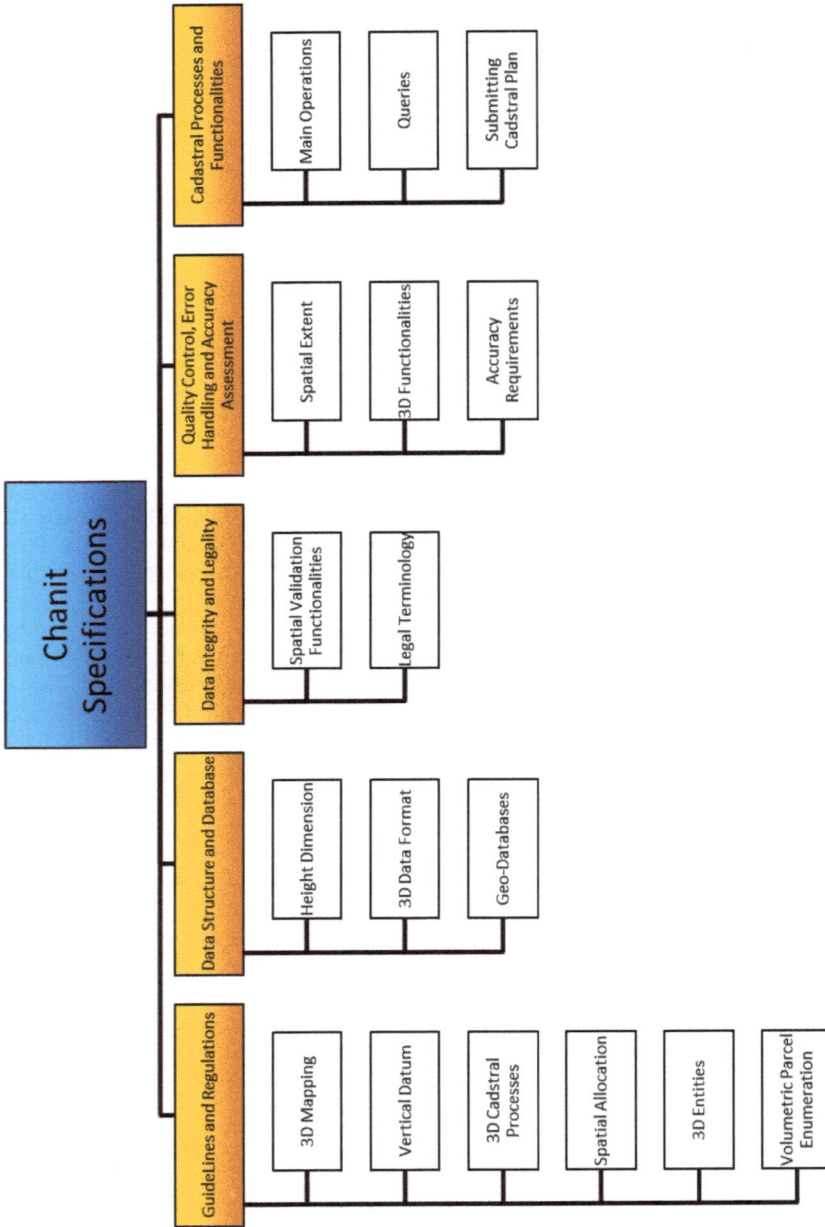

Figure 1. Relevant topics in CHANIT specifications analyzed in this study (extended with respect to 3D cadaster).

3.1. Cadastral and Geometrical Processes

1. Main Operations: according to Jaljolie et al. [2], cadastre management systems should offer these main operations: (1) 3D data collection and organization; (2) Visualization and navigation in the 3D environment; and, (3) 3D analysis, editing, and querying. However, for performing these operations, the technical framework needs to be determined in advance, including data structure, database, software, and hardware. Jaljolie et al. [2] also states that a useful 3D cadastre management system is required to efficiently process the operations and functionalities for a computerized handling of 3D cadastre RRRs, among them: (1) Insertion of a new 3D object (3D volumetric parcel); (2) Visualization of 3D objects (via search criteria); and, (3) Area analysis for plan and design. The first process is presented in Section 4.1 in detail with respect to the basic system functionalities. For providing efficient services, while archiving land RRRs in different zoning plans, cadastral processes include diverse functions for varying purposes. Some of which support taxation, property valuation, registering mortgages for future objects (and other fiscal operations). Other functions aim to enable efficient conveyancing, to support land use planning and land distribution processes. Well-built functions enable executing changes (derived from new/past land arrangements, such as subdivision/split, consolidation/union, transfer between lots, expropriation). Operating functions should be compatible with the correct spatial units, i.e., enabling survey, measure, visualize and store the property in convenient spatial units.

2. Queries: Similarly to functions, it is possible to expand and upgrade the existing 2D queries to 3D cadastral systems. The implemented 3D queries should answer the user's questions and fulfill his demands, which are analogous, in principle, to queries generally activated in 2D systems. For instance: calculating length and area (and now—volume) of parcels/buildings, calculating position (coordinates, datum), identify relationships of land parcels within an area of interest (ID, name, ownership, history, tax, value, ...), 2D and 3D parcels, lots and objects—on- and sub-surface.

3.2. Functionalities—Data Integrity and Legality

This topic deals with the topological rules and geometrical validation of 2D and 3D parcels. Many clauses in CHANIT specifications emphasize the importance of 2D data integrity. The expanded specifications suitable for 3D support are designed to ensure 3D data integrity also, with an emphasis on vertical information, from different perspectives:

1. Spatial Validation Functionalities: fulfilling the integrity requirements, it is important to build a chain of validation functionalities that ensure the veracity of the 3D data. The input of these new functionalities consist of 3D entities (e.g., structures, volumetric parcels, blocks), and the output can be: "valid", in a case that there is no conflict of topologic and geometric rules, or "error", otherwise. Several of these functionalities are listed and detailed in [2].

In the process of insertion a new 3D object/volumetric parcel (see Section 4.1), several geometric and topologic functions and validations should take place, before it is authenticated and inserted into the geodatabase, e.g., receives a system ID. The checklist includes format validity, "safe" distance validity and geometric validity. Format validity verifies that the 3D object topology is valid. The distance between two adjacent objects (above or below the ground) should be no less than the minimum distance according to the existing definition, that ensures, among others, environment protection, maintaining the stability of physical structures, and preventing negative mutual influence between objects. Usually, law and regulations determine the range of physical separation distance between objects. Keeping "safe" distances among 3D objects (especially when it comes to handling objects with no planar geometry, such as tunnels with curved facades, depicted in Figure 2) is one of the most important recommendations that was made by SOI's 3D cadastre committee (recommendations that will probably be formulated into the 3D cadastre treatment and application). Geometric validity verifies that there is no partial or full intersection between the newly inserted volumetric parcel and the existing 3D objects.

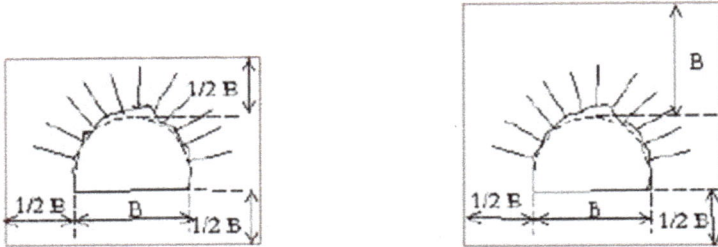

Figure 2. 1/2 B is the "safe distance" for a tunnel with an existing building above it (**left**); B is the "safe distance" for a tunnel with no existing building above it (**right**). Buffer function is used for verifying "safe distance" (source: SOI).

2. Legal Terminology: specifications should include cadastral guidelines for managing the 3D legal reality (RRRs—rights, restrictions, and responsibilities). Preliminarily, relevant terms should be officially and clearly defined, such as spatial parcel, volumetric parcel, "safe" distance, spatial physical object, etc. Following the basic existing definitions, clear regulations regarding the different aspects of 3D mapping and measurements, recording and storing, editing and submission, should be announced to enable planning institutions and surveyors to work properly and uniformly. For this purpose, new guidelines should be set, and old ones should be edited and updated.

3.3. Data Structure and Database

Although functional 3D databases (geodatabase) exist, enabling the storing, querying and representing of spatial geometric objects, they usually are not fit for managing 3D cadastral-objects; they need to be enhanced so that they would provide sufficient tools for handling complex 3D cadastral topological data models [34]. Geodatabases have yet to be developed and expanded, while according to Breunig and Zlatanova [35]: "the integration of 2D and 3D data models and the development of dimension-independent topological and geometric data models" is of a big importance.

CHANIT specifications provide a number of advantages, mainly with respect to the uniform contents and configurations of all cadastral maps that prevent file incompatibilities (e.g., incompatibility between CAD and SRV (A format used for cadastral data transfer from external sources to SOI: http://mapi.gov.il/ProfessionalInfo/DocLib/srv-v2.pdf (in Hebrew)) files), enabling easy transition from CAD environment to GIS environment. In principle, it is possible to profit from the same advantages in 3D reality-supporting specifications. However, the definition of a configuration for 3D cadastral maps, the type of files that will be used to construct and display the 3D cadastral reality, and the appropriate software for managing 3D properties, need to be defined. These issues are related directly to the data structure and database.

CHANIT specifications define the data structure suitable for 2D parcels and blocks. Accordingly, specifications should be updated for 3D reality management, such that a 3D data structure is required to be clearly defined in accordance with the following points:

1. Height Dimension: first critical step is required to add the z-value (height dimension) to the existing coordinates in the operative database. According to previous recommendations made in Felus et al. [6]: ... " So, for 3D parcels, it may be tempting to use relative height w.r.t. Earth surface. However, as the Earth surface may change over time (due to natural or man-related reasons) this is not a stable reference, and it is therefore advised to have at least absolute height

in coordinates of 3D parcels, and maintain and use Earth surface (height) description as separate registration". Such that we recommend that each point in the database will receive an absolute orthometric height value.

2. 3D Data Format: the format required for creating volumetric parcels should be determined. Different approaches are prospective for defining a 3D data structure, including presenting a 3D spatial parcel as an extrusion of a 2D parcel, a 3D spatial parcel as a solid polyhedral body, or a 3D spatial parcel as a combination of primitives (points, lines, and facades). To enable a relatively simple and fast transition from 2D to 3D reality, and vice versa, the format recommended for creating 3D volumetric parcels should conform to the data-structure that exists and is implemented today in the CHANIT specifications, i.e., points, lines, and polygons.

3. GeoDatabase: current databases necessitate to be amended and configured to be suitable for storing 3D cadastral data. A proper workspace for defining the various classes, features, and constraints of 3D entities is Python, as it allows creating vector layers in Shape File format (ESRI's shp), currently used in SOI systems. The spatial entities will be implemented into the previously defined classes, while the storage of the entities will be performed by ESRI's ArcMap and City Engine. Since open-source GIS software support Shape File format, working with alternative GIS software (e.g., QGIS, OpenGIS) should be considered at a later stage. Geodatabases should provide the framework to define the geometry and topology of nature-formed and man-made objects in a unified way [35]. In fact, for building and representing complex 2D/3D objects, it is necessary for the 3D cadastre management system to provide basic topological elements, for example, node, edge, face, and body, or geometric elements: points, line segments, triangles, tetrahedrons, and collections thereof to represent objects. In brief, the data-structure and database would significantly influence the development of the system, the way it is managed and the structure of the functions.

4. Implementation

Following the methodology and findings detailed in Section 3, we aim to provide practical recommendations for augmenting and promoting the existing operative 2D cadastral system to be suitable for comprehensive 3D land management. Accordingly, functions and processes are detailed here. Updated and additional classes, properties and methods, necessary for representing and handling 3D objects, are also suggested.

4.1. Functions and Processes

The basic system functionalities were identified and detailed based on a review of requirements that should serve all processes and terms of a 3D cadastral information system as mentioned in Section 3.2. The main processes were detailed in Jaljolie et al. [2], where as an example, in this section the process of insertion a new 3D object (3D volumetric parcel) is described in detail. In general, when inserting a new 3D volumetric parcel, several geometric and topologic functions and validations should take place, before the new 3D volumetric parcel is authenticated and inserted into the geodatabase, e.g., receives a system ID. These are depicted in the workflow in Figure 3, with a graphic example in Figures 4 and 5. The stages are as follows, where after a 3D object is inserted, it passes through a detailed checklist, which outputs 'true' in the case that the object fulfills all the requirements and integrity rules, or 'false'—in case it does not.

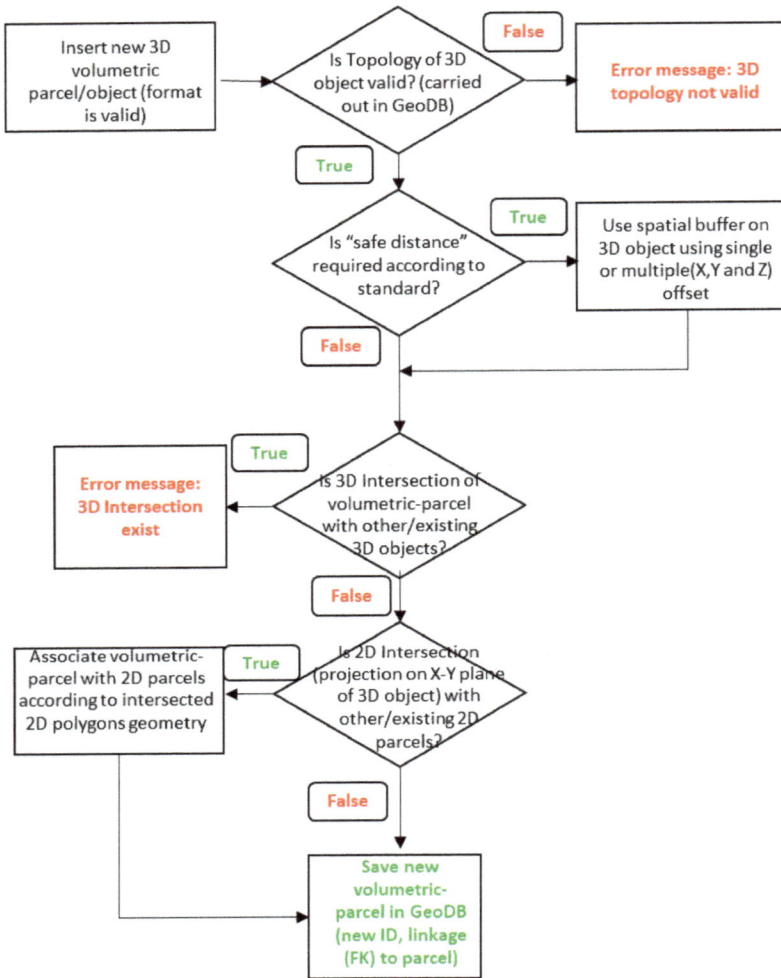

Figure 3. Insertion of a new 3D object (3D volumetric parcel) workflow.

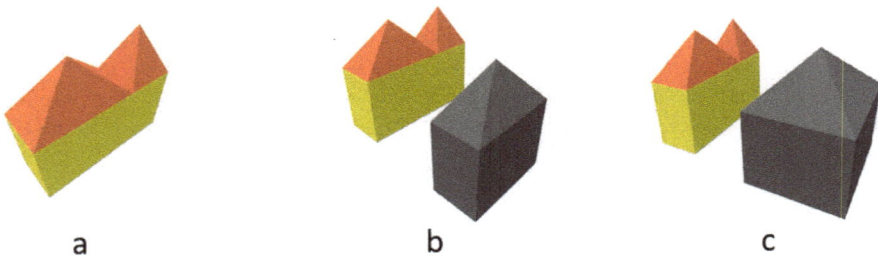

Figure 4. (**a**) Two parcels exist in the database; (**b**) A new volumetric parcel (in grey) is inserted into the database; (**c**) Applying buffer and intersection functions produce not spatial violations, geodatabase is updated.

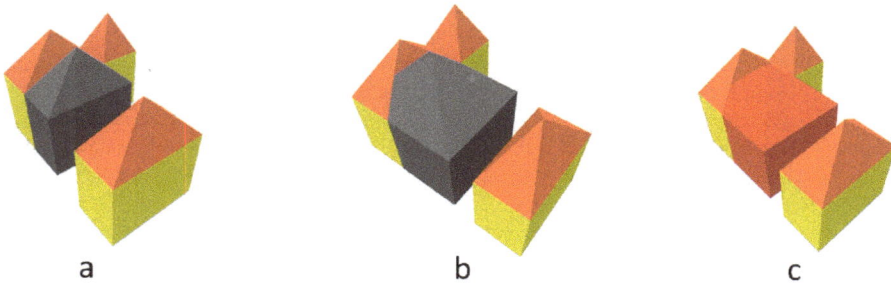

Figure 5. (**a**) A new volumetric parcel is inserted into the database (in grey) (**b**) A buffer is applied; (**c**) Applying the intersection function produces spatial violations, prompting error message.

4.1.1. Format Validity

In the first stage, the system checks whether the topology stored for the 3D object is valid with respect to the one used in the GeoDatabase. In case the topology is not valid an error message will be presented, otherwise the process proceeds to the next stage.

4.1.2. "Safe" Distance Validity

This spatial validity verifies whether the distances between the new input 3D object and other neighboring objects (objects exist in its near space) are legal. The output depends on the system's allowed "safe" distance definition, which is the minimum distance that should be preserved between any adjacent physical objects (see Section 3.2). For ensuring "safe" distances, and examining the proximity to neighboring 2D and 3D objects, the use of functions Buffer and Extrusion is made. According to the recommendations of SOI's 3D cadastre committee, enlargement and reduction of 3D objects are intended for the examination of 3D objects and their correspondence with cadastral conditions on the one hand, together with the possibility to join such two neighboring 3D bodies into a single 3D object—on the other. Using the Buffer function is available here in different implementations:

- Using multiple offsets: one offset in the horizontal plane ("sideways"), and another offset in the vertical plane ("Height"/Extrude). It enables choosing vertical and horizontal buffers separately and independently. Working in this mode is considered for a 3D object when preserving "safe" distance is requisite in only one plane: either horizontally or vertically. or in case different "safe" distances exist for the different planes.
- Using a single offset (XYZ): this function enlarges a 3D object both vertically and horizontally in the same factor.

If the distances between the new inserted 3D object and the existing neighboring objects in the system do not deviate from the minimal required "safe" distances (as required by the law and regulations), the object proceeds to the next validation stage.

4.1.3. Geometric Validity

This stage uses the Intersection function: it checks whether the newly inserted object (or its "buffer") intersects an existing object (or its "buffer"). In this case, an insertion of a 3D object, the intersection of 3D objects is applied first. In case the new inserted 3D volumetric parcel intersects existing 3D objects (whether partially or fully, as in contain), the system returns an error message, and the new 3D volumetric parcel is not added to the database. In case there exists no intersection among 3D objects, the next validation is concerned with the projection on the X-Y plane of the new 3D object and existing 2D objects (parcels). If there exists no intersection with other 2D objects, e.g., it is fully contained, the new 3D volumetric parcel is saved into the geodatabase, i.e., receives an ID, and the process of inserting a new 3D parcel ends. Otherwise, the inserted 3D volumetric parcel is associated with the 2D parcels according to the intersection of the 2D polygons geometry. According to [6], however, "To each of the 3D sub-parcels the same right and party should be attached, both initially, but also in future transactions (e.g., a tunnel that is sold to a company). The split process is necessary only in intermediate stages: "It is better to allow 3D parcels crossing many surface parcels. They could be created in one transaction involving all surface parcels, each selling a part of their property, to create a single 3D subsurface parcel to which the right and party can be attached (for the tunnel). So far, the historic reflections on the sub-parcel concept" [6].

The split action, which is required for indexing, is done by the Split function, which consists of four sub-functions; each is responsible for a slightly different operation. Only two are relevant here: (1) split of 3D objects in relation to existing/neighbouring 2D objects. The input is the new inserted 3D object and one (or more) 2D objects describing the limits of the 2D parcels and lots existing in the 3D parcel surroundings (above/below the 3D volumetric parcel). The function suggests the split of the 3D object on vertical planes determined by X-Y coordinates of the 2D objects. The output is two—or more—3D objects (multiple polyhedrons) created by splitting of the original 3D object. (2) Split of 3D objects as a function of geometric/cadastral constraints. The input is the new inserted 3D object, the geometric constraints function and/or the cadastral threshold values function. The output of the process will usually be composed of two—or more—3D objects derived from splitting up of the original 3D object. Examples of applying split of 3D objects as a function of geometric constraints could be splitting a 3D cadastral parcel on a horizontal or vertical plane; parallelism or perpendicularity between faces of the 3D object are maintained. Minimal object volume and minimal area of faces are examples of applying split of 3D objects as a function of cadastral constraints. To summarize, this step enables the split of a 3D parcel in accordance with both the cadastral aspect and topological aspect, as they are manifested in the projection of the 3D parcel on a 2D plane.

Following the Split operation, the process of inserting a new 3D volumetric parcel ends by saving the new 3D object (volumetric parcel) into the system's geodatabase, i.e., it receives an ID, including all required indexing. This process can be referred to as the process of "Insertion of a 3D plan" if implemented several times until all 3D volumetric parcels that exist in the 3D plan are validated and inserted into the system.

Figures 4 and 5 depict two graphic examples of the process described earlier. Figure 4 represents a valid process of inserting a new volumetric parcel into the existing database. After checking the topology of the newly inserted parcel, a buffer is created according to the required safe distance. Then, the system checks if the new parcel (including its buffer) intersects any existing parcel in the database. Because there is no intersection, the parcel is saved into the database. Figure 5 describes an invalid process, where the extent of the added volumetric parcel intersects an existing 3D parcel, generating a system error message; the new parcel is deleted, and the database is not updated.

4.2. Data Structure and Database

4.2.1. Classes

The main classes existing today in the operative 2D cadastral system, as well as the required new ones, are depicted in Figure 6. In red, new fields and methods relevant for 3D representation in existing classes. For representing 3D data, it is important to define two new classes, depicted in Figures 6 and 7: CLS_3DVolumetricParcels (CLS denotes class) and CLS_3DPolygons, where CLS_3DPolygons represents facades, floors, and roofs. 3D volumetric parcels consist of floor, roof, and facades, while facades, floor, and roof could be defined by the coordinates of their corners (CLS_Point), and by their borders (CLS_Line).

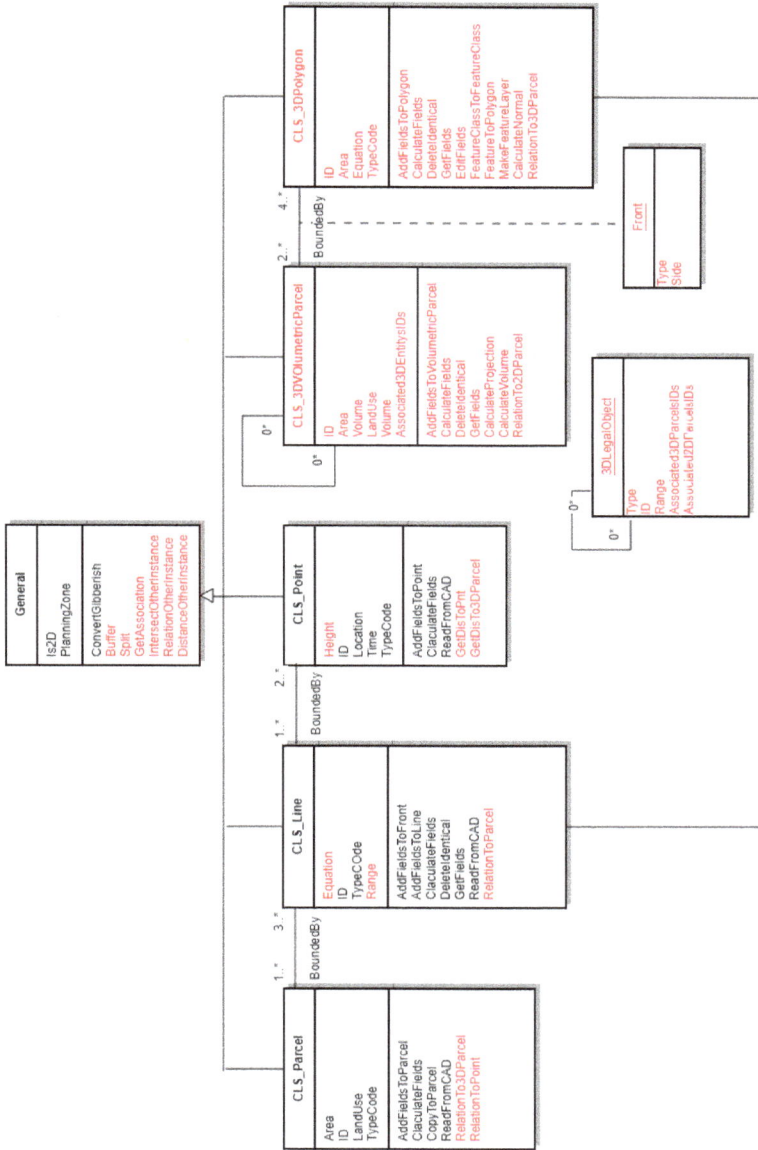

Figure 6. Main classes in SOI's operative cadastral 2D system, integrated with new classes and fields that should be added to the cadastral system to serve the 3D reality.

The existing 2D classes should have additional fields (that can be optional in case of pure 2D data), used for representing 3D classes and objects (appears in red in Figure 6). Principle new fields, such as "Height", should be added to class CLS_Point. CLS_Line indicates straight or curved line segment and should include a new field "Equation" that represents the equation function of the line in space required to make it possible to detect whether the line is vertical, horizontal or diagonal (essential for the practicality of functionalities—see Section 4.1). Adding the "Range" field (i.e., $\{x,y,z\}_{min}$ $\{x,y,z\}_{max}$) in CLS_Line to indicate the range of the coordinates can be helpful for detecting the extent (and location) of a spatial line.

The main classes (entities) for representing the land properties are point (CLS_Point), line (CLS_Lines) and parcel (CLS_Parcel). Since mutation plans are submitted in CAD format, additional classes are used for reading the input data from the CAD files and writing them into the geodatabases and tables via the defined processes (CLS_CadReader, CLS_GDBRead, CLS_GDB_writer, and CLS_InProcessTable, depicted in Figure 7). CLS_Incoding serves for converting the layers' codes that appear in the CAD files to a comprehensible language suitable for storing in the geodatabases. The review of CHANIT emphasizes the importance of different data formats for efficient usage, which makes these classes of big importance.

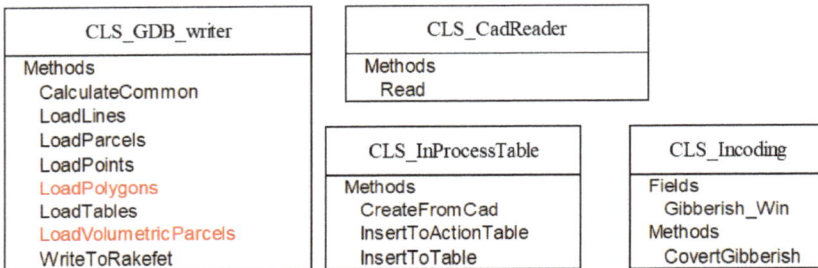

CLS_GDB_writer
Methods
CalculateCommon
LoadLines
LoadParcels
LoadPoints
LoadPolygons
LoadTables
LoadVolumetricParcels
WriteToRakefet

CLS_CadReader
Methods
Read

CLS_InProcessTable
Methods
CreateFromCad
InsertToActionTable
InsertToTable

CLS_Incoding
Fields
Gibberish_Win
Methods
CovertGibberish

Figure 7. Classes used for reading the input data from CAD files and writing them into the geodatabase and tables.

Hierarchical associations are created for better handling the different 3D entities: CLS_3DVolumetricParcel is associated to CLS_3DPolygon, which is associated to CLS_Line, which is associated to CLS_Point. This is necessary because a volumetric parcel consists of 3D polygons, where polygons consist of lines—which is consistent with the definitions of SOI's existing 2D land management system, practical for a smooth transition from 2D o 3D.

4.2.2. Fields and Methods

The spatial objects that are considered for registration in deeds are tunnels, foundations and construction poles, sub-ground buildings, parking lots, and infrastructure. All these object types would fall under the category of 3DLegalObject (Figure 6). Every type of entity, whether it is a Line, a Parcel, a Point, a 3DVolumetricParcel, and a 3DPolygon, should have a code for the effective description of the system, such that the relevant class type should be added as a new field named "TypeCode". For 3DVolumetricParcels, "TypeCode" declares the objects' type (characteristics), such as a tunnel, foundations, construction shaft, sub-ground building, parking lot—to name a few. The class 3DLegalObject may be composed of more than one 3DVolumetricParcel classes.

According to the review of CHANIT specifications, the planning zones are the valid range of the vertical mapping that should be stated in the regulations in accordance with planning zones. Determining the legal range of vertical mapping depends on the specific mutation planning zone, which may be object type dependent. That means that the valid range of the vertical mapping should be stated in the regulations in accordance with the planning zones and the spatial objects' types so that surveyors will be able to measure objects clearly and uniformly. This implies that it is necessary to add an association to planning zone (can be implied via spatial overlay). This new field should appear in all classes as "PlanningZone", also declared in the "General" table, which includes some of the common fields and methods that will be part of all participating classes. "Is2D" field holds the Boolean value "true" in case the object has a constant height value (flat), or "false" in case the object is not planar.

The review of CHANIT recommends allowing the allocation of several land uses for any 2D parcel in accordance with height, inferring adding the fields "Height" and "LandUse", which have a big importance. An array is a possible structure for the "LandUse" field, the same as a building could have several land uses associated with the horizontal and vertical location. The field "Height" is practically the fundamental field required for establishing a 3D management system since it extends the third dimension and the related important information.

The routines of edit and submission should be announced in the regulations. For practically editing the spatial objects, the methods "EditFields" and "GetFields" should be inserted and added to all classes in the data structure, so that the user can edit fields in a case that a real change is made. The numbering and naming of the 3D spatial parcels should be declared. Numbering is represented by the "ID" field in the data structure. Since the system aims to handle 3D mutation plans, "ID" field should be uniformly set in accordance with the planning zone and the data type of each object.

"LoadPolygons" method in the "CLS_GDB_writer" class is intended for loading facades, roofs, and floors, which are all 3D polygons, where "LoadVolumetricParcels" aims at loading 3D volumetric parcels for creating "volume tables" into the national cadastral database. These methods enable writing the relevant 3D objects into the geodatabase and tables. CLS_Parcel is associated with the 3D volumetric parcels that exist in the 3D entity that contains the volume above and below the 2D "CLS_parcel", which is necessary for efficient processing in the system (e.g., search), as it associates between a 2D parcel and the 3D objects that are above or below it. Classes in Figure 7 are important since CHANIT specifications demand the submission of CAD files, where certain topological rules are validated and maintained. These rules need to be extended to multi-dimensional reality with an emphasis on 3D parcels. One clause in the specifications requires assuring that 2D parcels consist of closed polylines; this clause will be expanded, as follows: the volumetric parcels will consist of closed facades. The rules, related to 3D management, should be applied, including (among others): a border must not have self-intersection, and no overlaps between neighbouring parcels should exist. Stoter and Zlatanova [33], for example, suggested a method for determining the topological relationship between two 3D objects.

As depicted in Figure 8, it should be defined to which volumetric parcel the roof, floor and facades belong to, as well as stating to which 2D parcels a volumetric parcel is related to. CLS_3DPolygon is associated to two CLS_3DVolumetricParcels (i.e., left/right adjacent volumetric parcels or top/bottom). In case there is no bounded 3D parcel at one side of a face, then the adjacent corresponding CLS_3DVolumetricParcels would refer to the outside world/space (which can then be traced to the relevant column(s) of space defined by the 2D surface parcels).

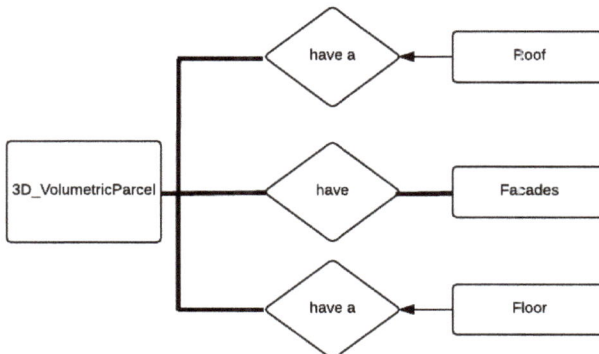

Figure 8. A 3D_VolumetricParcel must have a Roof, Facades, and Floor. Every Floor and Roof belongs to a 3D volumetric parcel.

5. Conclusions and Future Work

The objective of this research paper is to present recommendations concerning the complete and computerized sets of classes and functionalities required for the management and handling of 3D cadastral objects in a comprehensive 3D land management system in Israel. The rationale and motivation were based on physical and jurisdictional aspects related to the configuration and guidelines made by the SOI by adopting the currently operative 2D cadastral system based on CHANIT specification. Constructing proper and comprehensive classes, properties, methods, and functionalities is an indispensable part of building a good and reliable system. This paper started by identifying and categorizing the existing database and functionality gaps for establishing a 3D land management system implementation in Israel, and, accordingly, outlined the classes, methods and some of the functionalities necessary, required to extend and augment the existing 2D system rather than establishing a new cadastre system. The suggested functionalities and classes belong to two categories: one that is related to the handling of objects and data in the geodatabase (storage and converting between different data types), and the second that refers to the necessary geometric and topologic processes for data management.

So far, research in the field mainly concentrated on the topology of 3D cadastral objects, focusing on 3D spatial relationship models, with almost no discussion on the functions needed for the handling of reliable and correct operations. We believe that implementing these functionalities is vital to allow the establishment of these systems. Our next step is to follow these recommendations and construct an operational 3D geodatabase and data-structure required for handling 2D and 3D objects, integrating these functionalities, while implementing the different cadastral processes. Establishing this will contribute to the practice of good governance, in accordance with the definitions, guidelines and accuracy requirements made by the SOI. The research will also include requisites for functionalities validation and examining their workflow in various conditions and different situations within a system, to be adopted and implemented by the SOI in the near future.

Future work will also consider the creation of drawings and sketches, and the visualization of 3D mutation plans. Looking for a visual representation of 3D mutation plans, it would be necessary to comprehensively investigate the current software used for 2D representation, the symbology, layers, syntax rules, and the manner of converting data to drawings. An investigation of these aspects will indicate on updates and modifications required, so that vertical details, standard data, spatial objects, and information would be clearly represented in future cadastral systems. Next step will include the structuring of a small-scale prototype and the implementation of a pilot for checking the reliability of the data structure, database, and functions.

Acknowledgments: The authors would like to thank the Survey of Israel for their generous support.

Author Contributions: The methodology was developed by Ruba Jaljolie, Peter van Oosterom and Sagi Dalyot, with implementations, case studies and analyses carried out by Ruba Jaljolie.

Conflicts of Interest: The authors declare no conflict of interest.

References

1. Aien, A.; Kalantari, M.; Rajabifard, A.; Williamson, I.P.; Shojaei, D. Developing and testing a 3D cadastral data model: A case study in Australia. In *ISPRS Annals of the Photogrammetry, Remote Sensing and Spatial Information Sciences: XXII ISPRS Congress*; ISPRS: Paris, France, 2012.
2. Jaljolie, R.; Van Oosterom, P.; Dalyot, S. Systematic analysis of functionalities for the Israeli 3D cadastre. In Proceedings of the 5th International FIG 3D Cadastre Workshop, Athens, Greece, 18–20 October 2016.
3. Shoshani, U.; Benhamu, M.; Goshen, E.; Denekamp, S.; Bar, R. A multi layers 3D cadastre in Israel: A research and development project recommendation. In Proceedings of the FIG Working Week 2005 and GSDI-8, Cairo, Egypt, 16–21 April 2005.
4. Caine, A. Spatial rights legislation in Israel—A 3D approach. In Proceedings of the FIG Working Week 2009, Eilat, Israel, 3–8 May 2009; p. 14.

5. Sandberg, H. Developments in 3D horizontal land sub-division in Israel: Legal and urban planning aspects. In Proceedings of the Abstract Book Planning, Law and Property Rights (PLPR) 2014 Conference, Technion, Haifa, Israel, 10–14 February 2014; p. 75.

6. Felus, Y.; Barzani, S.; Caine, A.; Blumkine, N.; van Oosterom, P. Steps towards 3D cadastre and ISO 19152 (LADM) in Israel. In Proceedings of the 4th International Workshop on 3D Cadastres, Dubai, UAE, 9–11 November 2014.

7. Kazar, B.M.; Kothuri, R.; van Oosterom, P.; Ravada, S. On valid and invalid three-dimensional geometries. In *Advances in 3D Geoinformation Systems*; Springer: Berlin/Heidelberg, Germany, 2008; Chapter 2, pp. 19–46.

8. Stoter, J.E.; Van Oosterom, P.J.M. Incorporating 3D geo-objects into a 2D geo-DBMS. In Proceedings of the ASPRS/ACSM Annual Conference, Washington, DC, USA, 22–26 April 2002.

9. ORACLE. *Oracle Spatial User's Guide and Reference Release 9.0.1 Part Number a88805-01*; Technical Report; ORACLE: Redwood City, CA, USA, 2002.

10. Groger, G.; Kolbe, H.; Nagel, C.; Hafele, K. *OGC City Geography Markup Language (CityGML), Encoding Standard, Version 2.0.0.*; OGC: Wayland, MA, USA, 2012.

11. Lee, Y.T.; Fen, F. 3D reconstruction of polyhedral objects from single parallel projections using cubic corner. *Comput. Aided Des.* **2011**, *43*, 1025–1034. [CrossRef]

12. Ying, S.; Guo, R.; Li, L.; Van Oosterom, P.; Stoter, J. Construction of 3D volumetric objects for a 3D cadastral system. *Trans. GIS* **2015**, *19*, 758–779. [CrossRef]

13. Karki, S.; Thompson, R.; McDougall, K. Development of validation rules to support digital lodgement of 3D cadastral plans. *Comput. Environ. Urban Syst.* **2013**, *40*, 34–45. [CrossRef]

14. Shojaei, D.; Kalantari, M.; Bishop, I.D.; Rajabifard, A.; Aien, A. Visualization requirements for 3D cadastral systems. *Comput. Environ. Urban Syst.* **2013**, *41*, 39–54. [CrossRef]

15. Aien, A.; Rajabifard, A.; Kalantari, M.; Williamson, I. Aspects of 3D Cadastre—A case study in Victoria. In Proceedings of the FIG Working Week, Marrakech, Morocco, 18–22 May 2011; p. 15.

16. Döner, F.; Thompson, R.; Stoter, J.E.; Lemmen, C.H.J.; Ploeger, H.D.; van Oosterom, P.; Zlatanova, S. 4D cadastres: First analysis of Legal, organizational, and technical impact—With a case study on utility networks. *Land Use Policy* **2010**, *27*, 1068–1081. [CrossRef]

17. Eriksson, G.; Jansson, L. Strata titles are introduced in Sweden. In Proceedings of the XXIV International FIG Congress, Sydney, Australia, 11–16 April 2010; p. 13.

18. Guo, R.; Ying, S.; Li, L.; Luo, P.; Van Oosterom, P. A multi-jurisdiction case study of 3D cadastre in Shenzhen, China as experiment using the LADM. In Proceedings of the 2nd International Workshop on 3D Cadastres, Delft, The Netherlands, 16–18 November 2011; pp. 31–50.

19. Karki, S.; McDougall, K.; Thompson, R. An overview of 3D cadastre from a physical land parcel and a legal property object perspective. In Proceedings of the XXIV International FIG Congress, Sydney, Australia, 11–16 April 2010; p. 13.

20. Paulsson, J. 3D Property Rights—An Analysis of Key Factors Based on International Experience. Ph.D Thesis, Royal Institute of Technology (KTH), Stockholm, Sweden, 2007; p. 351.

21. Pouliot, J.; Roy, T.; Fouquet-Asselin, G.; Desgroseilliers, J. 3D cadastre in the province of Quebec: A first experiment for the construction of a volumetric representation. In Proceedings of the 5th International 3D GeoInfo Conference, Berlin, Germany, 3–4 November 2010; p. 15.

22. Rahman, A.A.; Hua, T.C.; Van Oosterom, P. Embedding 3D into multipurpose cadastre. In Proceedings of the FIG Working Week, Marrakech, Morocco, 18–22 May 2011; p. 20.

23. Stoter, J.; Ploeger, H.; van Oosterom, P. 3D cadastre in the Netherlands: Developments and international applicability. *Comput. Environ. Urban Syst.* **2013**, *40*, 56–67. [CrossRef]

24. Ho, S.; Rajabifard, A. Towards 3D-enabled urban land administration: Strategic lessons from the BIM initiative in Singapore. *Land Use Policy* **2016**, *57*, 1–10. [CrossRef]

25. Drobež, P.; Fras, M.K.; Ferlan, M.; Lisec, A. Transition from 2D to 3D real property cadastre: The case of the Slovenian cadastre. *Comput. Environ. Urban Syst.* **2017**, *62*, 125–135. [CrossRef]

26. Kim, S.; Heo, J. Development of 3D underground cadastral data model in Korea: Based on land administration domain model. *Land Use Policy* **2017**, *60*, 123–138. [CrossRef]

27. Janečka, K.; Souček, P. A country profile of the Czech Republic based on an LADM for the development of a 3D cadastre. *ISPRS Int. J. Geo-Inf.* **2017**, *6*, 143. [CrossRef]

28. Kitsakis, D.; Dimopoulou, E. Addressing public law restrictions within a 3D cadastral context. *ISPRS Int. J. Geo-Inf.* **2017**, *6*, 182. [CrossRef]

29. Atazadeh, B.; Kalantari, M.; Rajabifard, A.; Ho, S.; Champion, T. Extending a BIM-based data model to support 3D digital management of complex ownership spaces. *Int. J. Geogr. Inf. Sci.* **2017**, *31*, 499–522. [CrossRef]

30. Van Oosterom, P.; Stoter, J.; Ploeger, H.; Thompson, R.; Karki, S. World-wide inventory of the status of 3D cadastres in 2010 and expectations for 2014. In Proceedings of the FIG Working Week 2011, Marrakech, Morocco, 18–22 May 2011; p. 21.

31. Van Oosterom, P. Research and development in 3D cadastres. *Comput. Environ. Urban Syst.* **2013**, *40*, 1–6. [CrossRef]

32. Vandysheva, N.; Sapelnikov, S.; van Oosterom, P.; de Vries, M.; Spiering, B.; Wouters, R.; Hoogeveen, A.; Penkov, V. The 3D cadastre prototype and pilot in the Russian Federation. In Proceedings of the FIG Working Week 2012, Rome, Italy, 6–10 May 2012; p. 16.

33. Stoter, J.; Zlatanova, J. Visualisation and editing of 3D objects organised in a DBMS. In *Proceedings of EuroSDR Commission 5 Workshop on Visualisation and Rendering*; EuroSDR: Cambridge, UK, 2003.

34. Zhao, Z.; Guo, R.; Li, L.; Ying, S. Topological relationship identification in 3D cadastre. In Proceedings of the 3rd International Workshop on 3D Cadastres: Developments and Practices, Shenzhen, China, 25–26 October 2012; pp. 25–26.

35. Breunig, M.; Zlatanova, S. 3D geo-database research: Retrospective and future directions. *Comput. Geosci.* **2011**, *37*, 791–803. [CrossRef]

MDPI

St. Alban-Anlage 66

4052 Basel

Switzerland

Tel. +41 61 683 77 34

Fax +41 61 302 89 18

www.mdpi.com

ISPRS International Journal of Geo-Information Editorial Office

E-mail: ijgi@mdpi.com

www.mdpi.com/journal/ijgi

www.ingramcontent.com/pod-product-compliance
Lightning Source LLC
Chambersburg PA
CBHW051717210326
41597CB00032B/5517